ATOMIC ASSISTANCE

A VOLUME IN THE SERIES

CORNELL STUDIES IN SECURITY AFFAIRS

EDITED BY ROBERT J. ART, ROBERT JERVIS, AND STEPHEN M. WALT

A list of titles in this series is available at www.cornellpress.cornell.edu.

ATOMIC ASSISTANCE

HOW "ATOMS FOR PEACE" PROGRAMS CAUSE NUCLEAR INSECURITY

MATTHEW FUHRMANN

Cornell University Press
Ithaca and London

First published 2012 by Cornell University Press
First printing, Cornell Paperbacks, 2012
Printed in the United States of America

Library of Congress Cataloging-in-Publication Data

Fuhrmann, Matthew, 1980–
    Atomic assistance : how "atoms for peace" programs cause nuclear insecurity / Matthew Fuhrmann.
        p. cm. — (Cornell studies in security affairs)
    Includes bibliographical references and index.
    ISBN 978-0-8014-5090-7 (cloth : alk. paper)
    ISBN 978-0-8014-7811-6 (pbk. : alk. paper)
    1. Nuclear nonproliferation—International cooperation.
    2. Nuclear industry—International cooperation.   3. Technology transfer—International cooperation.   4. Technical assistance—International cooperation.   5. Security, International.
    I. Title.   II. Series: Cornell studies in security affairs.
    JZ5675.F85 2012
    327.1'747—dc23   2012005396

Cornell University Press strives to use environmentally responsible suppliers and materials to the fullest extent possible in the publishing of its books. Such materials include vegetable-based, low-VOC inks and acid-free papers that are recycled, totally chlorine-free, or partly composed of nonwood fibers. For further information, visit our website at www.cornellpress.cornell.edu.

Cloth printing   10 9 8 7 6 5 4 3 2 1
Paperback printing   10 9 8 7 6 5 4 3 2 1

Contents

Tables and Figures

TABLES

FIGURES

Preface and Acknowledgments

The genesis of this book dates back to 2004–5 when I was interested in exploring how international commerce enabled states to augment their military capabilities. I began to investigate how countries acquired military technology. It appeared that in many cases countries did not pursue the most obvious strategy: the procurement of complete weapons systems from other states. Nor did they seem to turn to black markets to the same degree that mainstream media coverage seemed to imply. On the contrary, I uncovered many cases where countries legally purchased subcomponents on the open market and built weapons systems indigenously. This seemed to happen with both conventional military technology and weapons of mass destruction (i.e., chemical, biological, and nuclear weapons).

Most of the components of major weapons systems are dual use in nature, meaning they have commercial applications in addition to their military uses. For example, a valve that is widely used in the semiconductor industry plays a key role in producing highly enriched uranium, which is a critical ingredient for producing nuclear weapons.

Iraq's nuclear weapons program provided one example of the challenges posed by this so-called dual-use dilemma. Consider the case of triggered spark gaps. This item is used by hospitals in machines called lithotripters to break up kidney stones. On the other hand, it can also be used as a trigger in nuclear bombs. Saddam Hussein reportedly procured several lithotripters with extra spark gaps—ostensibly for use in Iraqi hospitals. Yet many people feared that his intent was to use them for military purposes. In the end, Iraq did not build the bomb (in part because the inspections regime that followed the 1991 Persian Gulf War made it difficult to reconstitute the weapons program), but this case underscores that it can be difficult to discern states' intentions when they procure dual use commodities.

The more I looked, the more examples like this I found.

This raised some fascinating questions. How do supplier countries cope with the dual-use dilemma? What explains patterns of dual-use trade? How do governments regulate firms that produce commodities and technology that have both civilian and military purposes? To what extent does the acquisition of technology for civilian purposes contribute to the diffusion of military power? I began to explore these and other related questions. They interested me not only because they had significant policy implications but also because I thought they could speak to some central issues in international relations such as the efficacy of international institutions and the role of trust in state-to-state interactions.

My initial strategy was to produce a global database on dual-use trade. I wanted to account for *all* such commerce that could contribute to military capabilities. I started with the United States, which was one of the largest producers of dual-use commodities. To my surprise, data on American dual-use exports of the nature I wanted were not publicly available at the time. With the help of a colleague, I obtained data from the U.S. Department of Commerce on licensed dual-use exports from 1991 to 2001. However I was unable to obtain similar data for other supplier countries such as China, France, and Russia. A different approach was needed to compile the type of cross-national dataset that I had in mind.

I decided to focus explicitly on peaceful nuclear assistance—the transfer of nuclear technology, materials, and know-how for civilian purposes. There were no existing precanned datasets on global nuclear commerce, but most of the necessary information appeared to be in the public domain. Data considerations aside, I found nuclear trade to be the most interesting type of dual-use commerce. It was here that the dual-use dilemma was the most vexing. From the perspective of suppliers, civilian nuclear assistance is one type of economic statecraft that they could employ to enhance their influence. And from the standpoint of recipients, peaceful nuclear programs could partially satisfy growing energy needs and address the problem of global climate change. Yet there were dangers associated with civilian nuclear assistance. In providing this aid suppliers risked spreading the most destructive weapon known to mankind. Under what conditions would supplier countries take this risk? Did their gambles backfire by inadvertently contributing to nuclear weapons proliferation?

I embarked on a journey to find out. This book represents the culmination of that expedition.

A number of institutions and individuals helped to make this book possible. At the University of Georgia, I benefited tremendously from helpful friends and mentors. My two principal advisors, Gary Bertsch and Jaroslav Tir, provided unwavering support and together taught me that systematic scholarly research and policy relevance are not mutually exclusive. Jeff Berejikian and Doug Stinnett also supplied training and guidance that proved to be very helpful in writing this book. I also want to thank the others in my cohort at UGA, especially Bryan Early. Bryan provided sage advice about this project from its initial conception through the final revisions. We had countless conversations over the years about this book. Each one of them helped me improve the project.

The Center for International Trade and Security (CITS), which was under the direction of Bertsch at the time, provided me with funding and support while I was at UGA. My colleagues at the Center shaped my early thinking about peaceful nuclear cooperation by reminding me that nuclear proliferation cannot be understood without a deep appreciation of international commerce. I am thankful for the feedback and encouragement I received from everyone at CITS, especially Mike Beck, Seema Gahlaut, Jim Holmes, Scott Jones, Julia Khersonsky, Igor Khripunov, Dmitriy Nikonov, Anupam Srivastava, and Richard Young.

I spent 2007–8 as a research fellow at Harvard University's Belfer Center for Science and International Affairs. The fellowship at Belfer proved to be tremendously helpful in completing this book. I am forever indebted to Matt Bunn, John Holdren, Marty Malin, and Steve Miller for the opportunity to spend eighteen months at the Belfer Center. All four were terrific mentors and they provided extensive feedback on my research. The other fellows at Belfer also helped me improve this book in countless ways, often by commenting on draft chapters of the manuscript. For feedback and advice, I would like to particularly thank Hassan Abbas, Kristin Bakke, Kayhan Barzegar, Emma Belcher, Tom Bielefeld, Malfrid Braut-Hegghammer, Jon Caverley, Erica Chenoweth, Eric Dahl, Alex Downes, Sarah Kreps, Matt Kroenig, Jon Monten, Vipin Narang, Wendy Pearlman, Negeen Pegahi, Phil Potter, Matthew Sharp, Paul Staniland, and Dominic Tierney. Susan Lynch and Neal Doyle also helped in countless ways by making sure that the proverbial train ran on time.

Colleagues at the University of South Carolina, where I was an assistant professor from 2009 to 2011, also provided valuable feedback on my work and on the book publishing process more generally. Katherine Barbieri, Kirk Randazzo, and Harvey Starr were particularly helpful in that regard.

A Stanton Nuclear Security Fellowship at the Council on Foreign Relations in Washington, D.C., during 2010–11 helped me put the finishing touches on the book. I thank Michael Levi, Jim Lindsay, Paul Stares, and Micah Zenko for the chance to spend the year at the Council. I benefited tremendously from their guidance and advice on how to make my work more accessible to a policy audience. I also thank Victoria Alekhine and Janine Hill for making sure that everything went smoothly during my year at CFR.

I submitted the final copy of the book after joining the political science department at Texas A&M University. I am grateful for all of the support that I received from my colleagues at TAMU, particularly Quan Li, who helped me sharpen my arguments from the very early stages of the project. Benjamin Tkach also provided helpful editorial and research assistance in the final stages of the project.

Other colleagues commented on various chapters of this book along the way and/or offered helpful advice: Graham Allison, Victor Asal, Kyle Beardsley, Bill Boettcher, Duane Bratt, David Carter, Xenia Dormandy, Erik Gartzke, Chris Gelpi, Charlie Glaser, Mike Glosny, Joe Grieco, Aaron Hoffman, Mike Horowitz, Walt Kato, Jeff Legro, Andrew Long, Sean Lynn-Jones, Alex Montgomery, John Mueller, Paul Nelson, T. V. Paul, Brian Pollins, Bill Potter, Dan Reiter, Scott Sagan, Karthika Sasikumar, Todd Sechser, Etel Solingen, Adam Stulberg, Chris Way, and Alex Weisiger. I appreciate the time they took to help me strengthen all parts of the book.

James Keeley deserves special thanks for sharing his list of bilateral nuclear cooperation agreements, which was critical for my data collection efforts. So do Alex Downes and Todd Sechser for participating in multiple workshops to discuss the ways in which this book could be improved.

Roger Haydon at Cornell University Press diligently ushered the manuscript through the review process and provided helpful guidance and support. I wish to thank Roger, the editors of the Cornell Studies in Security Affairs, and the entire team at the Press for their efforts. Two anonymous reviewers also provided detailed feedback on the full manuscript. The end result is a much improved book and for that I am grateful.

I presented portions of this book in research seminars at Duke University, Georgia Tech, Harvard University, North Carolina State University, Princeton University, the Savannah River National Laboratory, the University of Pennsylvania, the University of South Carolina, and the University of Virginia. I thank all of the participants in those workshops for useful and constructive feedback.

While I was fortunate to benefit from tremendous feedback, those who commented on the book are certainly not responsible for any of its weaknesses. Of course, any remaining errors are my own. And my apologies to anyone I inadvertently left out.

Select portions of this book were previously published in peer-reviewed journals. Parts of chapter 2 were first published in Matthew Fuhrmann, "Taking a Walk on the Supply Side: The Determinants of Civilian Nuclear Cooperation," *Journal of Conflict Resolution* 53, no. 2 (2009): 181–208. Portions of chapters 7 and 8 appeared in Matthew Fuhrmann, "Spreading Temptation: Proliferation and Peaceful Nuclear Cooperation Agreements," *International Security* 34, no. 1 (2009): 7–41. I thank MIT Press and Sage Publications for permission to reprint excerpts from those articles in the book.

On a personal note, I would like to thank my parents, Cindy and Chris, and my siblings, Andrew and Kristin, and Kristin's husband, Jason, for their support while I was writing this book. I am also grateful for the support from my wife's family: Adam, Brian, Evan, Jane, and Monica. My wife, Lauren, read the complete manuscript and saved me from several errors. More importantly, she made every word of this book easier to write by bringing joy into every day. It is mostly for this reason that I dedicate the book to her.

Abbreviations

AEB	Atomic Energy Board (South Africa)
AEC	Atomic Energy Commission (United States and India)
AEOI	Atomic Energy Organization of Iran
AP	Additional Protocol
ASEAN	Association of Southeast Asian Nations
CANDU	Canada Deuterium Uranium
CEA	French Atomic Energy Commission (Commissariat à l'énergie atomique)
CENTO	Central Treaty Organization
CIR	Canada-India Reactor
CIRUS	Canada-India-Reactor-United States
CNEN	National Committee for Nuclear Energy (Italy)
CoCom	Coordinating Committee on Multilateral Export Controls
COW	Correlates of War
EURATOM	European Atomic Energy Community
FFEP	Fordow Fuel Enrichment Plant
GDP	gross domestic product
HEU	highly enriched uranium
IAEA	International Atomic Energy Agency
KANUPP	Karachi Nuclear Power Plant
LEU	low enriched uranium
MAD	mutually assured destruction
MID	militarized interstate dispute
MOX	mixed oxide
MW	megawatt
NAM	Nonaligned Movement

NATO	North Atlantic Treaty Organization
NCA	nuclear cooperation agreement
NNWS	nonnuclear weapons state
NNPA	Nuclear Non-Proliferation Act
NPT	Nuclear Nonproliferation Treaty
NSC	National Security Council
NSG	Nuclear Suppliers Group
NWS	nuclear weapons state
NWFZ	Nuclear Weapon Free Zone
PAEC	Pakistan Atomic Energy Commission
PHWR	pressurized heavy water reactor
PINSTECH	Pakistan Institute of Nuclear Science and Technology
PSI	Proliferation Security Initiative
PTA	preferential trade agreement
PWR	pressurized water reactor
RAPP	Rajasthan Atomic Power Plant
TNRC	Tajoura Nuclear Research Center
UAE	United Arab Emirates
UF_6	uranium hexafluoride

ATOMIC ASSISTANCE

Introduction

Unintended Consequences in International Politics

> The nations with the most developed peaceful [nuclear] programs will be nearest to a military bomb capability. It is therefore possible for a nation to proceed a considerable distance toward a bomb capability, to achieve an advanced state of nuclear "pregnancy."
>
> —Richard Rosecrance, Memo to Secretary of State Dean Rusk, May 28, 1968

In the late 1950s a South African nuclear scientist named J. Wynand de Villiers traveled to the United States to visit Argonne National Laboratory—a hub of America's atomic research at the time—which was located about 25 miles southwest of Chicago.[1] He had been invited by the U.S. government to receive training in the peacetime applications of nuclear energy. In the spirit of "atoms for peace," Washington hoped that de Villiers would use the knowledge he obtained to help South Africa experience the benefits of atomic power. Once he returned home, de Villiers did just that. In the 1960s, South Africa established the Pelindaba Nuclear Research Center, where it constructed reactors with assistance from the United States and began research on uranium enrichment so that it could indigenously produce fuel for these facilities. By the early 1970s, it had a well-developed civil nuclear infrastructure.[2] South Africa was benefitting from the peaceful uses of nuclear energy, just as the United States hoped that it would.

But atoms for peace became atoms for war. Realizing that the same technology and materials that South Africa acquired for peaceful purposes could also be used to build nuclear weapons, Prime Minister John Vorster authorized a nuclear bomb program in 1974.[3] J. Wynand de Villiers, the same scientist who had received training for "peaceful" purposes in the United

States fifteen years earlier, headed the nuclear explosives project.[4] Benefiting from the knowledge he had obtained abroad, de Villiers delivered the South African bomb. He helped the country build its first nuclear weapon in 1979—just five years after Vorster began the bomb program. South Africa developed six bombs before dismantling the weapons program in the early 1990s.

This short story illustrates why Richard Rosecrance expressed concerns about "nuclear pregnancy" more than forty years ago. Nuclear technology, materials, and know-how are dual use in nature, meaning they have both peaceful and military applications. As Swedish Nobel Prize–winning physicist Hannes Alfvén put it, the peaceful atom and the military atom are "Siamese twins."[5] For instance, a nuclear reactor intended for electricity production can be used to make plutonium for bombs if a country also has the means to separate it from spent nuclear fuel. And as the South African case underscores, knowledge in nuclear engineering and other relevant fields can be applied to peaceful ends, but it can also be used to overcome scientific challenges associated with the production of nuclear weapons.

Despite this dual-use dilemma, countries have regularly engaged in peaceful nuclear cooperation, which is defined as the state-authorized transfer of technology, materials, or know-how intended to help the recipient country develop, successfully operate, or expand a civil nuclear program.[6] In a historic address before the UN General Assembly in December 1953, President Dwight D. Eisenhower called on countries to "serve the peaceful pursuits of mankind" by providing "abundant electrical energy in the power-starved areas of the world."[7] This speech helped set the stage for the nuclear marketplace to take off over the next several decades. The United States, for instance, provided research reactors and enriched uranium to countries such as Iran, Pakistan, and Japan in the 1950s and 1960s. France, Brazil, and Italy supplied Iraq with nuclear facilities and materials in the 1970s. And the Soviet Union helped Argentina, Cuba, and Libya get their civilian nuclear programs off the ground in the 1980s.

The nuclear marketplace became especially active in the 2000s as countries looked for ways to combat high oil prices, enhance their energy security, and address the problem of global climate change. President Barack Obama underscored the utility of nuclear power in today's world when he said in February 2010, "To meet our growing energy needs and prevent the worst consequences of climate change, we'll need to increase our supply of nuclear power. It's that simple."[8] Outside the United States, dozens of countries that do not currently operate nuclear power plants are exploring nuclear energy development as part of a movement commonly known as the "nuclear renaissance." The nuclear energy aspirants are as diverse as Belarus, Bolivia, Mongolia, Myanmar, Nigeria, Sudan, and Venezuela.[9] In the Middle East, twelve countries are exploring nuclear energy and many of them have already received pledges of support from supplier countries.[10] A South Korean consortium, for example, won a landmark bid in December 2009 to build nuclear power plants in the United Arab Emirates (UAE) over the next decade.[11]

The future of nuclear energy remains uncertain in the aftermath of the March 11, 2011, earthquake and tsunami in Japan that led to the release of radioactive materials at the Fukushima Daiichi nuclear power plant. Just as the accidents at Three Mile Island in 1979 and Chernobyl in 1986 slowed global nuclear energy development, the Fukushima disaster will likely make some countries reluctant to rely on nuclear power for their electricity needs. The Japanese accident will probably have the largest effect on democratic governments, which are more susceptible to the negative public reactions that usually ensue from major nuclear disasters.[12] Yet nuclear power will most likely remain a viable option for many countries in the future, especially if concerns about the effects of climate change remain salient. Chile, for example, signed a deal with the United States in the immediate aftermath of Fukushima to receive assistance in developing civilian nuclear power.[13] Many states in the Persian Gulf also appear to be moving forward with plans for nuclear energy development in the wake of the Japanese accident.[14]

The book explores the rich history of peaceful nuclear cooperation. It addresses three main questions: Why do nuclear suppliers provide peaceful nuclear assistance to other countries? Does peaceful nuclear assistance raise the likelihood of nuclear weapons proliferation? Have international institutions influenced the nuclear marketplace and effectively separated the peaceful and military uses of the atom? Although these questions center around one particular topic, the book is more generally about the use of economic statecraft to achieve foreign policy objectives and the ways that tools of international influence can have unintended consequences.

It is fashionable to spotlight the Pakistani-based A. Q. Khan network— which shelled out nuclear technology to help Iran, Libya, and North Korea build nuclear weapons—and other cases of deliberate proliferation assistance.[15] The book departs from the prevailing trend. It is principally concerned with atoms for peace and how they unintentionally become atoms for war.

Why Peaceful Nuclear Assistance Matters

Why should anyone care about a book on peaceful nuclear assistance? The most obvious answer to this pointed question is that civilian nuclear aid matters in the real world. It matters, in part, because governments use it as a means to enhance their influence in international politics. In this respect, peaceful nuclear cooperation is not unlike other tools of economic statecraft such as preferential trade agreements (PTAs), economic sanctions, or foreign aid. History is rife with examples of countries using these and other economic policies to bolster their national security. During the cold war, for instance, the Coordinating Committee on Multilateral Export Controls (CoCom)—essentially the economic arm of the North Atlantic Treaty Organization (NATO)—restricted trade with the Soviet Union and its allies. Through CoCom, the United States and its Western partners pursued a strategy of economic containment against the Soviets aimed at weakening Moscow's military capabilities.[16] More recently, the United States has used PTAs to support the global war on terror and restricted trade to proliferators such as Iran and North Korea in an attempt to curtail their nuclear weapons programs.[17]

There is a bit of a divergence between the frequency with which policymakers employ economic tools of statecraft and the degree to which these instruments are studied by political scientists. In 1985, David Baldwin wrote in his seminal book, *Economic Statecraft*, "The study of economic instruments of foreign policy has been neglected relative to the study of other policy tools."[18] This statement still rings true today. The books about military instruments of foreign policy outnumber the books about economic statecraft. Yet scholars are increasingly aware that international commercial activities have implications for international security.[19] This book joins a small but growing number of others that emphasize the ways that economic statecraft can promote geopolitical interests.[20]

Although peaceful nuclear cooperation shares some similarities with other economic instruments, it deserves special treatment for at least two reasons. First, policymakers believe that civilian nuclear assistance can *transform* bilateral relationships. For example, the controversial civil nuclear deal between the United States and India—which emerged following a joint statement issued by President George W. Bush and Indian prime minister Manmohan Singh in July 2005—is the single most important initiative aimed at improving Indo-American relations in the twenty-first century. A strategic partnership with India would not be possible, U.S. officials believe, in the absence of civilian nuclear assistance.[21] The utility of peaceful nuclear assistance as a means to manage relationships stems partially from the prestige associated with nuclear technology. The notion that nuclear energy brings cachet is widespread in international politics. Pham Duy Hien, a scientific advisor to Vietnam's agency for nuclear safety, underscored this when he said, "Everybody knows the respect a country receives once it owns nuclear power."[22] The military potential of nuclear power also adds to its appeal—even when countries have no intention of building nuclear weapons. By offering something that is valued, supplier countries such as the United States are able to signal favorable intentions and, they believe, substantially increase the likelihood of future bilateral cooperation.[23] Although peaceful nuclear assistance is just one tool in the foreign policy toolkit, it is a powerful instrument that policymakers have at their disposal.

Second, the proliferation potential of nuclear technology makes atomic assistance a unique tool of economic statecraft. It is simultaneously helpful and potentially dangerous for international security. Scholars such as John Mueller have argued that sanctions and other economic tools can have unintended humanitarian and political side effects.[24] Yet no other instrument of economic statecraft risks the inadvertent spread of the most destructive weapon known to mankind. Accordingly, nuclear cooperation presents supplier governments with a Faustian bargain. They can promote their interests by exporting nuclear technology, materials, and knowledge—but only if they are willing to accept the risk that their cooperation might contribute to proliferation. Uncovering the conditions under which countries are willing to accept this trade-off can add to our understanding of risk taking in international politics.

The book shows that peaceful nuclear assistance ultimately raises the risk of nuclear proliferation. It therefore allows us to obtain a more complete understanding of how and why nuclear weapons spread. Debates persist about the strategic and political effects of nuclear weapons.[25] It is fair to say, however, that many policymakers believe that the spread of nuclear weapons constitutes a significant threat to international peace and security. These threats are especially acute in the aftermath of North Korea's nuclear tests in 2006 and 2009 and Iran's alleged pursuit of the bomb. Obama made this clear in a speech he delivered in Prague on April 5, 2009, when he said that measures to curb the future spread of nuclear weapons are "fundamental to the security of our nations and to the peace of the world."[26]

At the same time, we are in the midst of a possible renaissance in nuclear power, as I previously indicated. Now, perhaps more than ever, policymakers in the United States and elsewhere who are concerned about proliferation need to understand the connection between civilian and military nuclear programs.

Yet our present understanding of civilian nuclear assistance is limited. Early work called attention to the perverse connections between the peaceful and military uses of nuclear technology. Albert Wohlstetter and his colleagues famously warned in the 1970s that the distinction between "safe" and "dangerous" nuclear activities was becoming increasingly blurred.[27] Their groundbreaking work had a major effect on the way scholars thought about peaceful nuclear assistance and contributed to a change in U.S. nonproliferation policy. Despite the efforts of Wohlstetter and others,[28] critical gaps remain in our knowledge. This is, in part, because much of the recent research on nuclear proliferation has emphasized the demand side of the proliferation equation.[29] That is, it focuses largely on states' strategic, economic, or psychological incentives to develop nuclear weapons while downplaying the importance of capabilities and technology diffusion. This book is one of the few in recent memory to systematically explore the supply side of nuclear proliferation.[30] It is unique in that it emphasizes the proliferation potential of peaceful nuclear assistance—as opposed to indigenously acquired nuclear capabilities or deliberate proliferation assistance.[31] Although military assistance in developing nuclear weapons may be important, it is exceedingly rare, having occurred on only eight occasions from 1945 to 2008.[32]

Yet another reason to study civilian nuclear assistance is that it can inform debates about the efficacy of international institutions. One of the central questions in the field is whether treaties and institutions can constrain the behavior of powerful countries and, if so, under what conditions.[33] Analyzing peaceful nuclear cooperation provides a unique opportunity to address this question. I am able to evaluate critical components of the nuclear nonproliferation regime—perhaps the most important regime in the area of international security—that have rarely been subjected to rigorous empirical analysis.[34] In contrast to the current trend in scholarship, my final conclusion about international institutions is mostly negative.

Finally, this book provides an opportunity to build bridges between energy policy and security studies. Governments frequently highlight the connection between energy and national security. Ashton Carter, the undersecretary of defense for acquisition, technology, and logistics, said in May 2009, for example, "As you look out over the scenarios and the sources of conflict and the sources of threat to the United States, you see one after another that is driven by energy or in which energy is an important consideration."[35] Yet our understanding of *how* energy influences security is incomplete.[36] This book aims to change this in some small way by providing one concrete example of how decisions on energy production affect national and international security. The nexus between peaceful nuclear assistance and nuclear proliferation is not always what policymakers have in mind when they talk about the strategic effects of energy policy. Nevertheless, this book lends some credence to the linkage between energy and security.

The Conventional Wisdom

The questions addressed in the book have standard answers. The conventional explanation for why countries provide atomic assistance has to do with promoting nonproliferation. Simply put, suppliers share nuclear technology, materials, and know-how to limit the risk that the recipient country will build nuclear weapons. This notion dates back to Eisenhower's Atoms for Peace proposal. Eisenhower viewed his plan as an arms control measure.[37] He believed that sharing nuclear technology and know-how would reduce the likelihood of proliferation because foreign suppliers could obtain assurances from the recipient country that any assistance it provided would be used exclusively for peaceful purposes. The supplier could also potentially gain a degree of control over the recipient's activities by enhancing its dependence on external technology. On the other hand, if suppliers embargoed nuclear sales, they might encourage other countries to develop technologies indigenously and they would sacrifice leverage. Under such circumstances, countries would have fewer assurances that facilities would be used exclusively for civil purposes.[38]

The 1968 nuclear Nonproliferation Treaty (NPT) reinforced this conception of atomic assistance. This treaty divides countries into two groups: (1) nuclear weapons states (NWS); and (2) nonnuclear weapons states (NNWS). The NWS—China, France, Russia, the United Kingdom, and the United States—are permitted to possess the bomb under the NPT because they conducted nuclear tests prior to January 1, 1967. All other countries in the world pledge to forgo the development of nuclear weapons. To verify that NNWS keep this commitment, the International Atomic Energy Agency (IAEA) institutes a system of verification known as safeguards.[39] A critical "grand bargain" of the NPT is that NNWS are entitled to assistance in developing civilian nuclear programs.[40] This trade-off sent a clear message to states considering ratification. Countries that enter the treaty and play by the rules will be rewarded with the free-flow of nuclear technology for peaceful purposes. On the other hand, states that do not to commit to the NPT will be denied access to foreign nuclear assistance. Since nuclear suppliers have incentives to preserve this bargain, the logic of this argument predicts that they will offer aid to NPT members and restrict cooperation with nonmembers—especially those that pursue nuclear weapons.

According to this perspective, peaceful nuclear cooperation does not significantly raise the risk of nuclear weapons proliferation. If anything, suppliers' initial objectives are realized and nuclear aid lowers the risk that the recipient state will initiate a nuclear weapons program. Analysts recognize that atomic assistance led to nuclear proliferation in a few high-profile cases. Most notably, India tested a nuclear explosive device in 1974 using a research reactor supplied by Canada and nuclear materials exported by the United States exclusively for peaceful purposes. Yet this event represents a rare anomaly; virtually all nuclear cooperation is innocuous or even positive from a nonproliferation standpoint. Proponents of this view often point out that countries such as Germany and Japan accumulated large amounts of atomic assistance in the postwar period but did not pursue nuclear weapons.[41]

Improvements made to the nonproliferation regime over time have reinforced this perception of peaceful nuclear cooperation. India's 1974 nuclear test led to the establishment of the Nuclear Suppliers Group (NSG), a loose affiliation of suppliers designed to harmonize national export control policies and make it more difficult for countries with weak nonproliferation credentials to secure atomic aid. The NSG, according to the conventional wisdom, reined in suppliers and discouraged cavalier exports that maximized proliferation risks. Moreover, the IAEA safeguards regime has been strengthened since the early days of the nuclear age and this has made it more difficult for recipient states to draw on civilian technology and know-how for military purposes. Proliferation-prone atomic assistance, to the extent that it ever existed, is an artifact of an era when there were few institutional checks on the behavior of suppliers and recipients in the nuclear marketplace. Regime proponents argue that today the NSG, NPT, and IAEA provide policymakers with confidence that the looming renaissance in nuclear power can unfold without contributing to the spread of nuclear weapons.

The implication of this argument is that atomic energy assistance is not risky as long as suppliers play by the rules. The only cause for concern is aid that is explicitly intended to promote the spread of nuclear weapons, which is prohibited by the NPT.

The arguments outlined above have shaped policy for nearly sixty years. But they are incomplete—and potentially dangerous. This book will show that concerns about nonproliferation generally do not drive suppliers to provide nuclear assistance. NPT members are no more likely than nonmembers to receive peaceful nuclear assistance, even accounting for states' interest in nuclear energy and whether the supplier is part of the treaty. Countries that commit to the NPT are not even more likely than nonmembers to receive training and technical assistance or to import "minor" technologies related to nuclear research. Suppliers do not systematically deny atomic assistance to NNWS that pursue nuclear weapons and in some cases proliferators are actually more likely to receive aid. And the NSG has had a fairly limited effect on the way that suppliers conduct their business.

Countries that receive peaceful nuclear assistance are more likely to initiate nuclear weapons programs and acquire the bomb than those that do not receive aid—especially if they also face external security threats. The type of assistance makes a difference, but not in the way that the conventional wisdom suggests. Aid intended to help the recipient develop a nuclear power program raises the likelihood of nuclear weapons pursuit and bomb production. The link between nuclear energy assistance and proliferation is robust even when suppliers explicitly withhold enrichment and reprocessing technologies. More limited types of assistance are not as proliferation-prone, however. Joint research and technical exchanges raise the risk of weapons pursuit but not bomb acquisition. Similarly, transfers relevant to research programs play little role in enabling bomb production, although they contribute to the pursuit of nuclear weapons if the recipient experiences subsequent interstate conflict.

Military assistance facilitates the acquisition of nuclear weapons, but it is unrelated to nuclear weapons pursuit since it would be highly unusual for a country to provide proliferation aid to a state that was not already trying to build the bomb. Yet peaceful nuclear cooperation also enables bomb acquisition, even when accounting for transfers explicitly intended to help the recipient state proliferate. There is some evidence that military assistance is insufficient to enable weapons production if the recipient state does not receive adequate levels of civilian nuclear aid. Libya, for instance, failed to become a nuclear power despite having a weapons program that spanned three decades and getting military assistance explicitly designed to permit bomb production. A modest amount of civilian nuclear assistance came to Libya from the Soviet Union but the weapons program ultimately failed, in part, because Tripoli lacked the indigenous knowledge base that often results from more substantial peaceful nuclear cooperation.[42]

International institutions have been less effective than many think in minimizing the dangers of peaceful nuclear cooperation. The connection between atomic assistance and nuclear weapons program onset remains statistically and substantively significant throughout the nuclear age. Moreover, there is a limit to the constraining power of the NPT. Some tests indicate that states ratifying the treaty are less likely to initiate a nuclear weapons program if they have not received atomic assistance. But this relationship washes away when countries accumulate aid. A close examination of historical cases shows that the NPT did not play a critical role in nuclear decision making, even when we would have expected the treaty to make a big difference. For example, Japan would probably be nonnuclear today if the NPT had never been created as long as all other factors remained constant. Syria's nuclear activities over the last decade underscore that states can circumvent the rules and regulations that currently govern the safeguards regime. Indeed, IAEA safeguards have not deterred many proliferators from drawing on civilian nuclear programs to augment their military capabilities. The agency also has a poor track record of identifying violations when they occur to provide timely warning to the international community. The bottom line is that the perils of civilian nuclear cooperation persist today despite the strengthening of the nonproliferation regime over time.

The Arguments in Brief

Why, then, do countries provide peaceful nuclear assistance? This is puzzling given that atomic energy assistance raises the danger that nuclear weapons will spread to additional countries.

Nuclear suppliers are not naïve. Nor are they conniving. Governments understand that atomic assistance could inadvertently augment a nuclear weapons program and they choose to play a dangerous game anyway.

Countries use peaceful nuclear assistance to manage relationships with allies and adversaries in ways that promote their political and strategic interests.[43] In an anarchic international system where the prospect of nuclear war looms large, countries use peaceful nuclear aid discriminately. They provide it for very specific reasons: to strengthen their allies and alliances; to develop closer relationships with enemies of enemies; and—if the supplier is a democracy—to strengthen democracies and relationships with democracies. States also offer atomic aid to enhance their energy security by bartering nuclear technology for oil with petrol-producing countries but only when oil prices are extremely high. Oil-for-nuclear technology swaps were common in the immediate aftermath of the oil crisis during the 1970s and have reemerged more recently as oil prices rose above $100 a barrel for the first time in 2008.

Preventing proliferation is a major goal for nuclear suppliers such as the United States, Russia, and Canada. But, if they can achieve one or more of these benefits, the payoff from aiding a civil nuclear program is such that states will take a calculated risk that their exports will not contribute to the spread of nuclear weapons.

This strategy backfires by unintentionally raising the risk of nuclear proliferation. Why? The recipient does not necessarily trick the supplier into providing assistance with talk of peace. This can happen but it is not the norm. When assistance is given, the recipient typically desires a civil nuclear program and does not have ambitions to build nuclear weapons. But the recipient's calculus can change as it develops the potential foundation on which to construct a military program. The infrastructure and knowledge developed for peaceful purposes provides states with a greater degree of confidence that the bomb could be built in a relatively short amount of time. The technical advances arising from nuclear cooperation also empower scientists and other members of atomic energy commissions, some of whom have an interest in nuclear explosives. These outcomes make countries more likely to initiate bomb programs, compared to the probability of weapons pursuit if they did not benefit from assistance. The South African case discussed above shows that atomic assistance sometimes increases the risk of proliferation even if a state lacks compelling strategic incentives to proliferate.

Changes in a state's security environment over time significantly influence whether nuclear assistance triggers a decision to build nuclear weapons. Regimes come and go. New enemies emerge. When a crisis arises, countries that have accumulated peaceful nuclear assistance are more likely to try and build the bomb, relative to states that do not receive aid. Iran, for example, received atomic aid from the United States beginning in the late 1950s but it did not initiate a weapons program until the middle of its protracted war with Iraq in the 1980s. Pakistan began a civilian nuclear program innocuously in the 1950s but decided to draw on civilian assistance to build nuclear weapons in 1972 after suffering a humiliating defeat in a war with India. The combination of strategic threats and peaceful nuclear assistance creates conditions in countries that maximize the likelihood of nuclear weapons program initiation.

Once states initiate weapons programs, they typically seek atomic assistance to exploit the dual-use dilemma. Occasionally they get it. Exporting countries do not look for states with nuclear weapons programs and systematically line up to assist them. Should they find themselves in a situation, however, where they want to foster closer ties with a state for strategic reasons and that country happens to be pursuing the bomb, concerns about proliferation may not be sufficient to deter nuclear cooperation.[44] Although suppliers do not shy away from assisting states with active nuclear weapons programs under certain conditions, they generally withhold assistance to proliferators when they cannot extract strategic benefits from cooperating. The net effect is that states pursuing the bomb are often no more or less likely to benefit from peaceful nuclear cooperation.[45]

Atomic assistance ultimately enhances the probability that states will acquire nuclear weapons, compared to those that do not benefit from aid. Assistance provided prior to the onset of the bomb program ends up being the most helpful in facilitating proliferation. India, for instance, drew largely on dual-use technology and know-how provided before it pursued the bomb to conduct its first nuclear test. Pakistan and North Korea were able to construct bomb-related facilities in part because they had first received peaceful foreign assistance that established an indigenous technical base.

Readers should bear in mind two important caveats. First, the arguments and evidence presented in this book are based on probabilistic assessments. My argument for peaceful nuclear assistance does not explain every case where countries transferred nuclear technology, materials, or know-how. One can find cases that are not explained by my argument—and I examine several of them in the book. Yet, in general, a desire to manage politico-strategic relationships is a powerful motive for providing aid. Similarly, every case of civilian nuclear assistance does not lead to proliferation. The key point is that, on average, states that receive peaceful nuclear assistance are more likely to pursue and acquire nuclear weapons than those that do not get aid. It is also important to remember that the type of peaceful assistance makes a difference, as I previously discussed.

Second, this book's arguments are not the only ones that matter for understanding the causes and consequences of peaceful nuclear cooperation. Economic motives such as generating hard currency sometimes play a role when it comes to explaining the behavior of suppliers in the nuclear marketplace. That said, financial considerations usually matter less than a desire to manage politico-strategic relationships. In terms of the argument for nuclear proliferation advanced in this book, peaceful nuclear assistance is just one factor among a handful of others that shapes the probability that the bomb will spread to other countries. Proliferation is a multicausal phenomenon and no one argument can completely explain it.

With these caveats in mind, the book's conclusions come at a critical juncture. A nuclear energy renaissance may be on the horizon while the threats posed by nuclear proliferation continue to loom. And Western elites do not appear to fully appreciate that transactions conducted according to the rules of the nuclear marketplace can be problematic from a proliferation standpoint. The book suggests that policymakers may want to rethink some of their policies on nonproliferation and peaceful nuclear cooperation. If history is any indication, continuing to promote civilian nuclear programs abroad could be a recipe for the further spread of nuclear weapons.

Chapter 1

Definitions and Patterns of Peaceful Nuclear Cooperation

Since the initial drive by the United States to share technology and knowledge for peaceful purposes in the 1950s, civilian nuclear cooperation has occurred regularly. Nevertheless, it remains poorly understood and has rarely received scholarly attention. Some important questions must be addressed before analyzing the causes and strategic effects of nuclear cooperation. What is peaceful nuclear assistance? What are the different types of aid that nuclear suppliers can provide? How can we measure atomic assistance? What are the historical trends in civilian nuclear cooperation? How frequently have suppliers provided atomic aid, and with whom have they shared nuclear technology, materials, or knowledge?

Peaceful nuclear cooperation is the state-authorized transfer of nuclear facilities, technology, materials, or know-how from one country to another for civilian purposes. This definition captures transfers that enable the recipient country to develop, successfully operate, and expand a civil nuclear program. However, it excludes nuclear transactions that are not approved by the supplier country. Nonstate-sanctioned transfers could occur, for example, if nuclear technology or materials were stolen and sold to foreign clients.[1] Such cases are driven by the parochial interests of nonstate actors in the supplier country and it is very difficult if not impossible to systematically measure them. Moreover, because governments keep a tight leash on firms that export nuclear technologies, unauthorized exports of this nature occur relatively infrequently.

Civilian nuclear assistance may be intended to help the recipient state conduct research on the peaceful uses of nuclear energy or to produce electricity at nuclear power stations. There are six operational categories of peaceful nuclear cooperation: (1) safety; (2) intangible transfers; (3) nuclear materials; (4) research reactors; (5) power reactors; and (6) fuel cycle facilities.

Defining Peaceful Nuclear Cooperation

Nuclear Safety

For a civil nuclear program to function effectively, actions must be taken to prevent accidents involving nuclear or radiological materials and to minimize the consequences of mishaps in the event that they occur. These actions collectively encompass nuclear safety. Effectively safeguarding a nuclear program is crucial because nuclear reactors are susceptible to accidents, as Charles Perrow documents in his book *Normal Accidents*.[2] Just one accident, even a minor one, can affect a nuclear program. The March 1979 accident at the Three Mile Island nuclear power plant in Pennsylvania generated fear and anxiety among many Americans, restricting the growth of the U.S. nuclear program even though it did not result in the death of a single person.[3] The risk of nuclear accidents is especially salient today in light of the March 2011 meltdown at Japan's Fukushima Daiichi nuclear power plant.

Given the importance of nuclear safety, countries commonly cooperate in this area. States sometimes assist each other in safety inspections of reactors. In 1993 Canada aided Pakistan in a month-long safety inspection of the Karachi Nuclear Power Plant, which Ottawa built in the early 1970s.[4] A more frequent type of cooperation in this vein involves the sharing of research on nuclear safety. Over the last several decades, for instance, China has received access to safety-related research from France, Germany, Italy, Japan, and the United States.[5] In April 2008, the United States agreed to share research on safety with Israel, a move that attracted media attention because of Israel's refusal to sign the nuclear Nonproliferation Treaty.[6]

Intangibles

Training scientists or operators and conducting joint research and development on nuclear-related fields constitute intangible cooperation. A cadre of educated scientists and operators is vital to the functioning of a peaceful nuclear program. Countries with experience in nuclear matters often invite scientists and technicians from other states to visit their laboratories and receive training in nuclear physics and reactor operation. For instance, the United States, United Kingdom, and France all offered training to Iranian nuclear scientists prior to the 1979 Islamic Revolution.[7] Countries with well-developed nuclear energy programs sometimes cooperate with one another in conducting joint research and development of a new technology. Historically, many of the European countries—including France, Italy, Belgium, the United Kingdom, and Germany—cooperated to develop "breeder" reactors, which produce more plutonium than they consume. Japan and the United States have also conducted a substantial amount of joint research in areas such as liquid metal and breeder reactor development and fusion fuel processing.[8]

Nuclear Materials

Several materials play a prominent role in the nuclear marketplace. Among them are natural uranium, enriched uranium, and plutonium, which can be used to fuel reactors. Natural uranium, which contains less than 1 percent of the isotope U-235, is mined from the earth's crust, milled and processed into a chemical substance called yellowcake, and finally converted to a form usable in a reactor such as uranium metal or uranium dioxide.[9] Uranium in this form can be used to fuel some reactors. Enriched uranium contains a greater percentage of the isotope U-235. Many reactors operating in the world today require low enriched uranium (LEU). For use in a nuclear reactor, the uranium needs to be enriched to around 2–3 percent U-235.[10] Enriching uranium is a technical process that requires highly specialized facilities. A handful of developed states offer enrichment services whereby they import natural uranium (or use their indigenous supply), convert it to LEU, and export the enriched uranium for use in a reactor. The countries that have historically offered enrichment services include the United States, Britain, France, Germany, the Netherlands, and Russia.[11]

Some reactors use highly enriched uranium (HEU), which contains at least 20 percent of the isotope U-235. Note that HEU and "weapons-grade" uranium are not necessarily the same thing. The latter typically refers to uranium that is enriched to at least 90 percent U-235. Historically, the United States and the Soviet Union/Russia were the principal suppliers of HEU for civil purposes. Between 1950 and 2002, the United States exported more than twenty-five tons of HEU.[12] The United States exported HEU to Iran in the 1960s for use in a research reactor located in Tehran, for example.[13] The Soviet Union/Russia exported considerably less HEU—between 2.5 and 3.5 tons—during the same period. In more recent years, China has emerged as a significant supplier and has exported HEU fuel for research reactors in Nigeria, Ghana, Iran, Pakistan, and Syria.[14] As of 2003, more than thirty countries had received HEU as a result of civilian nuclear cooperation.[15]

Plutonium can also be used to fuel reactors but it does not naturally exist. Once in the reactor, uranium fuel creates a controlled nuclear chain reaction that releases neutrons. Spent fuel rods that are burned in a reactor contain new isotopes, including plutonium. To separate plutonium from other isotopes in the spent fuel, a procedure known as chemical "reprocessing" is necessary. Once separated, plutonium can be merged with uranium to form mixed oxide (MOX) reactor fuel. This process is commonly referred to as "recycling" because waste is converted to fuel, reducing the amount of spent material that needs to be stored. In the 1970s, the expansion of nuclear power coupled with the perceived shortages of uranium led to increased interest in the use of plutonium as a reactor fuel.[16] Ultimately, however, plutonium never became a prominent fuel source because recycling was economically inefficient and some major suppliers worried about the proliferation risks of its use in civil applications.[17]

Other materials relevant to a civil nuclear program include heavy water (water highly enriched with the hydrogen isotope deuterium), graphite, and thorium. Heavy water and graphite can be used as moderators in certain types of reactors to slow down the neutrons that are released when the nucleus of an atom is split. Reactors moderated by heavy water or graphite have the advantage of running on natural uranium, but the appropriate moderator must be present for the reactor to function properly. Of the two materials, heavy water is more widely employed; Canada, India, South Korea, China, Romania, Pakistan, and Argentina all possess heavy water moderated reactors.[18] Most of these countries are at least partially dependent on a heavy water supply from foreign sources.[19] Thorium is a naturally occurring material that can be used as reactor fuel because it is capable of breeding a fissile uranium isotope, U-233.

Research Reactors

Research reactors are used for training purposes or to produce isotopes that have medical applications and are often exported to countries that are just beginning nuclear programs.[20] They are smaller than power reactors and typically have a capacity of less than 100 MWt, meaning they have less than one-hundred-million watts of electric capacity.[21] As of May 2007 there were 283 research reactors operating in the world. As one might expect, some of these reactors were produced indigenously by states that have had highly developed nuclear infrastructures. Russia, for example, has sixty-two research reactors and the United States has fifty-four.[22] However, many of these reactors are located in developing countries, including Algeria, Iran, Jamaica, and Libya, and were supplied by foreign sources.

Power Reactors

Power reactors, which are used to produce electricity, comprise a significant portion of all civilian nuclear cooperation. They extract usable energy, typically by splitting the nucleus of an atom to produce a series of controlled nuclear reactions. There are a few different types of reactors used for power production. Two common types are the American-designed pressurized water reactor (PWR) and the Canadian-designed Canada Deuterium Uranium (CANDU) pressurized heavy water reactor (PHWR). According to the International Atomic Energy Agency, there were 439 nuclear power plants in operation at the end of 2007 in thirty countries.[23] In many countries, including France, Lithuania, and Ukraine, nuclear power provides a significant share of electricity production. Unless they are built in one of the major reactor suppliers such as the United States, Russia, France, or Canada, power reactors are typically constructed with some foreign assistance.

In addition to supplying complete nuclear research reactors, nuclear exporters sometimes provide various reactor subcomponents (e.g., reactor pressure vessels) to other countries.[24] These transfers also constitute civilian nuclear cooperation.

Fuel Cycle Facilities

As described above, countries often import fuel in a form that is ready to be loaded into a reactor. States seeking to develop more advanced civil nuclear programs often demand the capability to produce reactor fuel indigenously or with minimal dependence on foreign suppliers. To achieve this capacity, facilities related to the nuclear fuel cycle—the processes leading to the production of electricity from uranium in power reactors—must be constructed. Most of these facilities are technologically sophisticated and are usually built via civilian nuclear cooperation. In addition to reactors, components of the fuel cycle include (1) uranium mines; (2) facilities to convert solid uranium yellowcake to the gas uranium hexafluoride; (3) enrichment facilities that can increase the concentration of the isotope uranium-235;[25] (4) fuel fabrication facilities that transform enriched uranium into fuel rods; (5) facilities capable of storing or disposing spent fuel; and (6) reprocessing facilities capable of chemically separating plutonium from spent nuclear fuel. Additionally, heavy water production facilities are sometimes considered to be part of the fuel cycle given the significance of heavy water for certain reactors that run on natural uranium.

Transfers of fuel cycle facilities—or subcomponents of these facilities—occur, albeit less frequently than reactor exports. For example, in the 1970s Switzerland supplied a custom-made uranium conversion facility to Pakistan, while China assisted Iran in developing a similar facility in the 1990s.[26] The European countries involved in URENCO jointly developed uranium enrichment facilities and subsequently assisted other countries in building these complexes. URENCO, for instance, recently built an enrichment facility in the United States.[27]

What Peaceful Nuclear Assistance Is Not

My definition excludes assistance that is provided for military purposes. Military assistance includes transfers of complete nuclear warheads or bomb designs. It also includes dual-use technology or materials that are exported explicitly to help the recipient country build nuclear weapons. My conception of military aid is similar—but not identical—to what Matthew Kroenig has called "sensitive nuclear assistance."[28] Kroenig defines sensitive nuclear assistance as "the state-sponsored transfer of the key materials and technologies necessary for the construction of a nuclear weapons arsenal to a nonnuclear weapon state."[29] This aid takes three forms: (1) assistance in the design and construction of nuclear weapons; (2) the transfer of significant quantities of weapons-grade fissile material; or (3) aid in the construction of uranium enrichment or plutonium reprocessing facilities. The first form of this aid always constitutes military assistance. The latter two forms fall under my definition of military aid only if there is evidence that the supplier wanted to help the recipient build nuclear weapons. Kroenig identifies fourteen cases of sensitive nuclear assistance from 1951 to 2000. Seven of these cases meet my definition of military assistance and they are described below. I classify the others—including France's export of a reprocessing plant to Japan in the 1970s and Italy's sale of a radiochemistry lab to Iraq—as peaceful nuclear assistance.

Military assistance occurs infrequently, especially relative to peaceful aid. Between 1942 and 2008, there were eight cases where suppliers offered military assistance to a nonnuclear weapons state. In the 1950s France transferred a reactor and a reprocessing facility to Israel and the Soviet Union provided nuclear technology to China to support Beijing's bomb program.[30] After benefiting from military assistance, China provided similar aid to Algeria and Pakistan in the 1980s.[31] Pakistan provided military assistance to Iran beginning in the late 1980s and North Korea and Libya in the 1990s as part of an elaborate network that had at least tacit approval from the government.[32] Most recently, North Korea transferred a plutonium producing reactor to Syria to facilitate Damascus's ability to produce nuclear weapons.[33] This reactor was destroyed by an Israeli airstrike in September 2007, before construction had been completed.[34]

There are at least two additional cases where suppliers offered military aid to states that already possessed nuclear weapons.[35] In the 1950s the United States provided assistance to improve the safety and efficiency of Britain's nuclear arsenal.[36] Washington provided similar aid to bolster the French *force de frappe* (i.e., the French nuclear deterrent) beginning in the 1960s.[37]

While military assistance may be important it falls largely outside the scope of this book. My interest here is in explaining patterns of civilian nuclear assistance and in showing that such aid can have *unintended* consequences for international security. Note, however, that throughout this book I will analyze the similarities and differences between peaceful assistance and military aid. I will also distinguish enrichment and reprocessing assistance that is provided for peaceful purposes from other types of civilian aid since the former is thought to be the most closely associated with nuclear weapons development. As I will show in Part II of the book, civilian nuclear assistance increases the risk of nuclear weapons proliferation even when it does not involve the transfer of enrichment and reprocessing facilities.

Measuring Civilian Nuclear Cooperation

This book uses bilateral nuclear cooperation agreements (NCAs) to measure peaceful nuclear cooperation. These treaties authorize the exchange of nuclear technology, materials, or know-how and they are typically signed at high levels of government. They are the means by which countries regulate the nuclear marketplace; supplier states generally do not permit peaceful nuclear cooperation unless they have signed a treaty with the recipient country. Before exporting the Tehran Research Reactor, for instance, the United States inked an NCA with Iran, as I will discuss in chapter 4.

A typical NCA includes several general provisions: (1) authorizations of transfers related to nuclear facilities, technology, materials, or knowledge; (2) guarantees that technology transferred will not be used for any nuclear explosive device or for any research and development of explosive devices; (3) assurances that safeguards will be applied to all technology, materials, and know-how that is transferred; (4) prohibitions against transferring facilities or materials to unauthorized third parties; and (5) guarantees that adequate physical security is maintained for all nuclear materials and facilities transferred as part of the agreement. The nature of these provisions varies based on respective suppliers and recipients, but these terms appear in many treaties.

One potential drawback of using NCAs for my purposes is that the presence of an agreement does not always imply that a substantial degree of nuclear commerce is taking place.[38] Critics have claimed that this makes NCAs a "highly inaccurate measure of nuclear cooperation."[39] Yet this critique is overstated. Although there is not a perfect correlation between the number of NCAs a country signs and the amount of aid it receives, NCAs are a decent proxy for nuclear assistance.[40] The evidence presented in Part II of this book suggests that the vast majority of NCAs lead to actual assistance. Countries that participate in NCAs are more likely to want nuclear weapons and successfully build atomic bombs compared to countries that do not participate in NCAs. One could craft an alternative explanation for these observed correlations but I will show in chapters 7 and 8 that assistance stemming from treaties plays a key role in understanding nuclear proliferation dynamics. Moreover, agreements capture governments' incentives to provide aid even when they do not bear fruit, which is significant given that much of this book explains the behavior of nuclear suppliers.

Some have suggested that it would be preferable to measure nuclear assistance based on transfers of technology.[41] I opted against this approach partly because I was not confident that it was possible to obtain complete information about all civilian nuclear transfers that had occurred. Nuclear exports are often reported in publications such as *Nucleonics Week* and *Nuclear Fuel* but the degree to which the coverage is comprehensive was unclear—especially when it came to transfers other than nuclear power plants, plutonium reprocessing facilities, or uranium enrichment centers. Without question, it would have been a struggle to find information about transfers of intangibles in these trade journals. This was particularly problematic given that I was interested in the connection between a country's scientific knowledge base and its ability to produce nuclear weapons.

It would have been possible to devise a rough indicator of assistance based on nuclear technology transfers even if I could not obtain comprehensive data. Kroenig, for example, measures "nonsensitive nuclear assistance" simply by identifying whether a state has ever received foreign assistance in constructing a research reactor or a power reactor.[42] This could be a reasonable approach depending on one's purpose. In my case, it is inadequate because it misses all nonreactor assistance (which constitutes a major portion of all nuclear cooperation) and because it does not capture important variation in levels of civilian nuclear assistance that countries receive. A dichotomous measure of atomic assistance implies that countries such as India and Jamaica received the same amount of aid. In reality these two states received widely divergent levels of assistance.[43] NCAs allow me to capture the differences in the amount of atomic aid that states receive over time. This is important because I will argue in Part II that the amount of assistance matters when it comes to understanding the connection between nuclear cooperation and proliferation.

There are other important advantages associated with NCAs. As I hinted above, these treaties can be disaggregated in ways that add to our understanding of the causes and consequences of peaceful nuclear assistance. There are five types of NCAs. Safety agreements, which cover any aspect of nuclear safety discussed above and do not authorize other types of cooperation, represent the first type of NCA. The second type includes intangible agreements that are strictly limited to technical exchanges, training, and/or research and development. Nuclear material NCAs are a third type of agreement. These deals authorize the exchange of materials such as uranium or heavy water but they do not lead to other types of cooperation such as the construction of reactors. Whereas these first three types of agreements are generally limited to a single area, the latter two types cover multiple areas of atomic assistance.[44] Comprehensive research agreements are intended principally to transfer research reactors, but they also authorize the sale of nuclear fuel and promote training and technical exchanges. The final type of treaty, comprehensive power agreements, help states develop or expand nuclear power programs to produce electricity. These NCAs authorize the transfer of nuclear power reactors. They also encourage the transfer of fuel cycle facilities and nuclear materials, in addition to authorizing cooperation in intangible spheres. Comprehensive power agreements may also approve cooperation in nuclear safety or the transfer of research reactors. This typology of NCAs does not perfectly reflect areas of nuclear assistance since some agreements are multifaceted. Although this means that we need to be cautious when equating the NCA type with a particular area of assistance, we can make inferences about the kind of aid that matters based on my typology.

NCAs also provide a unique opportunity to analyze the role of treaty strength in international politics. Virginia Page Fortna and others have argued that stronger treaties are more effective than weaker ones when it comes to tackling problems such as the durability of peace following wars.[45] Are stronger NCAs likewise associated with more favorable outcomes? I can evaluate this question by coding the nonproliferation restrictions that suppliers place on recipient states. Some NCAs are stringent with respect to nonproliferation while others are relatively lax. I evaluate whether this matters in Part II of the book. This type of analysis would be much more difficult if I measured nuclear cooperation solely on the basis of technology exports.

Dataset on Nuclear Cooperation Agreements

For the book, I produced a dataset that includes NCAs signed between 1945 and 2000 based on a list of bilateral agreements compiled by James Keeley.[46] The unit of analysis is the directed dyad, meaning that agreements are included more than once if multiple countries agreed to supply nuclear technology, materials, or know-how. The dataset identifies the month and year that each agreement was signed and classifies each NCA as one of the five types identified above. It also codes each state party to an NCA as a supplier or recipient. Readers interested in more details about this dataset are encouraged to consult appendix 1.1, which discusses how I identified the universe of NCAs, explains how I coded the agreement type, and outlines how I identified supplier and recipient states for each treaty.[47]

Trends in Civilian Nuclear Cooperation

When we examine patterns in civilian nuclear cooperation, the first thing that stands out is the sheer frequency of peaceful nuclear assistance. Nearly fifteen hundred NCAs were concluded during the period I studied.[48] These NCAs represent around twenty-five hundred yearly commitments to supply nuclear technology.[49] The number of commitments is greater than the number of agreements because, as I noted above, some agreements call for

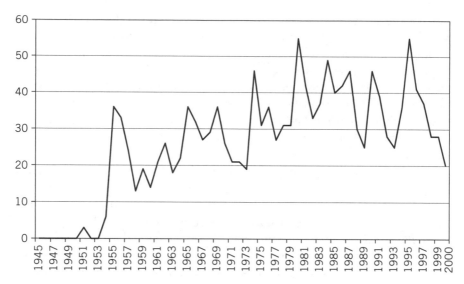

Figure 1.1 Number of NCAs signed per year, 1945–2000

multiple countries to supply nuclear technology or knowledge. This is significant because it indicates that peaceful nuclear cooperation occurs more frequently than is commonly assumed. Figure 1.1 plots the number of NCAs that have been signed over time. As the figure illustrates, the number of NCAs signed generally increased over time because more capable nuclear suppliers emerged and a greater number of countries developed civil nuclear programs. In the early days of civil cooperation, there were years when no agreements were concluded. The high points of cooperation occurred in 1980 and 1995 when fifty-five NCAs were signed. Since 1945, on average, twenty-six agreements have been signed each year. The average number of deals concluded since the nuclear marketplace began to flourish in the early 1970s is thirty-six.

Figure 1.2 disaggregates NCAs to explore whether some of the agreement types are more common than others. Interestingly, the figure illustrates that safety and intangible NCAs make up a large proportion of all agreements. Each of these two agreement types comprise around 30 percent of all NCAs signed between 1945 and 2000. We also see that comprehensive power NCAs are rather common, as they constitute nearly 25 percent of all treaties. Nuclear material and comprehensive research NCAs are the most limited types of agreements. Note, however, that these agreements are by no means rare. An average of two nuclear material NCAs have been signed each year since the beginning of the atomic age. During that same period, on average, nearly three comprehensive research deals were signed each year.

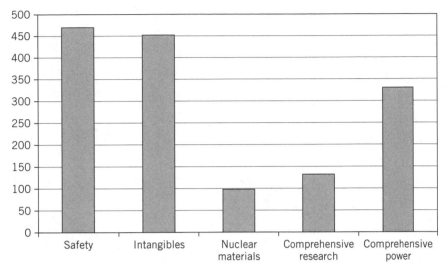

Figure 1.2 Number of NCAs signed by type, 1945–2000

Some may find two particular aspects of this breakdown surprising. The first is that safety NCAs are more common than treaties authorizing the exchange of nuclear technology (e.g., research or power reactors). The second is that there are 60 percent fewer comprehensive research NCAs than there are comprehensive power NCAs. The reason this may be unexpected is that the number of countries in the world today with operational research reactors is nearly twice as large as the number of states with power reactors producing electricity.

To shed further light on these trends, figure 1.3 plots the number of disaggregated NCAs signed between 1945 and 2000. The figure excludes nuclear material and intangible NCAs for ease of presentation, but note that the number of these deals signed over time is relatively steady. As the figure reveals, comprehensive research agreements were common in the 1950s, in the immediate aftermath of President Eisenhower's Atoms for Peace address. In 1955 states signed more than six times as many research NCAs as power NCAs. During this time, many countries were just beginning civilian nuclear programs and they wanted to build research reactors before moving on to the construction of more sophisticated power reactors. As more countries set their sights on nuclear power programs, the number of research NCAs declined, especially relative to the number of power NCAs. Countries continued to sign research NCAs, but typically only when a country that lacked preexisting infrastructure decided to begin a peaceful nuclear program.

Figure 1.3 also provides an interesting illustration of how cooperation in nuclear safety has evolved over time. Prior to the early 1970s, states rarely

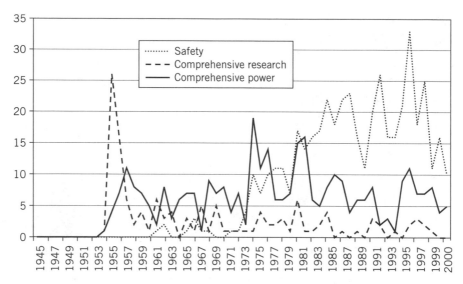

Figure 1.3 Number of NCAs signed per year, by type, 1945–2000

signed safety NCAs but by 1980 these deals represented the most common form of nuclear assistance. The spike in safety NCAs roughly corresponds to the nuclear accidents at Three Mile Island in the United States (1979) and Chernobyl in the Soviet Union (1986). This suggests that concerns about safety became especially salient after the occurrence of major nuclear accidents. This does not, however, tell the whole story. Note that the number of safety NCAs began to rise in 1973—six years before the Three Mile Island incident. Thus, states did not completely ignore nuclear safety prior to the first major accident at a nuclear power plant, but mishaps clearly accelerated the pace of international cooperation in this area. States continued to sign safety NCAs at a high rate more than a decade after Chernobyl, in part because they fear that an accident anywhere in the world would likely cripple their domestic nuclear program. It remains to be seen whether this perception is correct in the wake of the 2011 Japanese nuclear accident.

At this point, readers may wonder *where* peaceful nuclear cooperation occurs. Figure 1.4 illustrates the geographic dispersion of NCAs signed between 1945 and 2000.[50] In particular, it shows the percentage of all NCAs signed by states in the five major regions of the world: the Americas, Europe, Africa, the Middle East, and Asia.[51] The figure reveals that NCAs are not equally distributed around the world. European states signed 58 percent of all nuclear agreements during this period. This is not because a few states with large programs, such as France and Germany, dominate the nuclear marketplace; thirty-six European states (about 78 percent of the states in

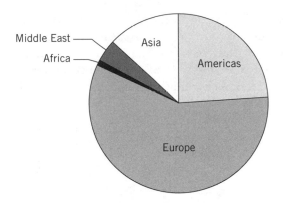

Figure 1.4 Regional dispersion of NCAs, 1945–2000

the region) signed NCAs. In terms of the other regions, states in the Americas signed about one-quarter of all NCAs while Asian countries concluded about 13 percent of all deals. The smallest proportion of peaceful nuclear cooperation occurred in the Middle East and Africa, where states signed 4 percent and 1 percent, respectively, of all NCAs.

Figure 1.5 illustrates the geographic dispersion of disaggregated NCAs signed between 1945 and 2000. Examining the dispersion of NCAs by type reveals a few interesting differences. European states signed large proportions of safety and intangible NCAs but the distribution across regions is more equal for material, research, and power NCAs. Also note that Middle Eastern countries signed a greater proportion of comprehensive research NCAs (nearly 20 percent) than any other type. Additionally, African states signed a higher proportion of nuclear material NCAs (about 4 percent) than other agreements, where they generally comprise 1 percent of the share.

Appendix 1.1: Dataset on Nuclear Cooperation Agreements

To analyze peaceful nuclear cooperation, I produced a dataset that identifies NCAs signed between 1945 and 2000. This appendix discusses the procedures employed to produce this dataset.

Identifying the Universe of NCAs

I began by consulting a list of bilateral NCAs signed between 1945 and 2003 compiled by James Keeley.[52] This list includes all agreements dealing with one of the areas of civilian nuclear cooperation described above (e.g., safety, intangibles, nuclear materials, research reactors, power reactors, and fuel cycle facilities).[53] Keeley's list contains the following information for each NCA: the title of the agreement, the state parties, the date the agreement was signed, and the source(s) used to justify the inclusion of the treaty.

Nearly eighteen hundred NCAs appear on Keeley's list. A small fraction of these NCAs do not meet my definition of atomic assistance. I excluded NCAs that appeared on the list for a few reasons. The first is if they dealt exclusively with nonproliferation assurances and did not actually authorize the supply of technology, materials, or know-how. A supplier country such

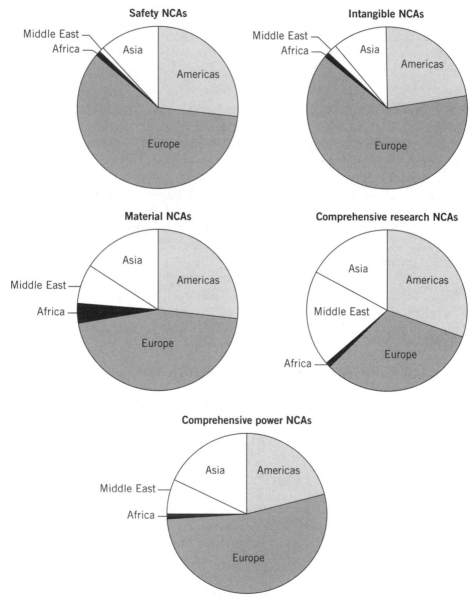

Figure 1.5 Regional dispersion of disaggregated NCAs, 1945–2000

as France, for example, might conclude an agreement with a recipient state that prevents the latter from using previously supplied technology for military purposes. Such a deal would not be included in my dataset. Similarly, I do not include agreements that are intended exclusively to help a country shut down a nuclear power reactor. I also exclude administrative agreements that do not explicitly authorize peaceful nuclear cooperation. For instance, joint statements that simply express a desire to conclude an NCA—such as the one issued by President George W. Bush and Indian prime minister Manmohan Singh in New Delhi in 2005—would not be included (the formal deal signed later would count). Neither would agreements that specify how nuclear transfers should be reported after they occur. Another reason I excluded NCAs is if there was not convincing evidence that an agreement was actually signed.[54] Finally, I do not include agreements that were intentionally or inadvertently included twice, which avoids the double counting of NCAs. Based on these criteria, my dataset includes 176 fewer NCAs than appear on Keeley's list.[55] The vast majority of these treaties are excluded because they only deal with nonproliferation assurances.

Identifying the Type of NCA

As I discussed earlier, there are five types of NCAs: (1) safety agreements; (2) intangible agreements; (3) nuclear material agreements; (4) comprehensive research agreements; and (5) comprehensive power agreements. The preferred method for identifying the treaty type was to obtain the full text of the agreement, which I was able to do for about 20 percent (300/1,500) of the NCAs. I obtained treaties through the United Nations Treaty Series, the Official Journal of the European Union, the International Atomic Energy Agency, and the atomic energy commissions of the respective state parties. If I had the full text available, I read the treaty and identified all of the areas of cooperation that it authorized. If I could not secure the full text of the NCA, I conducted research to identify the areas of cooperation. Of the roughly twelve hundred NCAs where the full text was unavailable, I was able to code more than one thousand (roughly 83 percent) based on this additional research.

I could not determine the relevant areas of cooperation for 198 NCAs (roughly 13 percent of all agreements)—even after an exhaustive search of relevant primary and secondary sources. I was able to verify that these agreements existed, but I could not explicitly determine which type of assistance they authorized. I coded these NCAs based on a few procedures designed to identify the *likely* areas of cooperation. For many of the missing NCAs, I had information on at least one other agreement that was signed by the same two countries. In such cases, I assumed that the NCA with incomplete information covered the same area(s) as the NCA with complete information. This is reasonable given that pairs of states tend to cooperate in the same areas over time. Of course, it is possible that the NCA covered a new area since cooperation could take on a "life cycle." In other words, assistance within a dyad might begin with the supply of a research reactor, progress to assistance in constructing power reactors, and eventually move on to nuclear safety. If I found no evidence that cooperation extended to other areas after conducing additional research, I assumed that the missing NCA was similar to other agreements signed within the same dyad.

For some of the missing NCAs, I did not have information on another agreement signed by the same two countries. To code these agreements, I determined the areas where the recipient state sought assistance at the time the agreement was signed. I assume that the agreement covered these areas as long as two conditions were met. The first is that the supplier had the technical capacity to provide aid in that area. Many of the suppliers included in my dataset have the capacity to offer assistance in all areas of atomic assistance. But there are some exceptions. For instance, Pakistan could not have supplied power reactors to Libya in the 1970s—even if Tripoli had wanted to develop a nuclear power program—because it lacked technical capacity to export technology in that area. The second is that there are no political considerations preventing the exporting state from offering assistance in that domain. The United States, for instance, does not export uranium enrichment facilities, plutonium reprocessing centers, or heavy water production facilities to nonnuclear weapon states because of prohibitions under domestic law.

Identifying Suppliers and Recipients

After identifying the type of assistance each NCA authorized I coded supplier and recipient countries.[56] Identifying suppliers and recipients is important since I am interested in studying exporting states' motivations for engaging in civilian assistance and the effects that receiving aid has on importers' proliferation choices. But it is not always easy to make a determination on this issue.[57] Sometimes the treaty text makes it clear that only one state is the supplier. As an NCA signed by the United States and Lebanon in 1955 states: "Lebanon desires to pursue a research and development program looking toward the realization of the peaceful . . . uses of atomic energy and desires to obtain assistance from the government of the United States of America." More often, however, an NCA is written in such a way to imply reciprocity of supply but in reality only one state is doing the supplying. In some instances, agreements written in this manner genuinely involve reciprocal nuclear assistance. This is often the case for deals involving United States and France, for instance. In other cases—for instance, an agreement involving the United States and Jamaica—the deal implies reciprocal supply when common sense dictates that this is not the case.

If the treaty text did not specify the supplier[s] or I could not obtain the text, I consulted the relevant historical literature to identify the exporting and importing countries. If these sources did not provide the information necessary to code suppliers, I assume that countries having the capacity to supply nuclear technology at the time the agreement was signed are exporters under the terms of the NCA. Capable suppliers include the "traditional exporters" and the "emerging suppliers" as defined by William Potter.[58] These countries are classified as suppliers beginning in the first year subsequent to 1950 that they acquire a nuclear engineering or uranium production capability. To determine when this occurred, I consult data on nuclear production capabilities compiled by Dong-Joon Jo and Erik Gartzke. These data include latent nuclear weapon production capability estimates for 192 countries between 1938 and 2002 based on seven components. Two of these components are uranium production—whether a country is known to have uranium deposits or produced uranium—and nuclear engineering capabilities.[59]

PART I

ATOMS FOR PEACE

Chapter 2

Economic Statecraft and Atoms for Peace

A Theory of Peaceful Nuclear Assistance

Why do countries provide peaceful nuclear assistance to other states? Suppliers use this type of foreign aid as a tool of economic statecraft to influence the behavior of their friends and adversaries. Civilian aid is strategically valuable in part because it strengthens the recipient country economically and bolsters the bilateral relationship between the supplier and importer. Some have argued that nuclear assistance can promote suppliers' strategic interests by helping recipient states build the bomb.[1] Although this may be true when it comes to military assistance, nuclear suppliers consciously seek to avoid this outcome when engaging in civilian nuclear cooperation—even when dealing with friendly nations.[2]

States use atomic aid to strengthen their bilateral relationships with recipient states and promote the development of those countries in ways that serve their politico-strategic interests. Because most states value receiving civilian nuclear assistance, suppliers can use it as a signaling device to convey valuable information about their foreign policy interests.[3] Aid helps nuclear exporters to (1) keep their allies and alliances strong; (2) constrain their enemies; and (3) prop up democracies in the international system. If suppliers can obtain one of these objectives by providing nuclear aid, they are much more willing accept or overlook the proliferation risks that accompany peaceful nuclear assistance. In the end, however, the gambits taken by nuclear exporters raise the risk that nuclear weapons will spread.

Many people believe that states provide atomic energy assistance to promote nonproliferation objectives, generate hard currency, or sustain their domestic nuclear industries. These arguments are intuitive and some of them are at least partially supported by the historical record. Yet my argument is more theoretically satisfying than these alternative explanations, and there is more evidence to support it.

The Strategic Value of Peaceful Nuclear Assistance

Strengthening the Recipient Country

There is a widespread belief that nuclear energy enhances a country's capabilities. Indeed, energy plays a critical role in facilitating economic growth and it is an important element of national power.[4] As Mohamed ElBaradei, the director general of the International Atomic Energy Agency highlighted in a September 2007 speech: "Energy is essential for development. Nearly every aspect of development—from reducing poverty to improving health care—requires reliable access to modern energy services."[5] Since nuclear power reactors improve a state's energy production capacity, they directly affect a country's capabilities. The Indian case illustrates this well. India experienced high annual growth rates of 8–9 percent during the late 2000s but energy shortages threatened to stymie this pace of development. To enhance its energy production capacity, India is turning to nuclear power. Although nuclear power accounts for only 4 percent of the country's energy, it plans to increase this by a factor of ten over the next twenty-five years.[6] If these plans materialize, nuclear power will play a significant role in facilitating Indian economic growth and promoting its material capabilities.

Power reactors also indirectly enhance a state's capabilities by freeing up resources. The resources saved by using nuclear power for domestic electricity production can, in turn, be used to augment a country's economic capacity or military capabilities.[7] For example, Mohammad Reza Pahlavi, the shah of Iran, desired nuclear power in the 1970s to produce electricity for domestic consumption so that oil could be saved for export or for other domestic uses. His logic was that the revenue generated from oil exports would enhance Iran's capabilities and influence in global politics. He argued on many occasions that oil is a "noble product that will be depleted one day" and that it is "a shame to burn the noble product for the production of energy to run factories and light houses" domestically.[8]

Strengthening the Bilateral Relationship

Scholars of international relations have long recognized that foreign assistance can strengthen relations between the supplier and recipient countries. Hans Morgenthau argued nearly fifty years ago, for instance, that foreign aid fosters closer political relationships by signaling favorable intentions and evoking a sense of gratitude that makes recipients more likely to cooperate with the supplier country in a wide variety of domains.[9] Assistance is especially valuable in strengthening bilateral relations if the asset being exchanged is valuable to the recipient and if it depends on the supplier country to obtain that asset.[10] Civilian nuclear cooperation meets these criteria; countries desire civil nuclear programs but are usually dependent on others to successfully develop them.

There are three principal reasons why countries value nuclear energy programs.[11] First, countries believe that they can stimulate economic development. Second, nuclear energy programs are matters of prestige and symbolize progress and technological modernity. Indeed, nuclear power has been described as the "flower of 20th century scientific achievement."[12] The perception that nuclear power and prestige are linked has drawn countries to fission since the beginning of the nuclear age. Following World War II, for example, Canada established its civil nuclear program in part because it desired the prestige associated with becoming a scientific power. In the words of Margaret Gowing, "atomic energy had helped to carve a new status for Canada in the post-war world. It had brought her to the top diplomatic tables and it had demonstrated and enhanced her underlying scientific, technological, and industrial strength."[13] As the nuclear age progressed, developing countries craved the enhanced standing that accompanied nuclear energy programs. Venezuelan president Hugo Chavez, for instance, has recently expressed an interest in building a nuclear reactor in part because he is interested in the prestige that comes with having a nuclear program.[14] Developed countries with rich histories in nuclear energy express similar sentiments today. For example, Boris Johnson, a former member of British Parliament and mayor of London, argued in March 2006 that the United Kingdom should expand its nuclear power program because "it would help to reinforce the crumbling science base of this country [and] . . . restore British activity and prestige in the physical sciences."[15]

Third, countries value nuclear power because it enhances energy security by promoting a stable supply of electricity. Most countries depend on fossil fuels—including coal, oil, and natural gas—to meet their energy needs. In many cases they do not possess indigenous reserves of these materials so they must import them. The 1973 Arab-Israeli War and the subsequent oil embargo highlighted that dependence on foreign suppliers could harm a country's interests. In particular, states realized that their ability to produce adequate energy could be jeopardized if political relations with oil-producing countries soured or if suppliers decided for other reasons to restrict the flow of petrol. These events would put oil-importing countries in a dangerous situation since their economic development and national security depend on adequate energy supply. Elites in many countries are well aware of this quandary. During World War I, for instance, Winston Churchill, the First Lord of the Admiralty, said that "safety and certainty in oil lie in variety and variety alone."[16] Chinese policymakers view energy security as a vital national security issue, and they have since the 1960s.[17] Historically, France has worried about this issue perhaps more than any other country. Former president Valery Giscard d'Estaing underscored this fear when he noted: "We buy our oil at the international price and in the event of a crisis or embargo, no country can protect itself all along."[18]

To guard against the possibility that oil or natural gas supplies could be disrupted, countries typically diversify their energy sources. Part of this strategy often includes developing or expanding nuclear power, which is seen as a relatively dependable alternative source of electricity. For example, the United Arab Emirates (UAE), which is the country most aggressively pursuing nuclear energy in the Middle East, issued a white paper in April 2008 highlighting that the development of nuclear power is essential to the country's future energy security.[19] Additionally, Belarusian president Aleksandr Lukashenko is currently pushing for the development of nuclear power plants because his country has become too dependent on energy imports, particularly Russian natural gas.[20] A recent study confirms the importance of energy security for nuclear power development, finding that states that are dependent on foreign sources to meet their energy needs are statistically more likely to build power reactors.[21]

Some have argued that countries value nuclear energy because of its connection to nuclear weapons. The idea is that states develop civilian programs because they realize that doing so will enhance their future ability to manufacture nuclear weapons. This concept is commonly referred to as "nuclear hedging."[22] I will show later in the book, however, that an interest in developing nuclear weapons is a less powerful driver of civilian nuclear development than many people think. With few exceptions, states do not want to build nuclear weapons at the time they first receive nuclear assistance (see chapter 7). Things might change once a state initiates a bomb program; they are likely to continue pursuing nuclear assistance after a proliferation decision. Yet civilian nuclear programs often do not begin with sinister intentions.

States value civil nuclear programs for the reasons articulated above but they are usually dependent on foreign suppliers to develop them. No state has developed or operated a civilian nuclear program using exclusively indigenous resources. Most countries simply do not have the capabilities required to construct reactors, fuel cycle facilities, or other infrastructure relevant for a civil nuclear program, as the cases analyzed later in the book will illustrate (see chapters 4–6). Countries that probably could build facilities indigenously given sufficient time and political commitment recognize that they can build a civil nuclear program much quicker if they receive foreign aid. Even the United States has received help from European countries in building uranium enrichment facilities, as I noted in chapter 1. Washington also depends on Canadian uranium to run many of its civilian nuclear power plants.

Discretion in the Nuclear Marketplace

Atomic assistance is an effective instrument of statecraft, but this does not imply that states shell it out indiscriminately. Exporters are aware of the dual-use dilemma; they are not naïve about the possible connection between peaceful and military nuclear programs. This has been true since the earliest days of civilian nuclear cooperation. President Eisenhower made this clear in his Atoms for Peace address when he poignantly said that the "miraculous inventiveness of man shall not be dedicated to his death, but consecrated to his life."[23] In several of the cases analyzed later in the book, suppliers knew that the assistance they were providing could aid the recipient country's pursuit of nuclear weapons. Canadian officials, for example, recognized as early as the mid-1950s that the reactors it provided to India *might* be leveraged to bolster a military program "without a prohibitive diversion of manpower resources."[24] This is not an outcome that Ottawa hoped for, but policymakers were aware that it was a possibility.

Cognizance of the relationship between the peaceful and military uses of the atom does not prevent countries from engaging in nuclear assistance. It does, however, cause them to use discretion in the nuclear marketplace. Except in cases of military assistance, suppliers do not want their exports to contribute to proliferation. Countries are keenly aware that the spread of nuclear weapons can threaten national and international security.[25] The United States, for example, has voiced concerns about the spread of nuclear weapons since the end of World War II. Although the United States is sometimes the most vocal about the threats stemming from proliferation, many less powerful countries (such as Brazil, New Zealand, and Sweden) also worry about the spread of the bomb.

Suppliers will accept the proliferation risks of atomic assistance—but typically under specific conditions. They will be more likely to supply nuclear technology, materials, or know-how if they believe that doing so will enhance their political and strategic interests. Expected strategic benefits can offset the possibility of proliferation, causing nuclear exporters to take a calculated gamble. This can even be true even if the recipient state is pursuing nuclear weapons.[26] In many circumstances, nonstrategic benefits (e.g., generating hard currency) will be insufficient to counter the perceived proliferation risks.

Allies, Adversaries, Democracy, and Civil Nuclear Assistance

So, what specific objectives motivate suppliers to engage in peaceful nuclear cooperation? I argue that supplier countries provide atomic assistance for three main politico-strategic reasons: (1) to keep their allies and alliances strong; (2) to constrain their enemies; and (3) to prop up democracies (if the supplier is also a democracy). Chapter 6 will show that suppliers also swap nuclear assistance for oil. These outcomes—which are possible because nuclear assistance strengthens the recipient country and the supplier's bilateral relationship with that country—provide suppliers with strategic benefits that offset the risk that their assistance *might* contribute to proliferation.

Atomic assistance is an effective tool states can use to achieve these strategic objectives. It is certainly not the only tool in the toolbox, however. Arms transfers, aid for economic development, and humanitarian assistance can be employed in the pursuit of similar ends.[27] Indeed, states sometimes ramp up other types of foreign assistance at the same time that they sign nuclear cooperation agreements. Thus, peaceful nuclear cooperation is sometimes one component of a broader strategy to promote a supplier's political and strategic interests.

Strengthening Allies and Alliances

States enter alliances in order to enhance their security. Alliances may be formed in order to establish a balance of power and constrain threatening states.[28] They may also represent attempts to bandwagon with more powerful states in security-enhancing ways.[29] Since they are useful instruments for balancing and bandwagoning, countries perceive that alliances deter third party aggression and promote peace.[30] Major powers sometimes seek alliances for other purposes such as managing weaker countries or obtaining autonomy benefits including the right to establish military bases or acquire overflight rights.[31] The United States, for instance, pursued an alliance with Iran in the late 1960s because it needed to operate an intelligence facility there to verify the Soviet Union's compliance with arms control treaties.[32]

The anticipated benefits stemming from alliances are vital to states' strategic interests. But the payoffs that countries expect when they form alliances do not always materialize. The absence of costly enforcement mechanisms, incentives to free ride, and changes in the strategic environment provide countries with incentives to abandon their alliance commitments.[33] Data on the dependability of allies in war underscore that allies are sometimes unreliable. Alan Sabrosky finds that allies fought together in only 27 percent of the cases and actually opposed one another 12 percent of the time in wars between 1816 and 1965.[34] In a more recent analysis, Brett Ashley Leeds, Andrew Long, and Sarah McLaughlin Mitchell find that alliances are relatively reliable but allies still do not aid one another in war roughly 25 percent of the time.[35] The 2003 Iraq War serves as a fitting reminder that allies do not blindly support military ventures. Many of the closest military allies of the United States vocally opposed military action to overthrow the regime of Saddam Hussein. A military alliance also does not guarantee cooperation in areas short of war. A country cannot assume that its allies will support its positions at the United Nations Security Council, ratify treaties that serve its strategic interests, share sensitive intelligence, restrict potentially lucrative trade with a third party, or work together to combat issues such as terrorism and weapons proliferation. All of this suggests that a military alliance alone does not guarantee that an ally will provide assistance in achieving strategic objectives.

Countries, therefore, have incentives to make sure that bilateral relationships with their allies remain strong even after a formal alliance is forged. By keeping intra-alliance relations strong, countries help to ensure that they actually experience the benefits that led them to create the alliance in the first place. Imagine an alliance where State A agrees to assist State B if it is attacked by State C. State B can increase the likelihood that State A will actually come to its aid if it works to continually strengthen bilateral relations with it. Likewise, State B will be more likely to cooperate with State A in a variety of other strategic domains if the partnership between the two countries is strong. And State B's cooperation will be more meaningful if its capabilities are enhanced. For example, the stronger State B is economically, the more likely it will effectively deter third party aggression.

Countries routinely use atomic assistance to strengthen alliance ties. American civil nuclear cooperation with Japan beginning in the late 1950s is a textbook example of using atomic assistance to strengthen an alliance. The United States depended on its alliance with Japan to balance against the Communist bloc, defend the western Pacific from Soviet aggression, and house American military bases essential to any operation conducted in the Far East. At the same time, declassified documents reveal that Washington recognized that it could not take its partnership with Tokyo for granted.[36] In 1958, the United States signed its first nuclear cooperation agreement with Japan to strengthen the alliance that the two countries had forged six years previously. By fostering Japanese economic growth via civil nuclear cooperation, the United States believed that it would strengthen Japan in ways that would reinforce the American-Japanese alliance and help ensure that Washington could depend on Tokyo to help attain strategic objectives.[37]

The value of atomic assistance in strengthening intra-alliance relations is partially contingent on restricting aid to adversaries. Lars Skalnes argues that attempts to bolster relations with allies by adopting discriminatory foreign economic policies can be undermined by granting enemies similar benefits because it sends mixed signals to both allies and adversaries.[38] Building on this logic, I argue that countries refrain from assisting enemies' nuclear programs to make nuclear assistance to allies more meaningful. Since suppliers restrict nuclear technology to some destinations, recipient countries view atomic assistance as a credible signal of intent to forge a strategic partnership.

Additionally, states want to avoid contributing to the capabilities of an adversary. They prefer that those they are fighting, or likely to fight in the future, are relatively weak because this increases their chances of victory. Countries also want to weaken their adversaries to increase their relative bargaining power and political influence. These considerations generally compel states to restrict foreign economic cooperation—including atomic assistance—with their enemies.[39] The classic example of this is the American-led restriction of nuclear assistance and other sensitive dual-use technologies to the Soviet Union and its allies during the cold war.[40]

Strengthening Relations with Enemies of Enemies

States have incentives to constrain the power of those they find threatening and often do so by cooperating with threatening states' enemies.[41] As I discussed above, State A and State B are likely to form an alliance if both actors are threatened by State C.[42] But countries that share enemies do not always formally ally. Today, for instance, the United States and India have a common enemy—China—but they do not have a defense pact or other formal alliance. Yet both countries are cooperating in a wide variety of areas, partially because they each fear China's rising influence in Asia. This type of strategic cooperation in the absence of a formal alliance is sometimes referred to as "soft balancing."[43]

Civilian nuclear cooperation is one type of soft balancing that countries can employ to counter the capabilities of potentially threatening states. The notion that nuclear assistance can be a means to constrain adversaries is not new. T. V. Paul, for instance, has argued that China's incentives to transfer nuclear technology to Pakistan "derive largely from Chinese concerns about the regional balance of power and are part of a Chinese effort to pursue a strategy of containment in its enduring rivalry with India."[44] Yet this and other related arguments imply that the strategic utility of nuclear assistance comes from helping countries build their first bomb or expand the size of their arsenals. My argument is that civilian nuclear assistance can undermine the influence of rivals for two reasons that have little to do with nuclear proliferation. First, it allows a supplier state to develop a closer relationship with the importing state, improving its ability to balance the threatening state's power. For example, India's civilian nuclear assistance to Vietnam beginning in the late 1990s was intended to forge an Indo-Vietnamese partnership to counter the rising influence of China in the region (see chapter 5). Second, civilian nuclear cooperation with a threatening state's enemy also constrains its power by making it more difficult for the threatening state to exert influence or commit aggression against its enemy (e.g., the recipient). As I will discuss in chapter 4, in providing nuclear aid to Iran in the 1970s, the United States strengthened it economically and politically, which made it more difficult for the Soviet Union—a common enemy in this case—to influence or attack Iran.[45]

States could use atomic assistance to constrain all of their enemies. Yet the likelihood that suppliers will employ nuclear cooperation to undermine their adversaries is conditional on the capabilities of the threatening state. Nuclear suppliers are especially likely to provide assistance to those states that are enemies of superpowers because states are most threatened by the strongest countries in the international system. Britain had rivalries with Iraq and the Soviet Union during the cold war.[46] My argument implies two things about the likelihood of atomic assistance under these circumstances. First, London was more likely to engage in nuclear cooperation with enemies of Baghdad and Moscow, compared to states that were not enemies of either state. Second, Britain was more likely to provide peaceful nuclear assistance to Soviet enemies relative to Iraqi enemies since it cared relatively more about limiting Moscow's influence.

Russia's recent nuclear cooperation with Venezuela is a fitting illustration of this argument.[47] In November 2008, Moscow pledged to supply the Latin American country with its first nuclear power plant during President Dmitry Medvedev's trip to Caracas. Moscow hopes to forge a closer relationship with Caracas aimed at countering the influence of the United States and nuclear cooperation represents an important component of this broader strategy. As one Latin American expert indicated, Russia's message to the United States in providing atomic assistance to Venezuela is clear: "We can exert influence in your backyard if you continue to exert influence in our backyard. If you don't take your missiles out of Poland and end NATO expansion we're going to increase our influence in Latin America and do things to provoke you."[48] Notice that the objective here is *not* to assist in the development of a military program. Indeed, Russia does not want Venezuela to build nuclear weapons. Moscow's intent is to constrain Washington by cooperating with one of its adversaries in a sensitive civilian area. The United States recognizes the peaceful intent of this cooperation—but it does not necessarily want one of its adversaries to possess technology that *could* be used to produce plutonium for nuclear weapons. Washington would be much more comfortable if Caracas did not possess what President Barack Obama's science advisor John Holdren has called an "attractive nuisance of a most dangerous kind."[49]

Propping Up Democracies

Scholars of international relations have long recognized that a country's regime type can affect international security. This is personified most prominently by the democratic peace theory, which holds that two democratic states are unlikely to fight each other because democracies share values and a respect for the rule of law and democratic leaders face institutional constraints that make conflict particularly costly.[50] The logic of this theory not only implies that democracies are less likely to engage in conflict but also that they are more likely to cooperate with one another. Indeed, democracies cooperate more than other pairs of states because they share similar interests and expect that accommodating behavior will be reciprocated.[51] Democratic states, therefore, have incentives to prop up other democracies because this puts them in a better position to achieve strategic objectives. A relatively weak democracy such as the Philippines, for example, can better assist the United States in countering terrorism or weapons proliferation if its capabilities are enhanced. This is in large part why the Princeton Project on National Security urges the United States to "sustain the military predominance of liberal democracies and encourage the development of . . . like-minded democracies."[52] Similarly, close bilateral ties between democracies enhance their ability to work together in achieving strategic objectives. This is exemplified by the 2006 National Security Strategy (NSS) of the United States, which states that "America's closest alliances and friendships are with countries with whom we share common values and principles. The more countries demonstrate that they . . . are committed to democratic principles, the closer and stronger their relationship with America is likely to be."[53] It is important to note that while shared values provide the foundation for strong relations between democracies they do not always guarantee a close relationship, just as an alliance does not ensure that an ally will always be willing to cooperate. Despite being the largest and second largest democracies in the world, for instance, India and the United States have only recently developed a close bilateral relationship. Sometimes, additional measures are necessary to develop strong relations—even among two democracies.

Strengthening a democracy can also limit the influence of nondemocracies. There are two reasons why this is the case. First, strengthening a democratic state makes it more difficult for a nondemocracy to assert itself as a regional hegemon. If India becomes stronger economically, for instance, it will be more difficult for China to emerge as the clear leader in Asia. Second, strengthening a democracy makes it more difficult for an autocratic country to exert influence against it, possibly weakening the democracy's commitment to the rule of law. During the cold war, the United States worried that the Soviet Union would pressure weak democracies in the developing world and attempt to convert them into communist states. Washington recognized that it could stymie this strategy by strengthening these countries, making them less susceptible to Soviet aggression or influence. Limiting the influence of nondemocratic states in these two respects serves a democratic state's strategic interests by constraining countries with which it could experience future conflict. Since it is more likely to fight a nondemocratic state, a democracy prefers these countries to be relatively weak.

The preceding logic suggests that democratic suppliers should be more likely to assist the civil nuclear programs of other democracies. In offering atomic assistance to other democracies, democratic suppliers strengthen recipient countries and their partnership with these countries in ways that promote their strategic interests. The U.S.-India deal, which I discuss in chapter 4, is a recent high-profile case that underscores the relationship between civilian nuclear cooperation and democracy. Virtually every senior U.S. decision maker involved in the deal—including President George W. Bush, Secretary of State Condoleezza Rice, and Undersecretary of State Nicholas Burns—has publicly justified atomic assistance to India on the grounds that it will transform relations with a democratic country in ways that enhance American security.

Summary of Hypotheses

The arguments I advanced above suggest five testable implications of peaceful nuclear assistance:

Hypothesis 3.1: Suppliers are more likely to provide peaceful nuclear assistance to their military allies than to nonallies.

Hypothesis 3.2: Suppliers are less likely to provide peaceful nuclear assistance to states with which they are engaged in militarized conflict than to states they are not fighting.

Hypothesis 3.3: Suppliers are more likely to provide peaceful nuclear assistance to enemies of enemies than to nonenemies of enemies.

Hypothesis 3.4: Suppliers are more likely to provide peaceful nuclear assistance to states that are enemies of the most powerful states in the international system than to nonsuperpower enemies.

Hypothesis 3.5: Democratic nuclear suppliers are more likely to offer peaceful nuclear assistance to democracies than to nondemocracies.

Alternative Explanations

Although there have been few attempts to explain civilian nuclear cooperation in a systematic fashion, there are possible alternative explanations that can be extracted from the case literature on atomic assistance and from more general literatures in political science and economics. Below, I describe three alternative explanations for peaceful nuclear assistance: that countries use atomic assistance to strengthen nonproliferation norms; that countries sell nuclear technology to make money; that countries offer nuclear assistance to sustain their domestic nuclear industries.

Promoting Nuclear Nonproliferation

The conventional wisdom for why countries provide peaceful nuclear assistance is that it promotes nonproliferation. This argument draws on the well-developed literature in political science on international norms—defined as "a standard of appropriate behavior for actors with a given identity."[54] Norms often emerge when powerful actors impose their preferences on others and threaten to censure those that fail to comply.[55] Once norms surface, they can shape state behavior in a variety of ways. States may follow norms because they believe it is the appropriate thing to do.[56] They may also comply out of fear that they will be punished for violating norms.[57] Countries might refrain from using chemical weapons in war, for instance, because they find them morally repugnant or fear that their use will bring about political, military, or economic costs.[58]

When President Eisenhower conceived of Atoms for Peace, he believed that the spread of nuclear technology for peaceful purposes could decrease the likelihood that countries would want nuclear weapons. His logic was in part that if the United States and other suppliers placed an embargo on nuclear assistance they would only encourage countries to build nuclear facilities indigenously.[59] Eisenhower's ideas were partially codified in the nuclear Nonproliferation Treaty, which entered into force in 1970. The NPT attempts to promote atomic assistance while limiting the prospect of further proliferation and working towards eventual nuclear disarmament. States that ratify the treaty make formal commitments not to manufacture or acquire nuclear weapons and institute measures to verify that these pledges are kept. In exchange, treaty signatories are entitled to nuclear assistance from those capable of offering it. With the backing of "norm entrepreneurs"—including the United States and the Soviet Union—the NPT helped establish the norms of the nuclear marketplace. Specifically, it defined the appropriate practice as assisting countries with strong commitments to nonproliferation that sign the NPT and denying civilian atomic aid to states that pursue nuclear weapons or flaunt the nonproliferation regime. These norms could affect the behavior of nuclear suppliers. They might refrain from assisting countries with weak nonproliferation records because they believe that doing so runs counter to the way that a responsible country should behave. Moreover, they may fear that they will be punished by members of the international community—especially the most powerful states—if they fail to comply with the perceived common practices. If this argument is correct, we would expect that nuclear suppliers would be more likely to assist members of the NPT and less likely to aid countries that are pursuing nuclear weapons.

Generating Foreign Exchange

The case study literature on atomic assistance often suggests that countries are motivated to generate foreign exchange. Etel Solingen posits that the need to make money motivated Brazilian nuclear exports beginning in the 1970s.[60] Likewise, Sara Tanis and Bennett Ramberg argue that Argentinean policymakers depended on atomic aid to produce much-needed hard currency.[61] Generating profits is also frequently highlighted as a motivation for nuclear cooperation in the mainstream media. Major newspapers around the world often report that suppliers such as Russia or North Korea export nuclear technology because they need the money. For example, news reports on Russian atomic aid to Iran in the mid-1990s continually suggested that Moscow's desire for hard-currency profit was the principal motivation for this cooperation.[62] One of the reasons why profit is routinely mentioned as a reason for nuclear assistance is that reactor exports can be quite lucrative. Canada's pledge to sell two reactors to China in 1996 represented a $4 billion contract and proposed American reactor exports to Iran in the 1970s would have generated $6.4 billion in revenue.[63]

Two related implications flow from this discussion. The first suggests that supplier countries shell out nuclear technology indiscriminately to any state that can afford it. This is the type of argument routinely made about Russia or North Korea. If this is true, we would expect that the strategic considerations I outlined above would have little effect on atomic assistance since they require states to restrict aid in cases where it could be lucrative. The second implication is that countries behave in ways that are economically efficient. There are strictly economic factors that are thought to influence foreign economic cooperation. It remains widely accepted that a state's capacity to supply exports as well as a state's demand for imports is directly related to its gross domestic product (GDP).[64] The distance between countries also affects nuclear commerce because of the transportation and transaction costs associated with trade. These costs are expected to rise as the distance between two countries increases.[65] Existing trade ties should also increase the efficiency of atomic assistance. If profit maximization is the goal, these economic factors should be especially salient in explaining atomic assistance.

Sustaining the Domestic Nuclear Industry

Economies of scale—the cost reductions that result from increased production—is a concept that has a rich tradition in economics dating back to Alfred Marshall's work in the late 1800s.[66] Paul Krugman and other proponents of "new trade theory" argue countries trade because of economies of scale.[67] It is economically efficient to concentrate production in a few locations so that high levels of production can be achieved in each location.[68]

This also means that a particular good is only produced by a few countries and exported everywhere else. A classic example of this is the North American auto industry. Both the United States and Canada produce cars and their subcomponents but each particular model or component is typically produced in only one country and exported to the other.[69]

Some scholars have argued that the problem of economies of scale influences atomic assistance. Specifically, they argue that nuclear suppliers are often constrained by small domestic markets for reactors and other nuclear technology so they provide foreign assistance in order to increase production and reduce costs. Doing so is sometimes a necessary strategy for the survival of domestic nuclear industries. Duane Bratt argues, for example, that Canadian reactor exports are necessary because there is not sufficient domestic demand to sustain a nuclear industry.[70] By exporting nuclear technology, Canada is able to reduce the production costs for the industry and make it economically viable. Claus Hofhansel and Erwin Hackel marshal similar logic to explain German nuclear assistance.[71] Hofhansel argues that Germany offers atomic aid to achieve economies of scale and compete with countries such as the United States while Hackel posits that German assistance is necessary to reduce the costs for the domestic nuclear industry. As these arguments indicate, sustaining the domestic nuclear industry is another plausible reason for nuclear assistance. If this argument is correct, we would expect that countries with less domestic demand for nuclear power would be more likely to offer atomic aid.[72]

Summary of Alternative Hypotheses

1. Countries use atomic assistance to strengthen nuclear nonproliferation norms.

1a. Countries that sign the NPT are more likely than countries that do not sign the NPT to receive atomic assistance.

1b. Countries that are pursuing nuclear weapons are less likely than countries that are not pursuing nuclear weapons to receive atomic assistance.

2. Countries sell nuclear technology to make money.

2a. Economically developed states are more likely to offer and receive atomic assistance.

2b. The likelihood of atomic assistance decreases as the distance between two countries increases.

2c. Countries with existing trade ties are more likely to engage in atomic assistance.

3. Countries offer nuclear assistance to sustain their domestic nuclear industries.

3a. Suppliers with lower domestic demand for nuclear energy are more likely to provide nuclear assistance than states with a high domestic demand for nuclear energy.

Methods and Cases

The arguments presented in this chapter suggest different rationales for peaceful nuclear assistance. In the next four chapters of this book, my goal is to empirically evaluate which argument offers the most explanatory power.

To do so, I use a "nested" research design that relies on both quantitative and qualitative methods.[73] In adopting this approach I attempt to benefit from the advantages that each method offers and inspire greater confidence in my core findings. Quantitative analysis offers several unique benefits. It allows researchers to minimize the risks of sample selection bias that can arise when only a small number of cases are analyzed. By providing the average effect of each variable on the dependent variable, statistical analysis can also tell us whether a given argument is salient across a large universe of cases and over time.[74] When using only qualitative analysis, it is often difficult to determine whether an argument is generalizable to a broader set of cases. Finally, statistical analysis is useful in determining the relative effects that variables have when controlling for possible confounding factors. For instance, it could tell us that Variable A doubles the likelihood of a particular outcome while Variable B triples the likelihood but does not have a significant effect once we account for Variable C.

Statistical analysis, of course, also has drawbacks. It generally evaluates whether two factors are correlated without assessing whether they are causally connected. Because it is based on average effects, statistical analysis may have little to say about potentially important outlying cases. It might be problematic, for example, to have a theory of war that explains most twentieth-century conflicts but does not account for World War I and World War II.[75] Another significant problem has to do with imprecise measurement. It is often difficult to perfectly quantify key concepts—especially intangible factors such as prestige. There are additional problems that researchers must grapple with when using statistical analysis to study rare events like nuclear proliferation, as I will discuss in part II of the book.

Qualitative analysis can address many of the shortcomings of statistical analysis, especially if researchers examine cases that fit the statistical analysis *and* outlying cases. By using process tracing to analyze cases where empirical predictions held true, researchers can do more to confirm whether the outcomes are attributable to the hypothesized mechanisms or some other explanation. Analysis of outlying cases can shed light on variables that are important but were left out of the initial statistical model.[76]

In chapter 3, I use statistical analysis to evaluate my argument and the alternative explanations. Then, based on the statistical results, I analyze two types of cases later in the book: (1) those that were successfully explained by the large-N analysis; and (2) those that deviate from the expected results. My goal in examining successfully predicted cases is to determine whether the causal logic driving my argument operates correctly. In examining outlying cases, I seek to determine whether there is a compelling explanation for atomic assistance that I initially failed to consider. Since civilian nuclear cooperation occurs so frequently, I am able to examine only a small sample of cases in each of these two categories. I use a few additional criteria to select cases among successfully predicted ones and outliers. First, I select cases that are historically significant. In addition to being theoretically and historically interesting, the most information is available on high-profile cases, which makes it easier to identify the underlying causal processes. Second, I chose cases that varied based on the characteristics of the supplier country and the time period. I examine cases in all periods of the nuclear age that involve major powers and nonmajor powers, major suppliers and emerging suppliers, and democracies and nondemocracies.

Chapter 3

The Historical Record

A First Cut

Countries use peaceful nuclear assistance as a means to enhance their political influence by managing their relationships with strategically important states. In particular, suppliers provide aid to: (1) keep their allies and alliances strong; (2) constrain their adversaries by cultivating closer ties with states that are vulnerable to influence or aggression from their enemies; and (3) prop up existing democracies (if the supplier is also a democracy).

In this chapter, I test my argument and the alternative explanations using statistical analysis and the dataset on peaceful nuclear cooperation that I described in chapter 1. The initial findings lend support to my theory. All of the variables operationalizing my argument are statistically correlated with the signing of nuclear cooperation agreements.

Although aspects of my theory explain the signing of all agreement types, it does not predict the emergence of all treaties equally well. For example, when attempting to enhance their influence, suppliers are more likely to employ agreements that transfer nuclear power plants and other advanced technology than limited treaties that only authorize the exchange of nuclear materials. India's 1974 nuclear test and the subsequent formation of the Nuclear Suppliers Group, an informal association of states designed to harmonize national export policies, are widely believed to have influenced the correlates of atomic assistance. Yet there is support for my theory both before and after the formation of the NSG and equally scant support for the competing explanations in both periods. Moreover, members of the NSG behave very similarly to non-members when it comes to the factors that motivate the signing of NCAs.

The competing explanations do not fare particularly well. The results undermine the nonproliferation argument because there is not a significant relationship between peaceful nuclear aid and membership in the nuclear Nonproliferation Treaty. Additionally, there is no evidence that suppliers systematically withhold assistance from countries that are pursuing nuclear weapons. I find some support for the argument that countries behave in ways that are economically efficient but there is little evidence in favor of the contention that countries engage in nuclear assistance to promote the growth of their own nuclear industries.

Dataset and Variables

To test the hypotheses on peaceful nuclear assistance outlined in the previous chapter, I produced a standard time-series cross-sectional dataset for the period 1950 to 2000 that consists of more than 150,000 observations. The unit of analysis is the directed dyad year.[1] The dataset includes twenty-three countries as capable nuclear suppliers and all countries in the international system as potential recipients of atomic assistance.

The dependent variable measures the state-authorized transfer of nuclear facilities, technology, materials, or know-how for peaceful purposes. It is based on the data on bilateral NCAs that are employed throughout the book. I code the dependent variable 1 if a suppler state and a recipient state sign an NCA in a particular year and 0 otherwise.[2] Of all the observations in the dataset, 2,532 (about 2 percent) experienced peaceful nuclear cooperation as I define it. Later in this chapter, I disaggregate the dependent variable to explore whether the motives for atomic assistance vary based on NCA type.

To test my argument and the competing explanations I code several independent variables, which I describe in more detail in appendix 3.1. I code variables measuring whether the supplier and recipient share a military alliance or are fighting in order to test my argument that states use atomic assistance to strengthen their allies and alliances. To test my argument that suppliers use nuclear aid to constrain their adversaries, I code variables measuring whether the supplier and recipient share a common enemy and whether the recipient is a superpower enemy. I test my democracy argument by coding a variable measuring whether the supplier and recipient are both democracies.[3]

Turning to the competing explanations, to test the nonproliferation argument I code variables measuring whether the recipient is pursuing nuclear weapons and has ratified the NPT. I test the profit argument by coding indicators of economic efficiency such as the gross domestic products of both countries, the distance between them, and the value of all exports from the supplier to the recipient. Finally, I code a variable measuring the number of domestic nuclear power plants per capita in the supplier state to test the economies of scale argument. The likelihood that a supplier engages in nuclear assistance should decline as its domestic demand for nuclear power increases.

The Determinants of Peaceful Nuclear Assistance

As a preliminary attempt to evaluate the hypotheses on atomic assistance I conduct a simple cross-tabulation analysis. Tables 3.1 to 3.5 present cross-tabulations that display the relationships among military alliances, conflict, shared enemies, superpower enemies, joint democracy, and peaceful nuclear assistance. These tables are revealing in several ways.

Table 3.1 illustrates that military allies engage in more nuclear cooperation than what we would expect from chance alone. Among the nonally observations, only 1 percent experienced nuclear cooperation while nearly 9 percent of the ally observations experienced atomic assistance. These figures indicate that suppliers are about 800 percent more likely to provide nuclear assistance to their allies than to their nonallies. Militarized conflict also has a significant effect on civilian nuclear assistance. Only 0.6 percent of the conflict observations experience nuclear cooperation while atomic assistance occurs in about 2 percent of the conflict-free observations, indicating that militarized disputes reduce the likelihood of nuclear commerce by 63 percent. Tables 3.3 and 3.4 indicate that having a shared enemy and being a superpower enemy substantially increases the likelihood of atomic assistance. Suppliers are 400 percent more likely to provide nuclear aid to states they share an enemy with and superpower enemies are 324 percent more likely than nonsuperpower enemies to receive nuclear assistance. Finally, table 3.5 shows that democratic suppliers are more likely to sign nuclear cooperation agreements with other democracies; joint democracy increases the probability of nuclear assistance by 493 percent.[4]

These simple cross-tabulations lend statistically significant support for my argument. They also clearly show that the relationships under examination are probabilistic—not deterministic. The explanatory variables raise (or lower in the case of conflict) the likelihood of peaceful nuclear assistance but they do not guarantee that atomic aid will occur. A military alliance, for instance, increases the likelihood of nuclear aid but allies do not engage in nuclear cooperation in the majority of the cases. Also note that the explanatory variables are not necessary conditions for peaceful nuclear cooperation. For example, democratic suppliers signed roughly 75 percent of all NCAs with other democracies but concluded a nontrivial number of deals (542) with nondemocracies. Even militarized conflict does not make it impossible for nuclear cooperation to occur; states signed four NCAs in the year following a dispute.

The next step is to determine whether these same relationships hold when controlling for confounding variables. I conducted multivariate statistical analysis designed to evaluate whether the historical record supports my argument when accounting for the other factors that are thought to influence

Table 3.1 Cross-tabulation of military alliances and civilian nuclear assistance, 1950–2000

Civilian nuclear assistance	Military alliance		
	No	Yes	Total
No	148,796 (98.97%)	10,164 (91.22%)	158,960 (98.43%)
Yes	1,554 (1.03%)	978 (8.78%)	2,532 (1.57%)
	150,350 (100%)	11,142 (100%)	161,492 (100%)

$\chi^2 = 4000\ p < .001$

Table 3.2 Cross-tabulation of conflict and civilian nuclear assistance, 1950–2000

Civilian nuclear assistance	Militarized conflict		
	No	Yes	Total
No	158,252 (98.43%)	708 (99.44%)	158,960 (98.43%)
Yes	2,528 (1.57%)	4 (0.56%)	2,532 (1.57%)
	160,780 (100%)	712 (100%)	161,492 (100%)

$\chi^2 = 4.69\ p = 0.030$

Table 3.3 Cross-tabulation of shared enemy and civilian nuclear assistance, 1950–2000

Civilian nuclear assistance	Shared enemy		
	No	Yes	Total
No	151,279 (98.70%)	7,681 (93.49%)	158,960 (98.43%)
Yes	1,997 (1.30%)	535 (6.51%)	2,532 (1.57%)
	153,276 (100%)	8,216 (100%)	161,492 (100%)

$\chi^2 = 1400\ p < .001$

Table 3.4 Cross-tabulation of superpower enemy and civilian nuclear assistance, 1950–2000

Civilian nuclear assistance	Superpower enemy		
	No	Yes	Total
No	142,674 (98.82%)	16,286 (95.13%)	158,960 (98.43%)
Yes	1,698 (1.18%)	834 (4.87%)	2,532 (1.57%)
	144,372 (100%)	17,120 (100%)	161,492 (100%)

$\chi^2 = 1400\ p < .001$

Table 3.5 Cross-tabulation of joint democracy and civilian nuclear assistance, 1950–2000

Civilian nuclear assistance	Joint democracy		
	No	*Yes*	*Total*
No	74,425 (99.28%)	32,522 (95.73%)	106,947 (98.17%)
Yes	542 (0.72%)	1,451 (4.27%)	1,993 (1.83%)
	74,967 (100%)	33,973 (100%)	108,940 (100%)

$\chi^2 = 1600 \ p < .001$

peaceful nuclear assistance. All five of the variables operationalizing my strategic argument are statistically related to nuclear cooperation agreements in the expected direction. Military alliances, shared enemies, superpower enemies, and joint democracy all raise the probability that two countries will engage in atomic assistance in a particular year while militarized conflict reduces this likelihood.

To illustrate these findings, I calculated the percentage change in the probability of civil nuclear cooperation produced by increases in each independent variable when all other factors are held constant.[5] Figure 3.1 shows that strategic considerations are highly salient in explaining nuclear assistance. A military alliance increases the likelihood of atomic aid by 237 percent while militarized conflict reduces this probability by 85 percent. Also notice that superpower enemies are 156 percent more likely than nonsuperpower enemies to receive nuclear aid. Having a shared enemy produces a more modest effect on nuclear cooperation; suppliers are 38 percent more likely to provide aid to enemies of enemies. We also see that democratic suppliers are 205 percent more likely to provide assistance to other democracies than nondemocracies. Collectively, these five strategic variables exert a large effect on atomic assistance. When all of these factors are present (or absent in the case of conflict), countries are 18,367 percent more likely to engage in civilian nuclear cooperation![6]

I turn to a few of the cases from the dataset to further illustrate the findings of the multivariate analysis. Table 3.6 lists the predicted probabilities of U.S. atomic assistance to members of the Association of Southeast Asian Nations (ASEAN) in 2000. It also displays the percentage increase in the probability that each country will receive U.S. aid, relative to the most unlikely recipient in the region. Since these countries generally have strong nonproliferation records and similar levels of development, the results illustrate nicely the salience of strategic factors in predicting whether nuclear assistance will occur. When there are no strategic incentives for atomic aid the likelihood that it will happen is very small. For instance, the probability of the United

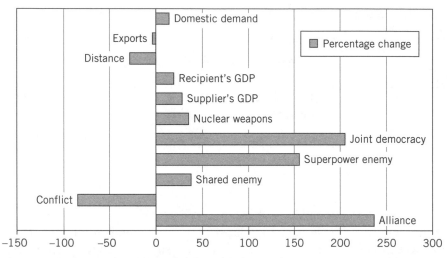

Figure 3.1 Percentage change in probability of civilian nuclear cooperation caused by each statistically significant independent variable
Note: Calculations are based on models 3 and 4.

States providing nuclear technology, materials, or knowledge to Burma or Cambodia in 2000 is 0.00867 and 0.00949, respectively. Even when countries express interests in developing nuclear power programs, suppliers are less likely to provide assistance when strategic incentives are absent. The likelihood of atomic assistance rises dramatically when the supplier has strategic incentives to offer it. The Philippines, which is a democracy and shares a common enemy and military alliance with the United States, is the most likely recipient of U.S. atomic assistance in Southeast Asia. The probability of U.S. nuclear aid to the Philippines in 2000 was 0.258, meaning that it is 3,145 percent more likely than Burma to receive U.S. assistance. Indonesia and Thailand are other likely recipients of U.S. nuclear aid in the region because they are democracies. They are 1,157 percent and 1,799 percent more likely than Burma to receive nuclear assistance from the United States, respectively.

The alternative explanations do not perform as well in the multivariate analysis. The quantitative evidence contradicts the nonproliferation argument. Surprisingly, NPT membership does not achieve conventional levels of statistical significance. One interpretation of this finding is that some states sign the NPT with no intention of developing a civil nuclear program. However, I replicated the analysis while only including recipient states that had at least one nuclear power plant in operation and I arrived at similar results. Another possibility is that NPT suppliers behave differently than non-NPT exporters since only the former are legally bound by the Article IV requirement to share technology, materials, and know-how with other parties of the

Table 3.6 Predicted probabilities of U.S. atomic assistance to ASEAN countries, 2000

Country	Predicted probability	Percentage increase (relative to Burma)	U.S. strategic incentives for nuclear assistance
Burma	.00795	–	None
Cambodia	.00843	6	None
Laos	.00856	8	None
Singapore	.00889	12	None
Vietnam	.0104	31	None
Malaysia	.0179	125	None
Brunei	.0229	188	None
Indonesia	.0999	1,157	Joint democracy
Thailand	.151	1,799	Joint democracy
Philippines	.258	3,145	Military alliance, Shared enemy, Joint democracy

Note: Calculations are based on model 4.

treaty. Although this conjecture is intuitive, it is not supported by the historical record. I find that suppliers that are part of the NPT are no more likely to sign NCAs with other members of the treaty, compared to nontreaty members. Moreover, the results show that NPT suppliers behave very similarly to non-NPT exporters; the five variables relevant to my argument similarly affect NPT and non-NPT suppliers.

Suppliers do not methodically restrict nuclear assistance to states that are pursuing nuclear weapons. The preliminary findings actually indicate that states with nuclear weapons programs are about 35 percent more likely to receive assistance in developing a civilian nuclear program, compared to states that are not pursuing nuclear weapons (see figure 3.1). Note, however, that this is not a robust finding. Most tests indicate that there is not a statistically significant relationship between signing NCAs and pursuing nuclear weapons. For example, when I limit the dataset to recipient states that have at least one nuclear plant in operation, there is not a significant relationship between these two variables. Either way, the findings are unsupportive of the nonproliferation argument.

There is some evidence in favor of the profit argument because the majority of variables measuring economic efficiency (e.g. supplier's GDP, recipient's GDP, and distance) have significant effects in the expected direction. However, this argument is not fully supported since existing trade linkages actually reduce the likelihood of nuclear assistance. Also note that these variables have very modest effects on civil nuclear cooperation. For instance, an increase in the supplier's GDP that roughly corresponds to the change in Germany's economic development from the late 1970s to the late 1990s raises the likelihood of nuclear cooperation by 28 percent; a similar increase the recipient's GDP raises the probability of assistance by only 19 percent. Substantial increases in the distance between the supplier and recipient (approximating Ethiopia's distance from the United States relative to Brazil's distance from the United States) lowers the probability of atomic assistance by 28 percent. And increasing the value of all exports from the supplier to the recipient lowers the probability of nuclear aid by a mere 4 percent. The increase in trade used to make this calculation roughly equates to the difference between U.S. exports to Iran in the mid-1990s and Spain in the late 1980s.

The statistical results contradict the export pressure argument since countries that have incentives to provide assistance in order to sustain their domestic nuclear industries are less likely to offer atomic aid. Substantively, increasing the number of domestic nuclear power plants per capita raises the likelihood of peaceful nuclear assistance by about 14 percent.[7]

Analysis of Disaggregated Peaceful Nuclear Cooperation

Recall from chapter 1 that there are five types of nuclear cooperation agreements: (1) safety NCAs; (2) intangible NCAs; (3) nuclear material NCAs; (4) comprehensive research NCAs; and (5) comprehensive power NCAs. The dependent variable employed above aggregated all of these nuclear cooperation agreements. Yet it is reasonable to wonder whether the determinants of civilian nuclear cooperation vary by treaty type. Suppliers might not view all treaties as equally effective in strengthening their bilateral relationships with recipient states. If this is true, the five variables testing my argument might be strongly associated with some types of NCAs but unrelated to others. It is also plausible that nonproliferation considerations influence the signing of some agreements even if concerns about proliferation do not drive the nuclear marketplace. For instance, suppliers might be willing to sell research reactors or nuclear materials to states that remain outside the NPT but deny these countries access to more advanced technology (e.g., nuclear power plants).

To explore these possibilities I code five dependent variables based on the disaggregated NCA typology and replicate the multivariate statistical analysis presented above.[8] The findings are generally consistent with the results reported in the preceding section. A military alliance is positively correlated with all five dependent variables. Conflict is also a strong predictor of all agreement types except nuclear material NCAs. Indeed, with few exceptions states do not sign treaties in the year following a militarized dispute.[9] The superpower enemy variable is statistically correlated with all of the disaggregated dependent variables. The two other variables testing my argument are correlated with most, but not all, treaty types. States with a common enemy are more likely to sign safety NCAs, intangible NCAs, research NCAs, and power NCAs but not nuclear material NCAs. Joint democracy raises the likelihood of signing safety, intangible, and power agreements but this variable is not correlated with material NCAs and comprehensive research NCAs.

What are the implications of these findings for my theory of peaceful nuclear cooperation? Elements of my argument explain the emergence of all five disaggregated agreements. My theory applies to some types of nuclear cooperation more than others, however. The argument performs best when it comes to safety NCAs, intangible NCAs, and comprehensive power NCAs; all five of the variables testing my argument are correlated with the signing of these three treaties. This is significant because these agreements make up 84 percent of all NCAs signed between 1950 and 2000. Moreover, the comprehensive power treaties are in many ways the most important because they are the broadest in scope. My theory also largely explains the emergence of comprehensive research NCAs. I do not find support for my argument that democracies sign these agreements to strengthen other democracies and their relationships with these countries but the results confirm the other elements of my argument. Nuclear material NCAs represent the one type of atomic assistance that my argument does not explain especially well; only military alliances are correlated with these treaties. But recall that these deals comprise less than 7 percent of all NCAs in the dataset and they are also limited in scope.

Turning to the control variables, nuclear weapons pursuit is negatively correlated with safety NCAs but positively correlated with material NCAs and comprehensive power NCAs, possibly because proliferators are more likely to seek the latter. This variable is statistically unrelated to intangible NCAs and comprehensive research NCAs. The findings regarding NPT membership similarly vary by agreement type. NPT members are more likely to sign safety NCAs but they are not more likely to sign the other four treaty types. These results indicate that the nonproliferation argument helps account for cooperation in nuclear safety but it is not salient in explaining any other type of atomic assistance. The proxies for the revenue maximization argument are generally correlated with all agreements except comprehensive research NCAs. Finally, domestic demand is correlated with all five dependent variables but in some cases it is positive (e.g., with safety NCAs) and in others it is negative (e.g., with research NCAs), consistent with theoretical expectations.

What about Military Assistance?

This book is about peaceful nuclear assistance, not aid provided to intentionally spread nuclear weapons. Yet readers may wonder whether the determinants of military assistance are similar to the correlates of peaceful nuclear assistance.[10] One needs to be exceedingly cautious when modeling an event that occurs as infrequently as military aid to develop nuclear weapons (which is even rarer than the acquisition of nuclear weapons). Nevertheless, existing statistical tools can shed some light on patters of proliferation assistance.[11]

I would expect the findings to vary wildly since military assistance is motivated by an entirely different calculus. The results support this expectation. None of the variables testing my argument achieve conventional levels of statistical significance. My argument does not work where it should not work. To understand the causes of military assistance, we would have to identify the conditions under which a supplier wants the recipient to build nuclear weapons. This is an important issue, but it falls outside the scope of this book.

So, what explains military assistance? The findings offer a preliminary answer to this question. States pursuing nuclear weapons are substantially more likely to receive aid, which is not surprising because a supplier is unlikely to offer this type of assistance to a state that is not trying to build the bomb. Somewhat unexpectedly, the results are supportive of economic explanations. Existing trade ties raises the likelihood of providing military assistance while distance is negatively correlated with this type of aid. There is also support for the export pressure argument; the likelihood of military assistance increases as the number of reactors per capita in the supplier state declines.[12]

The Nuclear Suppliers Group and Peaceful Nuclear Cooperation

Supplier states' incentives to engage in civilian nuclear cooperation may have evolved over time. In the 1950s, when President Eisenhower launched the Atoms for Peace program, there could have been some uncertainty whether countries would use civilian assistance to develop nuclear weapons. According to the conventional wisdom, this changed on May 18, 1974, when India exploded a nuclear device in the Rajasthan desert.[13] This event demonstrated that a developing country was capable of building nuclear weapons using technology supplied exclusively for peaceful purposes, as I will discuss further in chapter 7. By underscoring the proliferation dangers of atomic assistance, the Indian nuclear test could have altered the conditions under which supplier states shared nuclear technology, materials, and know-how.[14]

A group of seven countries established the NSG in the aftermath of India's nuclear test to harmonize national export policies and limit the proliferation potential of peaceful nuclear cooperation.[15] The NSG's guidelines, which were institutionalized in September 1977, required recipient states to provide nonproliferation assurances and accept safeguards designed to verify that nuclear facilities are used exclusively for peaceful purposes. The NSG's principles also encourage members to share information on their nuclear transactions and exercise restraint when exporting certain nuclear technology.[16] These guidelines are not legally binding on member states and the NSG is not set up to identify violations in the event that they occur. Nevertheless, the institution could discourage behavior that is risky from a nonproliferation standpoint by alleviating a classic dilemma. Suppliers may fear that if they control exports in the name of nonproliferation, the denied state would manage to secure aid from a different country. If this were to happen, the supplier would forgo the economic and strategic benefits that accompany nuclear transactions without obtaining a net nonproliferation benefit. To avoid this "sucker's payoff," countries might provide nuclear assistance despite obvious proliferation risks. The policy harmonization and transparency that emerges from the NSG could reduce suppliers' fears that they will be undercut if they refrain from providing assistance to a country with a weak nonproliferation record.

Two testable implications emerge from the preceding discussion. The first is that the factors causing atomic assistance vary in the periods prior to and following the Indian nuclear test. Specifically, countries should be more worried about nonproliferation considerations and less willing to use atomic aid to pursue strategic objectives after 1974. The second implication is that NSG members behave differently than non-NSG members when it comes to peaceful nuclear assistance.[17]

I conducted further statistical analysis to evaluate these propositions. The findings reveal that the correlates of peaceful nuclear cooperation are remarkably similar in the periods before and after 1974. Most significantly, all five of the variables relevant to my argument are statistically significant in the expected direction in these two time periods. The results also show that nonproliferation considerations had little effect on the nuclear marketplace both before and after 1974. States pursuing nuclear weapons were more likely than states that were not pursuing the bomb to receive assistance in both time periods. And NPT membership was statistically unrelated to peaceful nuclear cooperation from 1975 to 2000; this variable was significant and negative prior to 1974.[18]

NSG members and non-NSG members also behave very similarly when it comes to peaceful atomic assistance. There is empirical support for my argument regardless of suppliers' status with respect to the NSG. Moreover, there is equally scant support for the nonproliferation argument among both groups of suppliers. Non-NSG members are statistically less likely to assist NPT states than non-NPT states and more likely to cooperate with countries that are pursuing nuclear weapons than states that are not attempting to developing the bomb. Members of the NSG are similarly more likely to provide assistance to states with weapons programs compared to states that are not attempting to proliferate and no more likely to aid states that commit to the NPT than states that do not.

Despite these findings, it is plausible that the NSG has influenced the type of NCAs that countries sign. As I highlighted above, the organization encourages suppliers to exercise restraint when considering exports of reprocessing centers, enrichment facilities, and heavy water reactors. Seventy-six percent of all comprehensive power NCAs signed between 1950 and 2000 included a provision that explicitly prohibited the transfer of these facilities. If the NSG guidelines have influenced the export practices of suppliers, we would expect to find that members of the group were more likely than nonmembers to include this clause in the treaties that they signed. A simple cross-tabulation analysis shows, however, that this is not the case. The relationship between NSG membership and signing comprehensive power NCAs that place limits on the aforementioned transfers is statistically insignificant.[19] Indeed, non-NSG members signed roughly the same percentage of these limited NCAs as members of the group.[20] These findings lead to an unfortunate conclusion: the NSG has had a much more limited effect on the export behavior of nuclear suppliers than many scholars and policymakers believe.

Appendix 3.1: Data Analysis and Empirical Findings

Dataset

As I described above, I adopt a time-series cross-sectional data structure for the period 1950 to 2000. The unit of analysis is the directed dyad year. Included in the dataset are all nuclear suppliers and all potential recipient dyads in the international system. Nuclear suppliers are identified based on the procedures discussed in appendix 1.1. All states in the system are potential recipients of nuclear-related commodities. These spatial and temporal constraints leave 153,895 observations in the dataset.[21]

Independent Variables

My theory predicts that shared enemies, superpower enemies, military alliances, and joint democracy increase the likelihood of nuclear cooperation while militarized conflict reduces it. To determine whether states are military allies, I consulted version 3.0 of the Correlates of War (COW) Formal Alliance Data.[22] I include a dummy variable that equals 1 if the supplier and recipient share a formal defense pact in year t-1 and 0 otherwise. Data on militarized conflict are taken from version 3.0 of the COW Militarized Interstate Dispute (MID) dataset.[23] I include a dummy variable and code it 1 if the supplier and recipient states were involved in an ongoing MID in year t-1 and 0 otherwise.[24]

Data on shared enemies and superpower enemies were self-coded from a rivalry dataset compiled by James Klein, Gary Goertz, and Paul Diehl.[25] I create a dichotomous variable and code it 1 if two states are part of a rivalry with the same state in year t-1 and 0 if not. I create a second dichotomous variable and code it 1 if the importing state is a rival of either the Soviet Union between 1950 and 1991 or the United States between 1950 and 2000 in year t-1 and 0 otherwise.

Democracy data were obtained from the Polity IV dataset.[26] The Polity data use a twenty-one-point scale, with highly authoritarian states coded as -10 and highly democratic states coded as $+10$, to measure the relative openness of states' domestic instructions. I include a dummy variable and code it 1 if the recipient and supplier states both have a score of at least 6—which is the recommend cut point for a democracy—on the Polity scale in year t-1 and 0 in other cases.

Control Variables

I code the following indicators as proxies for the competing explanations:

- *Nuclear Weapons.* I create a dichotomous variable and code it 1 if the recipient state is pursuing nuclear weapons in year t-1 based on Sonali Singh and Christopher Way's proliferation data.[27] States are coded 0 once they produce at least one nuclear weapon (e.g., France after 1960) because at that point suppliers cannot prevent horizontal proliferation by withholding nuclear assistance.

- *NPT.* I include a dummy variable and code it 1 if the importing state ratified the NPT in year t-1 and 0 otherwise. To classify membership in this institution I consult a list compiled by the James Martin Center for Nonproliferation Studies.[28]

- *Supplier's GDP and Recipient's GDP.* I include variables measuring the GDP of the exporting state and the GDP of the importing state in year t-1.[29]

- *Distance.* I include a variable measuring the "great circle" distance between the capitals of states.[30]

- *Exports.* I include a variable measuring the total exports in thousands of U.S. dollars from the supplier to the recipient country in year t-1. I code this variable based on trade data compiled by Kristian Gleditsch.[31]

- *Domestic Demand.* I include a variable measuring the number of nuclear power reactors per capita in the supplier state. To calculate this variable I divide the number of power plants in the supplier country by its population and multiply this factor by 100 for the purposes of rescaling. I measure the number of working nuclear power reactors in the supplier country using data obtained from the International Atomic Energy Agency (IAEA).[32]

Method

Since the dependent variable is dichotomous, I use logistic regression analysis (logit) to estimate the relationships between the independent variables and peaceful nuclear assistance. I employ clustering across dyads to control for heteroskedastic error variance and use white robust estimation to correct the standard errors for spatial dependence. Additionally, I introduce a variable measuring the number of years since 1950 that lapse without a dyad signing an NCA and three cubic splines to control for temporal dependence in the dependent variable.[33]

Results

Table 3.7 displays the initial results of the multivariate statistical analysis that were previously illustrated in figure 3.1. Model 1 is a baseline model that only includes four variables operationalizing my strategic argument (joint democracy is excluded) along with controls for temporal dependence. Model 2 adds the democracy variable and limits the sample to democratic suppliers, which is why the number of observations declines compared to model 1.[34] Models 3 and 4 add the control variables to models 1 and 2, respectively. To appropriately test whether NPT membership influences atomic assistance I need to limit the sample temporally because the treaty did not enter in to force until 1970.[35] Thus, models 5 and 6 limit the estimation sample to observations from 1970 to 2000 and include the NPT variable. Recall that the calculations for figure 3.1 were based on models 3 and 4.

The five variables relevant to my argument are statistically significant in the expected direction across models 1–6. Thus, the findings supporting my theory do not appear to be sensitive to the inclusion of control variables, the regime type of the supplier, or the time period analyzed.[36] From a statistical standpoint the findings relevant to the economic arguments are also consistent across the models displayed in table 3.7. There are some differences when it comes to the nonproliferation argument, however. The relationship between signing nuclear cooperation agreements and NPT membership is statistically insignificant in model 5 but this variable is negative and statistically

Table 3.7 Determinants of peaceful nuclear cooperation, 1950–2000

	(1)	(2)	(3)	(4)	(5)	(6)
	Baseline	Baseline (democracy)	Full model	Full model (democracy)	NPT era	NPT era (democracy)
Strategic factors						
Alliance	1.643***	1.146***	1.235***	0.862***	0.996***	0.537***
	(0.084)	(0.099)	(0.091)	(0.108)	(0.094)	(0.108)
Conflict	−1.497***	−1.716*	−1.877***	−2.146**	−2.190***	−1.908*
	(0.516)	(1.043)	(0.551)	(1.072)	(0.790)	(1.109)
Shared enemy	0.719***	0.552***	0.326***	0.300***	0.254**	0.267**
	(0.101)	(0.107)	(0.105)	(0.111)	(0.109)	(0.121)
Superpower enemy	0.942***	0.916***	0.956***	0.967***	1.076***	1.199***
	(0.082)	(0.094)	(0.082)	(0.089)	(0.082)	(0.087)
Joint democracy		1.237***		1.130***		1.261***
		(0.095)		(0.093)		(0.103)
Nonproliferation						
Nuclear weapons			0.271*	0.092	0.273*	0.202
			(0.146)	(0.159)	(0.144)	(0.154)
NPT					0.016	−0.182*
					(0.086)	(0.099)

(*Continued*)

Table 3.7—*cont.*

	(1) Baseline	(2) Baseline (democracy)	(3) Full model	(4) Full model (democracy)	(5) NPT era	(6) NPT era (democracy)
Generating revenue						
Supplier's GDP			0.000***	0.000***	0.000***	0.000***
			(0.000)	(0.000)	(0.000)	(0.000)
Recipient's GDP			0.000***	0.000***	0.000***	0.000***
			(0.000)	(0.000)	(0.000)	(0.000)
Distance			−0.000***	−0.000***	−0.000***	−0.000***
			(0.000)	(0.000)	(0.000)	(0.000)
Exports			−0.000**	−0.000**	−0.000**	−0.000**
			(0.000)	(0.000)	(0.000)	(0.000)
Sustaining industry						
Domestic demand			5.219***	2.666***	4.086***	1.867**
			(0.901)	(0.978)	(0.884)	(0.925)
Temporal controls						
No NCA years	−0.222***	−0.229***	−0.174***	−0.183***	−0.240***	−0.241***
	(0.018)	(0.019)	(0.018)	(0.019)	(0.021)	(0.022)
Spline 1	−0.001***	−0.002***	−0.001***	−0.001***	−0.001***	−0.002***
	(0.000)	(0.000)	(0.000)	(0.000)	(0.000)	(0.000)

Spline 2	0.000**	0.001***	0.000	0.001***	0.000	0.001***
	(0.000)	(0.000)	(0.000)	(0.000)	(0.000)	(0.000)
Spline 3	0.000	−0.000	0.000	−0.000	0.000	−0.000
	(0.000)	(0.000)	(0.000)	(0.000)	(0.000)	(0.000)
Constant	−3.444***	−3.867***	−3.359***	−3.829***	−2.857***	−3.321***
	(0.086)	(0.100)	(0.103)	(0.114)	(0.126)	(0.133)
Observations	161492	108940	153895	102199	112893	79693

Notes: Robust standard errors in parentheses; * significant at 10%; ** significant at 5%; *** significant at 1%.

significant in model 6, indicating that democratic suppliers are actually less likely to provide nuclear assistance to states that enter the NPT. Recall that earlier in the chapter I performed tests to assess the robustness of the finding; in no case was the NPT significantly related to nuclear cooperation in the positive direction. The coefficient on the variable measuring whether the importing state is pursuing nuclear weapons is positive and statistically significant in models 3 and 5. Note, however, that this variable is indistinguishable from 0 in models 4 and 6. Altering the coding to include all states with nuclear weapons programs, regardless of whether they have produced the bomb or not, produces substantively similar results. Although proliferating countries seek out peaceful nuclear assistance, they are not necessarily more likely to receive it than states that are not pursuing nuclear weapons.

DISAGGREGATED NUCLEAR COOPERATION

Table 3.8 displays the results of the analysis with the disaggregated dependent variables. For each dependent variable, the first model tests my first four hypotheses and includes all suppliers in the sample. The second model is meant primarily to evaluate hypothesis 3.5; it adds the joint democracy variable and limits the sample to democratic suppliers. There are ten models reported in table 3.8.

Many of the findings remain consistent regardless of which dependent variable is employed, as I discussed earlier in the chapter. The alliance variable is statistically significant across models 7–16, indicating that allies are more likely to sign all five types of NCAs. The conflict variable is typically dropped from statistical models because it predicts failure perfectly.[37] In other words, states usually do not sign NCAs in the year after they experience a militarized dispute. Note that coefficients and standard errors for this variable are reported in three models; in two cases the variable is insignificant (models 10 and 11) and in one case it is correlated with nuclear cooperation (model 9). Technically, the results from model 10 indicate that democratic suppliers are no less likely to sign intangible NCAs with states that they fight than with states that they do not fight. This result is driven by a single agreement, however. India signed an intangible NCA with Sri Lanka in 1986 after experiencing a militarized interstate dispute in 1985; this is the only case of a democratic supplier signing one of these deals with a state it experienced a dispute with in the previous year. A cross-tabulation analysis reveals that 99.9 percent (887 of 888) of the cases of intangible assistance involving a democratic supplier followed peace in the previous year. The shared enemy variable is statistically significant in every case except models 11 and 12, indicating that it is unrelated to the signing of nuclear material NCAs. The superpower enemy variable is insignificant in model 8 but significant in all of the other models reported in table 3.8. This finding indicates that democratic suppliers are no more likely to sign comprehensive research NCAs with superpower enemies than nonsuperpower enemies. Yet when all

Table 3.8a Determinants of disaggregated peaceful nuclear cooperation, 1950–2000

	(7) Safety	(8) Safety	(9) Intangibles	(10) Intangibles	(11) Materials	(12) Materials
Strategic factors						
Alliance	1.041***	0.523***	1.611***	1.130***	0.843**	1.090**
	(0.138)	(0.150)	(0.115)	(0.136)	(0.352)	(0.454)
Conflict			-1.076*	-0.950	-0.492	
			(0.599)	(1.045)	(1.055)	
Shared enemy	0.437***	0.372**	0.239*	0.271**	-0.101	0.264
	(0.149)	(0.154)	(0.139)	(0.137)	(0.312)	(0.357)
Superpower enemy	1.022***	1.018***	1.144***	1.036***	0.823***	1.069***
	(0.123)	(0.127)	(0.098)	(0.101)	(0.272)	(0.336)
Joint democracy		1.603***		1.430***		0.001
		(0.148)		(0.129)		(0.332)
Nonproliferation						
Nuclear weapons	-0.573*	-0.115	0.111	-0.111	0.742*	0.716
	(0.343)	(0.338)	(0.227)	(0.255)	(0.385)	(0.475)
NPT	1.087***	0.797***	-0.047	-0.118	-0.756***	-0.665**
	(0.193)	(0.196)	(0.090)	(0.099)	(0.242)	(0.323)

(*Continued*)

Table 3.8a—cont.

	(7) Safety	(8) Safety	(9) Intangibles	(10) Intangibles	(11) Materials	(12) Materials
Generating revenue						
Supplier's GDP	0.000***	0.000***	0.000***	0.000***	0.000***	0.000**
	(0.000)	(0.000)	(0.000)	(0.000)	(0.000)	(0.000)
Recipient's GDP	0.000***	0.000***	0.000***	0.000***	0.000***	0.000***
	(0.000)	(0.000)	(0.000)	(0.000)	(0.000)	(0.000)
Distance	−0.000***	−0.000***	−0.000***	−0.000***	−0.000***	−0.000*
	(0.000)	(0.000)	(0.000)	(0.000)	(0.000)	(0.000)
Exports	−0.000**	−0.000**	−0.000**	−0.000***	−0.000*	−0.000**
	(0.000)	(0.000)	(0.000)	(0.000)	(0.000)	(0.000)
Sustaining industry						
Domestic demand	10.676***	7.455***	2.336*	−2.008	−6.864*	−4.428
	(1.213)	(1.288)	(1.252)	(1.384)	(3.764)	(4.739)
Temporal controls						
No NCA years	−0.192***	−0.173***	−0.158***	−0.167***	−0.137**	−0.100
	(0.027)	(0.028)	(0.022)	(0.024)	(0.064)	(0.081)
Spline 1	0.000	−0.000	−0.001***	−0.001***	−0.001*	−0.001
	(0.000)	(0.000)	(0.000)	(0.000)	(0.001)	(0.001)

	(1)	(2)	(3)	(4)	(5)	(6)
Spline 2	−0.001***	−0.001*	0.001***	0.001***	0.001	0.000
	(0.000)	(0.000)	(0.000)	(0.000)	(0.001)	(0.001)
Spline 3	0.001***	0.000***	−0.000*	−0.000**	−0.000	−0.000
	(0.000)	(0.000)	(0.000)	(0.000)	(0.000)	(0.000)
Constant	−5.031***	−5.567***	−4.331***	−4.764***	−5.856***	−6.490***
	(0.173)	(0.192)	(0.135)	(0.173)	(0.370)	(0.521)
Observations	153199	101870	153895	102199	153895	101870

Notes: Robust standard errors in parentheses; * significant at 10%; ** significant at 5%; *** significant at 1%. Conflict is dropped from some models because it predicts failure perfectly.

Table 3.8b Determinants of disaggregated peaceful nuclear cooperation, 1950–2000

	(13) Comprehensive research	(14) Comprehensive research	(15) Comprehensive power	(16) Comprehensive power
Strategic factors				
Alliance	2.043***	2.137***	1.090***	0.800***
	(0.220)	(0.261)	(0.162)	(0.194)
Conflict				
Shared enemy	0.808***	0.804***	0.380**	0.419**
	(0.287)	(0.310)	(0.155)	(0.174)
Superpower enemy	0.766***	0.399	0.936***	1.133***
	(0.289)	(0.301)	(0.151)	(0.182)
Joint democracy		0.142		0.738***
		(0.262)		(0.159)
Nonproliferation				
Nuclear weapons	−0.648	−1.065	0.828***	0.484**
	(0.595)	(1.029)	(0.198)	(0.241)
NPT	−0.557*	−0.882***	−0.385**	−0.719***
	(0.325)	(0.338)	(0.162)	(0.190)

Generating revenue				
Supplier's GDP	0.000	−0.000	0.000***	0.000***
	(0.000)	(0.000)	(0.000)	(0.000)
Recipient's GDP	−0.000	−0.000	0.000***	0.000***
	(0.000)	(0.000)	(0.000)	(0.000)
Distance	0.000	0.000***	−0.000***	0.000
	(0.000)	(0.000)	(0.000)	(0.000)
Exports	−0.000	−0.000	−0.000***	−0.000***
	(0.000)	(0.000)	(0.000)	(0.000)
Sustaining industry				
Domestic demand	−30.704***	−30.607***	−3.933*	−1.391
	(9.101)	(8.668)	(2.201)	(2.052)
Temporal controls				
No NCA years	0.018	0.055	0.018	0.041
	(0.051)	(0.060)	(0.034)	(0.044)
Spline 1	0.002**	0.002**	0.001**	0.001*
	(0.001)	(0.001)	(0.000)	(0.000)
Spline 2	−0.002***	−0.002**	−0.001***	−0.001*
	(0.001)	(0.001)	(0.000)	(0.000)

(Continued)

Table 3.8b—*cont.*

	(13)	(14)	(15)	(16)
	Comprehensive research	Comprehensive research	Comprehensive power	Comprehensive power
Spline 3	0.001***	0.001**	0.000***	0.000*
	(0.000)	(0.000)	(0.000)	(0.000)
Constant	−6.531***	−6.922***	−5.532***	−6.309***
	(0.288)	(0.324)	(0.207)	(0.247)
Observations	153199	101870	153199	101870

Notes: Robust standard errors in parentheses; * significant at 10%; ** significant at 5%; *** significant at 1%. Conflict is dropped from some models because it predicts failure perfectly.

suppliers are included in the sample, this variable is correlated with the signing of research agreements (see model 13). Finally, the joint democracy variable is statistically significant in models 8, 10, and 16 but insignificant in models 12 and 14, revealing that democratic suppliers are not more likely to sign material NCAs and research NCAs with other democracies than with nondemocracies.

FURTHER ANALYSIS OF THE NSG

Table 3.9 presents the findings of the statistical analysis designed to evaluate whether the Indian nuclear test in 1974 and the subsequent formation of the NSG influenced the correlates of peaceful nuclear cooperation. Models 17–20 include all of the independent variables but vary based on the observations that are included in the estimation sample. Model 17 includes observations from 1950 to 1974 (i.e., prior to the creation of the NSG), while model 18 includes post-1974 observations. Model 19 limits the sample to suppliers that are part of the NSG and model 20 includes only non-NSG suppliers.

All five variables testing the hypotheses are statistically significant in the expected direction, indicating that my argument applies to a diverse set of suppliers in distinct time periods. When it comes to the control variables, the results are generally consistent compared to the findings reported in table 3.7. None of the variables testing the nonproliferation argument behave as expected, while most of the variables testing the revenue maximization argument are statistically significant. Collectively, these findings suggest that NSG members and non-NSG members behave similarly when it comes to providing peaceful nuclear assistance and that India's 1974 nuclear test was less of a shock to the system than the conventional wisdom suggests.

ROBUSTNESS CHECKS

In this section, I present a series of robustness checks to evaluate the sensitivity of my findings.[38] To begin, I alter the size of the estimation sample. The initial tests relied on fairly lenient criteria to identify nuclear suppliers. I replicate the statistical analysis while adopting a stricter definition of a capable nuclear supplier that only includes states in the estimation sample when they can offer advanced nuclear technology.[39] It is also possible that economically underdeveloped states might never have the opportunity to receive atomic assistance. I reestimate the full model displayed in table 3.7 but only include recipients that have at least one nuclear power plant in the sample.

As table 3.10 illustrates, the five variables testing the hypotheses remain statistically significant in the expected direction. Many of the findings with respect to the control variables are also consistent, although nuclear weapons pursuit is statistically insignificant in models 21 and 22. Domestic demand is also insignificant in model 21; it was statistically significant in the models reported in table 3.7.

Table 3.9 The Nuclear Suppliers Group and peaceful nuclear cooperation, 1950–2000

	(17) Pre-NSG	(18) Post-NSG	(19) NSG supplier	(20) Non-NSG supplier
Strategic factors				
Alliance	1.204***	0.493***	0.335***	1.386***
	(0.121)	(0.104)	(0.125)	(0.104)
Conflict	−1.416*	−1.703**		−1.072*
	(0.723)	(0.784)		(0.626)
Shared enemy	0.325**	0.254**	0.271**	0.409***
	(0.152)	(0.109)	(0.120)	(0.136)
Superpower enemy	0.460***	1.117***	1.102***	0.833***
	(0.131)	(0.084)	(0.103)	(0.102)
Joint democracy	0.646***	1.152***	1.082***	0.784***
	(0.110)	(0.090)	(0.110)	(0.088)
Nonproliferation				
Nuclear weapons	0.299	0.383***	0.088	0.419**
	(0.235)	(0.136)	(0.183)	(0.166)
NPT	−0.529***	−0.160*	−0.055	−0.238***
	(0.142)	(0.097)	(0.127)	(0.087)

Generating revenue				
Supplier's GDP	0.000***	0.000***	0.000***	0.000***
	(0.000)	(0.000)	(0.000)	(0.000)
Recipient's GDP	0.000***	0.000***	0.000***	0.000***
	(0.000)	(0.000)	(0.000)	(0.000)
Distance	−0.000***	−0.000***	−0.000***	−0.000***
	(0.000)	(0.000)	(0.000)	(0.000)
Exports	0.000	−0.000**	−0.000**	−0.000
	(0.000)	(0.000)	(0.000)	(0.000)
Sustaining industry				
Domestic demand	3.360	−2.719**	0.750	−8.705**
	(2.232)	(1.121)	(0.943)	(3.898)
Temporal controls				
No NCA years	−0.063***	−0.267***	−0.232***	−0.025
	(0.021)	(0.025)	(0.022)	(0.029)
Spline 1	0.000	−0.002***	−0.001***	0.001
	(0.000)	(0.000)	(0.000)	(0.001)
Spline 2	−0.001*	0.001***	0.001**	−0.001
	(0.000)	(0.000)	(0.000)	(0.002)
Spline 3	0.000***	−0.000**	−0.000	0.001
	(0.000)	(0.000)	(0.000)	(0.002)
Constant	−4.150***	−2.673***	−3.065***	−4.071***
	(0.128)	(0.168)	(0.130)	(0.150)
Observations	105448	48317	96977	56918

Notes: Robust standard errors in parentheses; * significant at 10%; ** significant at 5%; *** significant at 1%.

Table 3.10 Determinants of peaceful nuclear cooperation with limited sample of suppliers and recipients, 1950–2000

	(21)	(22)
	Advanced suppliers	Advanced recipients
Strategic factors		
Alliance	0.593***	0.629***
	(0.099)	(0.097)
Conflict	−1.192**	−1.772**
	(0.572)	(0.792)
Shared Enemy	0.379***	0.208**
	(0.105)	(0.100)
Superpower enemy	0.952***	0.628***
	(0.088)	(0.085)
Joint democracy	0.952***	0.287***
	(0.082)	(0.086)
Nonproliferation		
Nuclear weapons	0.216	−0.158
	(0.138)	(0.177)
NPT	−0.090	−0.209**
	(0.070)	(0.082)
Generating revenue		
Supplier's GDP	0.000***	0.000***
	(0.000)	(0.000)
Recipient's GDP	0.000***	0.000***
	(0.000)	(0.000)
Distance	−0.000***	−0.000**
	(0.000)	(0.000)
Exports	−0.000***	−0.000**
	(0.000)	(0.000)
Sustaining industry		
Domestic demand	−1.437	4.066***
	(0.985)	(1.059)

(*Continued*)

Table 3.10—*cont.*

	(21)	(22)
	Advanced suppliers	Advanced recipients
Temporal controls		
No NCA Years	−0.207***	−0.210***
	(0.019)	(0.022)
Spline 1	−0.001***	−0.002***
	(0.000)	(0.000)
Spline 2	0.001**	0.001***
	(0.000)	(0.000)
Spline 3	−0.000	−0.000***
	(0.000)	(0.000)
Constant	−2.878***	−2.224***
	(0.123)	(0.156)
Observations	99005	17502

Notes: Robust standard errors in parentheses; * significant at 10%; ** significant at 5%; *** significant at 1%.

As a second test to evaluate the robustness of my results, I explore whether there are differences in the correlates of atomic assistance in the post–cold war era. Some might argue that the shift from a bipolar to unipolar international system after the cold war made countries less inclined to link atomic assistance to their strategic interests. Joanne Gowa, for example, finds that states are more likely to tie trade to the flag when the international system is bipolar because alliances are more stable.[40] With the collapse of the bipolar structure, nuclear commerce may be less influenced by international security because there is less of a strategic need to link trade and security in a unipolar world.[41] It is also possible that revelations following the 1991 Persian Gulf War changes patterns of peaceful nuclear cooperation. As I will discuss throughout Part II of the book, the international community learned after the American-led coalition expelled Iraq from Kuwait that Baghdad had built an expansive secret nuclear infrastructure throughout the 1980s. In response to Iraq's nuclear program, the NSG revised its guidelines and the IAEA revamped its system of safeguards by instituting the Additional Protocol. These events may have encouraged suppliers to exercise greater caution when providing nuclear assistance.

To test whether this is the case, I replicate the statistical analysis for the period from 1992 to 2000. The results are displayed in table 3.11. Model 23 excludes the joint democracy variable from the model and includes all

Table 3.11 Determinants of peaceful nuclear cooperation, 1992–2000

	(23)	(24)
Strategic factors		
Alliance	0.406***	0.066
	(0.149)	(0.155)
Conflict		
Shared enemy	0.297**	0.358**
	(0.150)	(0.157)
Superpower enemy	1.276***	2.174***
	(0.175)	(0.172)
Joint democracy		1.673***
		(0.161)
Nonproliferation		
Nuclear weapons	−1.561***	−2.058**
	(0.543)	(0.940)
NPT	0.224	−0.188
	(0.184)	(0.192)
Generating revenue		
Supplier's GDP	0.000***	0.000***
	(0.000)	(0.000)
Recipient's GDP	0.000***	0.000***
	(0.000)	(0.000)
Distance	−0.000	0.000
	(0.000)	(0.000)
Exports	−0.000**	−0.000*
	(0.000)	(0.000)
Sustaining industry		
Domestic demand	3.245**	3.101**
	(1.413)	(1.496)
Temporal controls		
No NCA Years	−0.194***	−0.152***
	(0.037)	(0.040)

(*Continued*)

Table 3.11—*cont.*

	(23)	(24)
Spline 1	−0.000	−0.001
	(0.001)	(0.001)
Spline 2	0.000	0.000
	(0.000)	(0.001)
Spline 3	−0.000	−0.000
	(0.000)	(0.000)
Constant	−3.367***	−4.548***
	(0.216)	(0.252)
Observations	35889	29226

Notes: Robust standard errors in parentheses; * significant at 10%; ** significant at 5%; *** significant at 1%. Conflict is dropped from both models because it predicts failure perfectly.

suppliers in the sample. Model 24 includes only democratic suppliers in the sample and adds the joint democracy variable. The results generally demonstrate that my argument is valid in the post–cold war era. This is striking because it indicates that strategic factors drive the nuclear marketplace even after the collapse of the bipolar international system and the shocks resulting from Iraq's nuclear program. There is only one difference with respect to the five variables testing the hypotheses. Military alliance is statistically insignificant in model 24, indicating that democratic suppliers are no more likely to sign NCAs with their allies in the period from 1992 to 2000 than they were during the cold war. Note, however, that the alliance variable is statistically significant in model 23, which suggests that nondemocratic suppliers continued to use atomic assistance as a means to strengthen their allies and alliances in this era.

The alternative explanations perform about as well (or as poorly) as they did previously. The one noteworthy difference is that nuclear weapons pursuit is negative and statistically significant in both models. The nonnuclear weapons states that were pursuing the bomb during this period—Iran, Iraq, Libya, and North Korea—were statistically less likely to receive nuclear assistance.[42] On the one hand, this implies that nonproliferation factors are salient in explaining atomic assistance in the post–cold war era. Yet surprisingly, the NPT variable remains statistically insignificant in both models. Only four states remain outside of the NPT today (India, Israel, North Korea, and Pakistan). Some states accumulated nuclear assistance throughout the 1990s, however, despite being non-NPT members at the time.[43]

As a final robustness check (not reported), I replicate the statistical analysis with additional control variables that might affect atomic assistance. These new variables are described briefly below.

- *Neighbor NCA.* Contagion effects could influence nuclear assistance. In particular, a state might be more likely to demand nuclear assistance when its neighbors receive aid because they learn from the experiences of states around them or as a result of relative gains considerations. To control for this possibility, I create a dummy variable and code it 1 if a state's neighbor imports nuclear technology in year t-1 and 0 otherwise.[44]

- *Energy Demand.* Countries with greater energy needs may be more likely to seek atomic assistance. To control for this, I include a variable measuring the ratio of the country's energy production capacity to its population. These data are obtained from the COW's National Military Capabilities dataset.[45]

- *Oil Price.* The price of oil might affect nuclear assistance because countries may be more likely to turn to nuclear energy when the alternatives become costly. To control for this, I include a variable measuring the price of a barrel of oil measured in U.S. dollars in year t-1. Data on oil prices come from the *Historical Statistics of the United States.*[46]

- *NSG.* Membership in the NSG might inspire greater confidence that states will not use peaceful nuclear assistance for military purposes. I include a variable measuring whether the importing state is part of the NSG in year t-1. To classify membership in the group I consult a list compiled by the James Martin Center for Nonproliferation Studies.[47]

I add these new controls one by one to the full statistical model. All five variables relevant to my argument remain statistically significant in the expected direction. Likewise, the variables testing the competing explanations perform similarly, although domestic demand does not achieve conventional levels of statistical significance in any of the models. Some of the new controls are statistically significant and their effects are usually in the expected direction. For instance, countries are more likely to receive atomic assistance when their neighbors sign NCAs and when they are members of the NSG. Energy demand and oil price are statistically significant in some models but the findings are sensitive to model specification.

Chapter 4

Nuclear Arms and Influence

Assisting India, Iran, and Libya

Do the causal processes driving my theory operate correctly in actual cases of civilian nuclear assistance? One way to answer this question is to qualitatively evaluate cases where my statistical model correctly predicted the occurrence of nuclear cooperation.[1] Such cases should yield two main pieces of evidence if my argument is correct. First, leaders and other senior decision makers should justify atomic assistance on the grounds that it will strengthen the recipient country and the supplier's bilateral relationship with that country. Second, government officials should indicate that this is part of a broader strategy to (1) keep their allies and alliances strong; (2) counter the influence of their adversaries; or (3) prop up existing democracies.

I analyze three successfully predicted cases in this chapter: U.S. assistance to Iran (1957–79); Soviet assistance to Libya (1975–86); and Canadian assistance to India (1955–77). To evaluate whether my argument explains a more recent, high-profile case, I also examine U.S. nuclear cooperation with India (2001–08).[2] Given the statistical results, different parts of my argument should explain the behavior of nuclear suppliers in these cases. In the United States–Iran case, we should see evidence that a desire to strengthen the military alliance and counter the influence of the Soviet Union motivated Washington to initiate atomic assistance to Tehran. Countering the influence of a shared adversary—the United States—should be salient in explaining the Soviet–Libya case. The Canada–India case ought to yield evidence that atomic aid occurred in part because India was a democracy. Finally, countering the influence of a common adversary, China, and strengthening an existing democracy should be salient in explaining the United States–India case.

The historical record supports these expectations. In each case, suppliers used atomic energy assistance to strengthen the recipient country and their bilateral relationship with that country in ways that promoted their political/ strategic interests. Yet there is considerably less evidence to support explanations rooted in nonproliferation or economics. Supplier states sometimes had suspicions that the recipient state was pursuing nuclear weapons or that it might initiate a bomb program in the future—but this did not prevent nuclear cooperation from taking place. This does not imply that exporters wanted the recipient to build nuclear weapons. It does show, however, that countries will take calculated gambles in the nuclear marketplace under the right circumstances. Economics mattered more than nonproliferation, but it was not the driving force in any of these cases.

U.S. Nuclear Cooperation with Iran, 1957–1979

U.S. civilian nuclear assistance to Iran began after the two states signed an agreement for cooperation in the peaceful uses of nuclear energy on March 5, 1957. This agreement stated that the United States would supply "information as to the design, construction, and operation of research reactors and their use as research development and engineering tools" as well as "appropriate equipment and services" to Iran.[3] It took a few years for nuclear cooperation to develop, in part because Iran had an insufficient electrical capacity and no national grid to accommodate even a single power reactor.[4] During the 1960s, the United States built a 5MWt water-moderated research reactor in Tehran and provided 5.5 kg of enriched uranium to fuel it.[5] In November 1967, the Tehran research reactor went critical.[6] During this period, the United States also provided Iran with "hot cells" and training for their use.[7]

After these initial transfers, the United States worked to deepen its nuclear cooperation with Iran. In March 1969, President Richard Nixon amended the 1957 agreement, extending it for an additional ten years. Plans for cooperation really began to accelerate in 1974 when the shah established the Atomic Energy Organization of Iran (AEOI) and pursued arrangements to generate twenty-three thousand MWe from nuclear power facilities. A State Department telegram on March 11, 1974, stated that Washington had an interest in "mov[ing] quickly" to "broaden ties with Iran," including cooperation in the development of nuclear breeder reactors. Officials in Washington "noted the priority the Shah gives to developing alternative means of energy production through nuclear power and agree this is the area in which we might most usefully begin on a specific program of cooperation and collaboration."[8] According to a declassified study on "Joint U.S.-Iranian Cooperation," the United States had an interest in becoming "a major source of the equipment as well as the technology used" in the burgeoning Iranian nuclear industry.[9] In May 1974, Dixie Ray, head of the U.S. Atomic Energy Commission (AEC) visited Tehran and established direct contact between the AEC and the Atomic Energy Organization of Iran. During Ray's visit, the United States agreed to supply enriched uranium for eight power reactors and considered plans for a "broad range of future cooperative activities" with Iran.[10] Iran also expressed interest in participating in a proposed commercial uranium enrichment facility that was planned to be built in the United States.

U.S. Secretary of State Henry Kissinger visited Iran from November 1–3, 1974. During Kissinger's visit, the two countries created a subcommission on nuclear energy—within a broader bilateral Economic Commission—to facilitate plans for intensified cooperation in this area.[11] This commission met in Washington on March 3–4, 1975, and announced Iran's intentions to purchase nuclear power plants from U.S. firms. This announcement was described as the "most dramatic feature of the meeting."[12] Henry Kissinger and Iranian finance minister Hushang Ansary signed a $15 billion trade agreement, which called for U.S. firms to supply eight nuclear power reactors in exchange for $6.4 billion.[13] In order for these transfers to take place, the United States needed to sign another nuclear cooperation agreement with Iran. The 1957 agreement, which had been extended until 1979, dealt only with nuclear research and could not be extended to cover nuclear power as well.[14]

The United States and Iran made progress toward signing such an agreement. In March 1975, President Gerald Ford called for a study looking at the implications of such an agreement. The major point of contention stymieing the signing of a nuclear agreement dealt with Iran's authority to reprocess plutonium.[15] This disagreement created tension between the two countries, and in August 1976 negotiations on a nuclear cooperation agreement were suspended.

Cooperation soon resumed, however. In April 1977, the two countries signed an agreement pledging to cooperate on technical matters related to nuclear safety.[16] And on August 8, 1977 talks between the two states regarding the export of power reactors resumed after Iran renounced its intention to build a reprocessing facility.[17] On January 1, 1978, President Jimmy Carter and the shah reached a tentative agreement on nuclear cooperation that would permit the aforementioned nuclear sales to take place. This agreement represented a compromise on reprocessing; it did not promise reprocessing rights for Iran but neither did it permit a U.S. veto over such rights. This provided Iran with "most favored nation" status as far as reprocessing is concerned.[18]

Just when the prospects for nuclear cooperation looked bright, troubles ensued. The United States grew increasingly troubled by the social unrest and political turmoil that existed in Iran.[19] Following the 1979 Islamic Revolution, the United States ended its supply of highly enriched uranium and terminated all nuclear cooperation with Iran.

U.S. assistance to Iran illustrates that nuclear suppliers play a dangerous game. In the end, atomic aid harmed U.S. security by indirectly—and inadvertently—contributing to Iran's nuclear weapons program.[20] For instance, scientists conducted covert experiments at the American-supplied Tehran research reactor that could help the Iranians build the bomb.[21]

Strengthening Iran and the American-Iranian Partnership

A review of declassified documents and other primary sources lends strong support to my argument that American nuclear assistance to Iran was intended to strengthen Tehran and solidify a strategic partnership between the two countries. American atomic aid was perceived to be important for Iran's development due to its growing energy needs and a diminishing oil supply. Gary Sick, who dealt with nonproliferation issues under Presidents Ford, Carter, and Reagan, noted that "the shah made a big convincing case that Iran was going to run out of gas and oil and they had a growing population and a rapidly increasing demand for energy."[22] As I will discuss further below, Washington intended its nuclear assistance to help Iran meet its rising energy demand and strengthen it economically. U.S. officials were also aware that atomic exports would enhance Iran's capabilities indirectly by freeing up resources. For instance, a National Security Council document that summarized the U.S. strategy for pursuing a nuclear cooperation agreement with Iran stated that receiving nuclear assistance from the United States will "free remaining oil reserves for export or conversion to petrochemicals," indirectly augmenting Iran's national power.[23]

Atomic energy cooperation was a "prominent centerpiece" of American attempts to strengthen its bilateral relationship with Iran.[24] In April 1974, prior to his visit to Tehran, Henry Kissinger called for the United States to cooperate with Iran in the peaceful uses of nuclear energy to provide "concrete evidence of our . . . interest in developing closer ties through specific concrete programs" and signal that we "attach highest value to a partnership" with Iran.[25] Similarly, a report produced by the U.S.-Iran Joint Commission on Economic Cooperation stated that U.S. nuclear exports to Iran would serve as a means to "engage the Iranians so intimately as to assure an enduring relationship under this or successor regimes."[26] And according to a State Department document, the purpose of intensified nuclear cooperation was to "reinforce our close and harmonious relations with Iran, with a view toward promoting a stable and enduring relationship."[27] Turning this logic on its head, the United States knew that failing to supply Iran with power reactors could have adverse effects on its strategic partnership. According to a declassified National Security Council document, a failure to cooperate with Iran in the peaceful uses of nuclear energy could have "serious short, as well as long-term, adverse effects in our relations, given the Shah's sensitivity towards U.S. attitudes and Iran's strong desires to be treated in a nondiscriminatory manner and as a nation that often has supported U.S. interests."[28]

There are two main reasons why U.S. officials wanted to use nuclear assistance to strengthen Tehran and enhance American-Iranian relations. First, the United States needed to ensure that its alliance with Iran remained strong since it depended on Tehran to meet important strategic objectives. And second, both countries were threatened by the Soviet Union and a strong American-Iranian partnership was needed to limit Moscow's influence.

STRENGTHENING THE AMERICAN-IRANIAN MILITARY ALLIANCE

Iran was an important part of the American alliance system during the early cold war period. In 1955, Great Britain, Iran, Iraq, Pakistan, Turkey, and the United States (as an associate member) formed the Baghdad Pact. This arrangement was renamed the Central Treaty Organization (CENTO) following the 1958 revolution in Iraq that resulted in Baghdad's withdrawal from the alliance.[29] In 1959, the United States bolstered its relationship with Iran by signing the Bilateral Defense Treaty on March 5, 1959. This treaty stipulated that the United States would come to the aid of Iran in case of aggression against it.[30]

The United States and Iran signed their first nuclear cooperation agreement just as the two states were becoming allies.[31] The nuclear agreement entered into force on April 27, 1959—just one month after the signing of the Bilateral Defense Treaty. The timing of these events seems to suggest a relationship between an alliance and the initiation of nuclear assistance. Indeed, U.S. nuclear cooperation with Iran was an attempt to strengthen the bilateral alliance, particularly in light of Washington's refusal to formally join the Baghdad Pact (i.e., to become a full, rather than associate, member). A declassified U.S. National Security Council document published on February 3, 1957, stated that technical assistance, including nuclear cooperation for peaceful purposes, is "important as a means of making the presence of the United States felt at all levels of the population throughout [Iran]."[32] This "presence" was perceived to be important in part because of Iran's "strategic location" between the Persian Gulf and the Soviet Union and because of its vast oil reserves.[33]

In the late 1960s and early 1970s, following the British withdrawal from the Persian Gulf, the U.S.-Iran alliance took on increased importance.[34] The United States depended on its alliance with Iran for several reasons. First, Washington relied on military intelligence facilities located in Iran that were "essential to the American capacity to monitor and analyze Soviet adherence to arms control agreements."[35] Second, the United States needed overflight rights allowing American aircraft to have access to the Indian Ocean and South Asia. The only alternatives to Iranian airspace were sensitive routes over Egypt or Israel. Third, the United States required access to Iran's oil, a point that I revisit in chapter 6. Finally, it was important for Washington that Iran play a "constructive regional role" to help limit Soviet influence in the region.[36]

Since the pursuit of U.S. strategic interests depended on the alliance with Iran, Washington worked to keep this partnership strong. It accomplished this, in part, by providing nuclear assistance. Henry Kissinger hinted at this when he justified U.S.-Iran nuclear cooperation by stating simply: "They were an allied country."[37] Charles Naas, who served as deputy U.S. ambassador to Iran in the 1970s, made the argument more directly when he suggested that nuclear assistance took place in part because the alliance with Iran "as a whole was very important."[38] It is telling that all U.S. nuclear cooperation with Iran ended following the collapse of the alliance in 1978–79. This lends support for my argument that states are willing to trade nuclear technology with their allies and reluctant to exchange such items with their enemies. In the post–cold war era—now that Iran is a nonally—the United States has exerted considerable effort to prevent Tehran from acquiring the same technology it was openly offering in the 1970s.

CONSTRAINING THE SOVIET UNION

Iran was a rival of the Soviet Union for a significant portion of the cold war period.[39] Tehran's concerns over Soviet domination date back to 1946 when Moscow refused to withdraw its military forces from northwestern Iran following the end of World War II. The Soviet Union eventually withdrew in May 1946, but this crisis raised concerns in Iran and the United States about possible Soviet control of the Persian Gulf.[40] A February 1957 U.S. Security Council document on "U.S. Policy Toward Iran" noted that "Iran remains concerned by the Soviet penetration of Afghanistan and exploitation of Arab disorder which threaten to outflank Iran, already exposed along a 1200-mile frontier with the USSR."[41] The shah's "nightmare scenario" was "Soviet envelopment of Iran through client forces in Iraq and Afghanistan."[42] Iran's rivalry with the Soviet Union ended following the 1979 Islamic Revolution. The United States had a more well-known rivalry with the Soviets during this same period that characterized the post–World War II international order.

The available evidence indicates that Washington wanted to strengthen Iran and bolster American-Iranian relations to constrain the influence of their common enemy—the Soviet Union. In the mid-1950s, the United States launched its policy of containment, which resulted in an extensive system of alliances designed to "restrain expansion of Soviet influence."[43] Building on the notion of containment, the "Eisenhower doctrine" asserted a U.S. interest in supporting the independence of countries in the Middle East. In March 1957, President Eisenhower received congressional approval to provide assistance to noncommunist Middle Eastern states threatened by Soviet aggression. During this period, the United States perceived that the best way to prevent Soviet penetration into Iran was to strengthen it economically.[44] This assistance was designed to "underwrite extensive modernization of the Iranian economy" so that it would be in a position to prevent Soviet aggression.[45] Assistance of all kinds, including cooperation in the peaceful uses of nuclear energy, was perceived to be an "important [pillar] supporting the Shah in his present paramount position."[46]

In the 1970s, U.S. interests in promoting strength and stability in Iran were especially salient. A declassified national security document from 1974 stated: "Our interests in Iran are substantial and are growing steadily. . . . It shares with us an interest in . . . limiting the influence of the Soviet Union. . . . Acting as a responsible regional power, Iran can help stabilize the area politically."[47] The sale of nuclear power reactors was part of a broader U.S. strategy to increase Iran's power and promote stability in the Persian Gulf, while mitigating Soviet influence in the region. In the mid-1970s it was believed that cooperation in the peaceful uses of nuclear energy could create "a framework and atmosphere" to discuss these types of strategic interests.[48] This sentiment continued through 1978, when President Carter and the shah reached an agreement on the sale of nuclear reactors to Iran. Indeed, Carter felt it was important to export nuclear power reactors to Iran to introduce further stability in the Persian Gulf.[49]

Alternative Explanations

NONPROLIFERATION

The empirical evidence suggests that nonproliferation-related considerations affected the terms of the arrangement negotiated by the United States and Iran, but that they had little bearing on the initial decision to transfer nuclear power reactors. The senior decision makers, especially those in the Nixon and Ford administrations, appear to have spent little time thinking about the proliferation consequences of U.S. nuclear exports to Iran.[50] For example, Henry Kissinger indicated that "we didn't address the question of them one day moving toward nuclear weapons."[51] The available evidence does reveal, however, that others at lower levels of government wanted to ensure that U.S. nuclear transfers did not contribute to nuclear proliferation. Fred Ikle, the director of the U.S. Arms Control and Disarmament Agency when the nuclear deal was being negotiated, noted that one of the key objectives of U.S. nuclear cooperation with Iran was to ensure that nuclear exports were used only for peaceful purposes.[52] There was some concern that Iran might use U.S. nuclear technology to pursue nuclear weapons. According to a U.S. Defense Department document, "the potential for instability and uncertain political situation in the Middle East [means that] the proposed agreement for nuclear cooperation could have serious national security implications in the future."[53] These concerns explain why the United States refused to give in to Iranian demands on reprocessing—at least initially.

That Iran was a member of the nuclear Nonproliferation Treaty did little to alleviate fears that U.S. exports might contribute to proliferation.

According to a U.S. National Security Council document, "The fact that Iran is a party to the NPT is a very positive element but the concern in our country over the possibility of proliferation is really extraordinary."[54] The United States recognized that refusing to provide Iran the right to reprocess spent nuclear fuel while allowing some non-NPT states to do so could pose political problems. The United States was concerned that "subjecting an NPT party like Iran to more rigorous controls" might be perceived as "undermining the NPT."[55] In addition to this concern, many U.S. officials were not convinced that Iran's status in the NPT provided sufficient assurances against future proliferation: "despite Iran's present benign attitude towards the NPT and non-proliferation, some are concerned over her possible longer-term nuclear weapon ambitions should others proliferate."[56]

The evidence suggests that concerns over whether Congress would approve the agreement helps explain why the United States wanted to negotiate an agreement with Iran that was strong on nonproliferation (e.g., why they did not want to give in on Iranian demands to reprocess spent fuel). Washington recognized that it had to "weigh any modification of our position [on reprocessing] carefully in the light of general public and Congressional concerns over proliferation."[57] If the agreement was perceived to be weak on nonproliferation, then the United States recognized that Congress would not approve it.[58]

In light of this evidence, it would be incorrect to conclude that the NPT was irrelevant for America's nuclear cooperation with Iran. At the same time, Tehran's NPT commitment did not play a major role in catalyzing atomic assistance in this particular instance.

ECONOMICS

There is some evidence that economic-related considerations were salient in explaining the decision to begin nuclear cooperation with Iran in 1957. For example, one of the goals of enhanced nuclear cooperation with Iran was "to obtain for the United States a major share in the additional business which Iran's oil wealth will generate."[59] As previously mentioned, the agreement forged in the mid-1970s would have generated $6.4 billion in revenue for U.S. companies. Declassified national security documents often mention this as a benefit of nuclear cooperation with Iran.[60]

Economic justifications for nuclear cooperation were not trivial but they were less salient than the political motives previously discussed. When economics was mentioned as a justification for cooperation, it was usually overshadowed by security-related considerations. For example, Kissinger argued that U.S. nuclear cooperation with Iran would provide economic benefits but emphasized that it would "underpin the broad political-military collaboration which we see in our interest."[61] Further, it was the change in the security environment that led to the collapse of U.S.-Iranian nuclear cooperation.

The United States could have gone through with plans to build nuclear reactors in Iran following the Islamic Revolution, generating large sums of money in the process, but it chose not to because of changes in the security relationship between the two countries.

Soviet Nuclear Cooperation with Libya, 1975–1986

The Soviet Union was Libya's principal supplier of civilian nuclear technology.[62] An agreement signed on May 5, 1975, initiated cooperation between these two countries. The terms of this deal authorized the Soviet Union to build a 10MW research reactor in Libya and assist in constructing the Tajoura Nuclear Research Center (TNRC).[63] In April 1981 the Soviet Union delivered the first shipment of highly enriched uranium (11.5kg) to fuel the Tajoura reactor and in August it became operational.[64] The TNRC became the focal point of Libya's covert nuclear weapons program until Tripoli abandoned these efforts in 2003. Suspicious activities that took place at this facility included undeclared work on uranium conversion, gas-centrifuge enrichment, and plutonium separation.[65]

While the TNRC was in the construction phase, the Soviet Union signed an additional nuclear cooperation agreement with Libya in December 1977, pledging to construct a 440MW nuclear power plant along Libya's Mediterranean coast.[66] Western sources indicate that the plans actually called for the export of two 440MW pressurized water reactors, the standard Soviet export reactor.[67] Libya ran into "major problems" with the Soviets in acquiring these reactors and nuclear trade moved slowly following the completion of the Tajoura project.[68] Although tangible progress was lacking, Soviet officials continued to indicate an interest in cooperating with Libya in the nuclear arena. During a visit to the Soviet Union in May 1981, Libyan leader Muammar Qaddafi discussed the nuclear reactor issue for the third time and reportedly made some progress toward concluding the deal.[69] And in May 1986 the Soviet Ambassador to Libya asserted that "we have an idea to cooperate in building a nuclear power station [in Libya]."[70] Despite these sentiments, the Soviet deal expired in late 1986 and the reactors were never delivered to Libya.[71]

Strengthening the Soviet-Libyan Partnership

Why did Soviet nuclear cooperation with Libya begin in 1975 and end in 1986? The empirical evidence indicates that Soviet atomic assistance to Libya was intended to forge a strategic partnership between the two countries. Prior to the 1970s, Moscow and Tripoli did not share warm relations. Libyan monarch Idris I, whose reign began in 1951, adopted a pro-Western posture.

And Colonel Muammar Qaddafi, who rose to power following a bloodless coup against Idris in 1969, initially kept his distance from the Soviet Union.[72] During the early years of Qaddafi's tenure, Moscow allied closely with Egypt and did not look to improve its relations with Tripoli. In 1972, however, Egyptian president Anwar Sadat expelled the Soviet Union from Egypt and Moscow looked to improve relations with Libya.[73] During a Libyan delegation's visit to Moscow in May 1974, Alexey Kosygin, premier of the Soviet Union, declared: "We hope to examine various aspects of the cooperation between our countries [and] exchange opinions relating to the prospects for expanding Soviet-Libyan ties. In our opinion, there are considerable opportunities for this."[74] Further, a joint communiqué issued following bilateral discussions in Tripoli stated that "the Soviet Union and the Libyan Arab Republic regard the development and deepening of friendship and cooperation between them as a highly important task of their foreign-policy activity."[75]

Atomic assistance was a crucial part of Moscow's strategy to improve relations with Tripoli. It provided a means for the Soviet Union to increase its influence in the region by drawing Tripoli closer.[76] Statements from Soviet leaders substantiate this assertion. For example, during bilateral discussions in May 1974, Kosygin stated that "we hope to examine various aspects of the cooperation between our countries. . . . we are prepared to come to an agreement that would make our cooperation more stable and long-term in nature. Coordinated actions in this area would undoubtedly facilitate the strengthening of mutual understanding and trust between our countries."[77] The cooperation that Kosygin referred to included cooperation in the nuclear area. Just months after making this statement, the Soviet Union signed its first nuclear cooperation agreement with Libya.

During the 1980s, the Soviet Union lost its interest in strengthening its bilateral relationship with Libya. Moscow became increasingly worried about Qaddafi's unpredictable behavior and his frequent conflicts with Arab neighbors. And Mikhail Gorbachev, who rose to power in 1985, did not want to risk other priorities by engaging Libya too closely.[78] In 1983 the Soviets refused Qaddafi's request for a formal treaty to strengthen relations between the two countries and attempted to keep the erratic Libyan leader "at arms length."[79] And Moscow failed to come to Libya's direct aid during two military confrontations with the United States in March and April 1986. Once Moscow no longer wanted to strengthen its partnership with Tripoli, all peaceful nuclear cooperation ended.

There is some evidence that Moscow desired closer relations with Libya because it wanted to forge a military alliance. Moscow was in need of a partner in North Africa after the expulsion from Egypt.[80] And Libya's strategic location made it a potentially important ally for the Soviet Union. Its position on the Mediterranean coast offered convenient ports of call for the Soviet navy and Libyan airfields could provide valuable support for Soviet actions in sub-Saharan states.[81] But the principal reason why Moscow valued a stronger partnership with Tripoli was to constrain the influence of the United States.

CONSTRAINING THE UNITED STATES

The Soviet Union and Libya were enemies of the United States during the period that nuclear cooperation took place.[82] Leaders of both countries routinely condemned American "imperialism."[83] The Soviet Union was engaged in a superpower rivalry with the United States while Libya and the United States terminated diplomatic relations and faced military confrontation in the 1980s. In May 1974, Qaddafi noted that Libya's friendship with the Soviet Union was based on a shared interest in opposing the United States. Libya desired to defend itself from the "U.S. diplomatic offensive beginning in the Middle East" while the Soviets wanted to gain a "strategic advantage" over its principal rival.[84]

This shared strategic interest provided the Soviets with strong incentives to develop closer relations with Libya. The threats these two states shared were alone insufficient to forge a partnership. Primary documents reveal that Soviet-Libyan nuclear cooperation was perceived as a vital means to strengthen bilateral relations and, in turn, counter the influence of the United States. During bilateral discussions in March 1972—which laid the groundwork for the 1975 nuclear agreement—the Soviet Union and Libya noted "the great importance of the friendship [between the two countries] for the success of the struggle . . . against imperialism."[85] Just prior to signing the 1975 agreement during a trip to Tripoli in May 1975, Kosygin stated that enhanced cooperation is necessary to succeed in the "struggle against the common enemy—[the United States]."[86] These sentiments were continually reinforced as the Soviet Union negotiated a second nuclear cooperation agreement with Libya and built the Tajoura reactor. In April 1981, Soviet leader Leonid Brezhnev stated that nuclear cooperation, and relations with Libya more generally, were vital in countering the influence of the United States: "[We are] good comrades and brothers-in-arms in the struggle for the rights and freedom of peoples, against imperialist oppression and aggression. . . . We in the Soviet Union appreciate the principled position that Libya takes on questions [of American imperialism] . . . you consistently oppose imperialist intrigues and encroachments on the rights of peoples."[87] Additionally, during discussions with the Libyans on cooperation in nuclear power in May 1982, Nikolai Tikhonov, chairman of the Soviet Council of Ministers, stated: "We see the unity of our two countries' goals in the anti-imperialist struggle . . . as the firm basis of our relations. It is [this] lofty goal that our cooperation ultimately serves."[88]

Moscow also believed that strengthening Libya could constrain the power and influence of the United States. By helping Libya develop a civil nuclear program, it hoped to enhance its energy production capacity and facilitate economic growth. This would make it more difficult for the common enemy (Washington) to exert influence or commit aggression against Tripoli.[89]

Following Gorbachev's rise to power in 1985, the Soviet Union lost much of its interest in constraining the United States. Gorbachev's "new thinking" in foreign policy sought to reduce cold war tensions and improve relations with Washington and the West.[90] The Soviet leader met with President Ronald Reagan for the first time in November 1985 and opened a dialogue aimed at warming U.S.-Soviet relations. The two leaders met again in Iceland in October 1986, were they famously called for the elimination of all nuclear weapons, and soon developed a close personal relationship.[91] This warming of Soviet-American relations goes a long way in explaining why the agreement to supply two power reactors to Libya fell through in 1986.[92]

Alternative Explanations

NONPROLIFERATION

Libya ratified the NPT on May 26, 1975, but it is widely known to have pursued nuclear weapons from shortly after the time Qaddafi rose to power until 2003. Qaddafi attempted to purchase nuclear weapons from China and India in the 1970s and reportedly was willing to pay $1 million in gold to anyone who could produce a complete warhead.[93] Despite this, the Soviet Union moved forward with plans to provide nuclear technology to Libya between 1975 and 1986.

The Soviet Union developed a reputation as a "responsible" nuclear supplier that held high nonproliferation standards.[94] Troubled by its contribution to the Chinese nuclear weapons program in the 1950s, the Soviet Union scaled back its nuclear exports in the 1960s and demanded that recipients of nuclear technology provide certain assurances. Moscow instituted a system of safeguards, insisted that all recipients of reactors obtain fuel from the Soviet Union and return spent fuel rods, and prohibited its Eastern European allies from obtaining uranium enrichment or plutonium reprocessing facilities.[95] It also generally demanded that recipients of nuclear technology sign the NPT.

The nuclear transfers to Libya are at odds with these priorities.[96] The Soviet Union's cooperation with Libya demonstrates its willingness to supply nuclear technology to a state with a weak nonproliferation record if doing so achieved important strategic objectives. The Soviet Union needed to forge a partnership with Libya—especially following its expulsion from Egypt in 1972—in part to constrain U.S. power and selling nuclear technology helped achieve this objective. This objective was perceived to be more important than Libya's weak nonproliferation commitments.

The Soviet Union was well aware of the proliferation risks that came with this nuclear cooperation and made some efforts to minimize them.

Libya's ratification of the NPT occurred just four days before it signed the first agreement with the Soviet Union, which led some observers to speculate that Moscow pressured Tripoli to ratify the treaty in exchange for nuclear assistance.[97] The Soviets also applied "stringent" safeguards to Libya to provide additional assurances that nuclear cooperation would not contribute to proliferation.[98] Given Libya's pursuit of nuclear weapons, however, the Soviet Union likely viewed its NPT commitment as a "hollow pledge."[99]

A final nonproliferation-related explanation for atomic assistance in this case is that Moscow wanted to "preempt the sale of nuclear technology . . . by less cautious suppliers."[100] The Soviet Union had previously suffered when other suppliers provided technology after it had refused to do so. For example, in the early 1970s the Soviet Union refrained from supplying a plutonium-producing reactor to Iraq only to see France and Italy assist Baghdad's civilian nuclear program (see chapter 5).[101] Not wanting this to happen again, Moscow may have supplied a reactor to Libya to prevent it from receiving proliferation-prone nuclear technology from Western European nuclear suppliers. In other words, the nuclear deal was conceived to make it more difficult for Libya to acquire nuclear weapons. The empirical evidence does not lend support for this argument. Qaddafi was determined to acquire nuclear technology and there is no indication that he stopped approaching other suppliers after the Tajoura facility was built.[102] Moreover, if the Soviets were supplying Libya so that Qaddafi would not receive assistance from less "responsible" suppliers, Moscow should have followed through with its plans to supply power plants, which of course is not what happened.

ECONOMICS

The evidence reveals that economic arguments are less salient in explaining nuclear cooperation with Libya than the Soviet Union's strategic interests. Some scholars suggest that the Soviet Union wanted to sell nuclear technology to Libya to illustrate that it was a viable supplier to developing nations.[103] The Soviet VVER-440 reactor was one of only two small reactors on the market at the time (the other was the Canadian CANDU) and Moscow may have been attempting to command a portion of the market where there was relatively little competition.[104] The problem with this explanation is that there were several developing countries that wanted nuclear power— including Iran, the Philippines, Cuba, and Egypt—to which the Soviet Union could have sold nuclear technology. This argument cannot explain why the Soviets sold to Libya when there were other ways they could have demonstrated their ability to be a viable supplier to developing nations.

A related economic rationale for the nuclear sales is a desire to maximize profits and gain access to Libya's hard-currency reserves.[105] This may have been a benefit of nuclear cooperation with Libya but it could not have been a primary motivating factor. There were many countries with better proliferation records than Libya that could have paid for Soviet nuclear technology.

Canadian Nuclear Cooperation with India, 1955–1976

India was an important partner for Canada in the aftermath of World War II. The Canadians were in a position to "assist the United States in parts of the world where American views by themselves were considered too extreme."[106] Ottawa was weak enough not to be viewed as an imperialist power but strong enough to strengthen India and Indo-Canadian ties in ways that were productive from a political standpoint.[107] Ultimately, as Escott Reid noted, Canada was able to engage India in ways that were more acceptable.[108]

Canadian-Indian nuclear cooperation began on April 28, 1956, when the two countries signed a nuclear cooperation agreement authorizing Canada to build a 40MW research reactor in India. This reactor, known as CIRUS (Canada-India-Reactor-United States), went critical in July 1960 as a result of civilian nuclear assistance provided by Canada and the United States.[109] CIRUS was intended to be a "teaching tool" to allow the Indians to build up their knowledge in nuclear engineering.[110] The Canadians placed very limited safeguards on the CIRUS transaction, stating only that "the Government of India will ensure that the reactor and any products resulting from its use will be employed for peaceful purposes only."[111] The absence of robust safeguards is not an indication of naïveté. The evidence presented below will underscore that suppliers often take risks that could be characterized as reckless from a proliferation standpoint if they are able to obtain political and (to a lesser extent) economic benefits.

The construction of a research reactor paved the way for Canada to assist India in the construction of power reactors. On December 16, 1963, Canada and India signed a second nuclear cooperation agreement. This agreement led to the construction of a 100MW Rajasthan Atomic Power Plant (RAPP-1). The terms of this deal called for the "free exchange of scientific and technical information . . . for the development of heavy water moderated reactor systems." It also enabled India to obtain "detailed design data, including plans and working drawings regarding the design and construction of nuclear power stations of the heavy water type."[112] Canada additionally agreed to provide one-half of the initial uranium fuel charge for the Rajasthan reactor.[113] Unlike the CIRUS arrangement, Canada was able to include fairly stringent safeguard provisions in the RAPP-1 deal. Most notably, the agreement allowed for Canadian inspections of the RAPP-1 reactor to ensure that it is being used only for peaceful purposes.[114] RAPP-1 was built by Indian scientists using designs obtained from Canada under the terms of this

NCA, not on a turnkey basis as many other reactors were.[115] This reactor
went critical on August 11, 1972.[116]

On December 16, 1966, Canada agreed to offer assistance in the design
and construction of a second nuclear power reactor at Rajasthan (RAPP-2).
The capacity of RAPP-2, 200MW, was twice as much as RAPP-1.[117] The
safeguards requirements attached to this agreement were more stringent
than the other two deals because they allowed inspectors from the IAEA to
verify that the reactor was not used for weapons-related purposes.[118]

Then, on May 18, 1974, India conducted what it termed a "peaceful"
nuclear test at Pokhran in the Rajasthan desert. Four days after the explosion, Canada temporarily suspended all nuclear assistance to India.[119] This
included suspension of the RAPP-2 reactor construction and the cancella-
tion of an export license covering $1.5 million in nuclear-related equipment
and materials.[120] After two years of negotiations, Canada permanently suspended all nuclear assistance to India on May 18, 1976.[121]

Canada's nuclear cooperation with India is significant because it enabled
New Delhi to acquire nuclear weapons, as I will discuss further in Part II of
this book. Canadian assistance was critical in two respects. First, the CIRUS
reactor provided a means to produce the plutonium necessary for use in a
nuclear weapon. In 1964, India had produced its first weapons-grade plutonium by reprocessing spent fuel from the Canadian-supplied CIRUS reactor.[122] Second, Ottawa's transfers of knowledge and technical expertise in
the 1950s and 1960s helped establish India's "self-sufficiency in reactor technology."[123] Once Canada terminated its nuclear relations with New Delhi in
1976, India's nuclear program suffered considerably.[124]

Strengthening India and Indo-Canadian Relations

Following World War II, Canada recognized the importance of developing
close relations with India, as previously noted. Indo-Canadian cooperation
during international crises in the 1950s—including the Korean War, the International Commission for Supervision and Control in Indochina, and the
Suez crisis—illustrated the value in forging close bilateral relations.[125] Prime
Minister Louis St. Laurent and External Affairs Minister Lester Pearson
"attached the highest importance to Canadian-Indian relations" and actively
worked to forge a strategic partnership.[126]

Atomic assistance was part of a strategy to keep the relationship between
Ottawa and New Delhi strong.[127] Canada wanted to provide India with the
CIRUS reactor in March 1955 because it "believed that such a move was of
considerable international political importance for the West in its attempts
to secure the loyalty of the developing world as the Cold War heated up."[128]
Jules Leger, the undersecretary of state for external affairs, argued just as

nuclear relations with India commenced that "politically, it would do more to strengthen our relations with India than anything I could think of."[129] Minister of External Affairs Lester Pearson echoed these sentiments when he noted on November 7, 1955 that nuclear cooperation with India had "a real political value in bringing us closer together."[130] Pearson also felt that nuclear commerce could help Canada "direct Indian energies into friendly channels—friendly, that is, to Canada and the West."[131]

In addition to strengthening its bilateral relationship with India, Canada strove to promote New Delhi's economic development. In January 1950, members of the Commonwealth gathered in Ceylon (now Sri Lanka) to consider extending aid to South and Southeast Asian countries.[132] What emerged from this conference became known as the Colombo Plan, a developmental aid program modeled after the Marshall Plan. Members of the Commonwealth—especially Canada—believed that it was crucial to assist the development of India and other developing countries in Asia.[133]

Peaceful nuclear assistance was a crucial part of Ottawa's strategy to promote New Delhi's economic development. Indeed, the nuclear technology Canada initially provided to India was donated as part of the Colombo Plan.[134] Pierre Trudeau, who served as Canadian prime minister from 1968 to 1984, argued that nuclear technology was "one of the most certain means" of industrializing developing countries and helping them "pass out from the medieval economic state."[135] Trudeau indicated that this belief had a tremendous effect on Canada's decision to initiate atomic assistance with India in the mid-1950s. By providing a reactor and transferring nuclear know-how, Canada hoped to help India in "overcoming its desperate problems of poverty."[136] Canadian officials directly responsible for beginning atomic assistance—including St. Laurent, Pearson, and Escott Reid, the high commissioner to India—shared these sentiments.[137]

Canadian interest in strengthening India and its bilateral relationship with New Delhi continued in the early 1960s, as India remained "a key nation in Canada's contacts with the non-aligned world."[138] As a result, Ottawa moved forward with plans to sell nuclear power reactors to India in December 1963. But by 1966 Canada perceived that it would accrue fewer strategic benefits by strengthening New Delhi and Indo-Canadian relations. As a result of disagreements about the war in Vietnam, India began to move away from Canada and the West and closer toward the Soviet Union.[139] Consequently, peaceful nuclear cooperation between the two countries lost some momentum. As Robert Bothwell notes, "the political rationale [for nuclear cooperation] was beginning to wear thin, as Indo-Canadian relations frayed in a number of areas, and the vision of the 1960s that an intercontinental bridge of friendship could be constructed . . . was becoming blurred."[140] Canada moved forward with the RAPP I and RAPP II projects because it was committed to see them through.[141] But it did so reluctantly, which explains in part why both reactors were completed far behind schedule. RAPP I was not completed until 1973 (four years behind schedule) and RAPP II was not finished until 1981 (nearly eight years behind schedule).

India's 1974 nuclear test had severe consequences for Indo-Canadian relations. Ottawa felt betrayed that India would use technology and materials provided exclusively for peaceful purposes to build the bomb. As Escott Reid noted, the nuclear explosion "had a catalytic effect on Indo-Canadian relations. It demonstrated that so far as Canada was concerned the special relationship had ceased to exist."[142] Not surprisingly, all bilateral nuclear cooperation came to a screeching halt once Ottawa lost interest in strengthening India and enhancing Indo-Canadian relations.

There are two reasons why Canada cared about improving its relationship with India and strengthening New Delhi beginning in the 1950s. The first was because India was a democracy. And the second was because India could help Canada counter the influence of the Soviet Union.

STRENGTHENING A DEMOCRATIC INDIA

Canada and India were two of the world's most vibrant democracies at the time atomic assistance commenced in the 1950s. One of the principal reasons why Ottawa cared about improving Indo-American relations and strengthening New Delhi is because India was a democracy. Canadian foreign policy during the period that Escott Reid dubs the "golden age"—which lasted from the end of World War II until the late 1950s—hinged on promoting democracy.[143] An important part of this strategy was to strengthen bilateral relations with existing democracies. Lester Pearson noted in April 1949: "The North Atlantic community is part of the world community and as we grow stronger to preserve peace, all free men grow stronger with us. The world today is too small, too interdependent, for even regional isolation."[144] With this statement, the Canadian foreign minister indicated a willingness to forge relationships with democratic countries because doing so would promote peace and stability. In this context, India was a vital country since so many of the "free men" were represented by New Delhi.[145] At the same time, Ottawa recognized that democracy did not guarantee the formation of a strong partnership with India or any other democracy. This was especially true because the Soviet Union had incentives to drive a wedge between liberal democratic countries. Other means were necessary, therefore to promote solid relations with other democracies. As Pearson stated in 1955, "It is essential . . . in the situation which we face, to maintain maximum unity among those nations who are working together to defend freedom to maintain peace. We certainly cannot afford to take this unity for granted [even though] all the members of our coalition are democracies with traditions of freedom and self rule."[146] In the Indian case, Ottawa used atomic assistance as the principal means to solidify its relationship with democratic India. If there was unity between Canada and India, Ottawa believed that peace and stability were more likely.[147]

Ottawa had an especially strong interest in ensuring that existing democracies, such as India, remained strong. If democratic states remained weak and economically underdeveloped, they might devolve into autocracies. This was an especially salient concern when it came to India. After achieving independence in 1947, economic conditions deteriorated in India and many feared that the lack of development would ultimately threaten the viability of India's democracy. For instance, in April 1949, Indian prime minister Jawaharlal Nehru stated that "democracy was . . . threatened by the present time from two directions—first, by a direct onslaught by communism; and secondly, by an internal weakening, largely due to unfavourable economic conditions."[148] Nehru repeated this message to the Canadian Parliament in October 1949: "The troubles and discontents of . . . the greater part of Asia are the result of obstructed freedom and dire poverty. The remedy is to accelerate the advent of freedom and to remove want."[149] Ottawa shared these sentiments. It worried that the Soviet Union would persuade New Delhi that its regime type was responsible for the economic woes and that this would lead to the collapse of India's democracy. Pearson was well aware that the Indian people "had achieved freedom without the immediate material improvement which they had . . . expected to follow freedom" and that the communists were "exploiting these and other grievances and would be able to create serious difficulties if economic conditions deteriorated."[150]

The collapse of India's democracy would have had significant, but perhaps not critical, implications for Canadian interests. Life in Canada may not have been drastically different if India devolved into a totalitarian dictatorship but Ottawa cared a great deal about preserving India's democracy. Lester Pearson argued after returning from a trip to India in 1950 that "there is no more important question in the world today [than whether] communist expansionism may now spill over into southeast Asia as well as into the Middle East."[151] Likewise, an official Canadian brief issued in the early 1950s stated that "the immediate problem of world peace is the containment of Soviet communism within its own borders."[152] These threats explain why Canada worked so hard to promote India's economic development. It believed that external aid, especially atomic assistance, was necessary to improve conditions in India. Ottawa knew that without its support, the "forces of totalitarian expansionism . . . cannot be checked."[153] As Pearson argued, assistance was "the best defence against communism" because "there was little value in preaching the virtues of the democratic way of life to starving people."[154]

There was an additional benefit of strengthening a democratic India. Ottawa knew that other countries would observe how democracy affected India. If it promoted economic growth and prosperity there, other states might have confidence that democracy would bring them similar benefits.

This would encourage existing democracies to stay the course—even if they experienced tough times—and could push autocratic states to democratize. Alternatively, if democracy failed in India, others might accept that it would not work in their states either. India, therefore, was an important example for the rest of the world. And with foreign support, it could be a positive model. Escott Reid illustrated this thinking when he wrote that "with adequate help from the more advanced countries India can, I am convinced, save itself by its exertions and save Asia and Africa by its example."[155]

COUNTERING SOVIET INFLUENCE

The main reason that Ottawa initiated atomic assistance to India was to strengthen an existing democracy. A secondary motive was to constrain the influence of the Soviet Union by strengthening New Delhi economically and fortifying Indo-Canadian relations. Ottawa was "acutely conscious of its membership in the American-led Western alliance against the Soviet Union and its allies."[156] Canada's commitments as part of this alliance affected its dealings with India. Following Soviet premier Nikita Khrushchev's historic visit to India in 1955, the country began to drift toward the Soviet Union.[157] But New Delhi and Moscow were not necessarily the closest of partners during the 1950s.[158] Ottawa believed, therefore, that it could make it more difficult to the Soviet Union to befriend India by developing a close bilateral relationship with New Delhi. Indeed, Prime Ministers Mackenzie King and Louis St. Laurent recognized that strong links between India and Canada would prevent New Delhi from "drifting into the embrace of the Soviet bloc."[159] They also knew that Moscow would be less able to exert influence on or commit aggression against India if New Delhi was more powerful.

Ottawa was in the ideal position to constrain Soviet influence by forging closer relations with India. U.S. intentions were viewed skeptically in India and Washington's attempts to entice New Delhi with nuclear technology would likely have been less successful—or at least this was the perception in Ottawa.[160] Canada recognized that nuclear transactions resulted in long-term partnerships between the supplier and recipient states and it did not want to see the Soviet Union develop stronger relations with other countries, especially developing countries.[161] It feared that the Soviet Union was prepared to offer atomic assistance to India.[162] By beginning nuclear cooperation in 1955 Ottawa hoped that it—not Moscow—would develop a closer relationship with India. Additionally, the Canadians hoped that the CIRUS transaction would limit Moscow's influence in Asia by counteracting the "positive impression" that the Soviet Union was helping China develop a "peaceful" nuclear program.[163] Ottawa cared so much about providing atomic assistance to New Delhi that it preferred to weaken nuclear safeguards demands than "allow the Soviets to use a nuclear reactor to establish closer ties with a developing country."[164]

Slowly, beginning in 1955, the "geopolitical interest" that originally motivated Canada to export nuclear technology to India began to fade because of New Delhi's closer alignment with the Soviet Union.[165] But Canada moved forward with nuclear cooperation in the 1960s and 1970s in part because it recognized that India remained strategically important and that a termination of nuclear trade at that point would put a serious strain on Indo-Canadian relations.[166] Constraining Soviet influence, therefore, is salient in explaining the onset of atomic assistance but less important in explaining the continuation of aid until 1974.

Alternative Explanations

NONPROLIFERATION

India has historically been one of the most ardent critics of the NPT because it allows five countries to possess nuclear weapons (the United States, the Soviet Union, Britain, France, and China) while prohibiting all other states from doing so, making the treaty discriminatory. New Delhi embarked on a concerted effort to acquire nuclear weapons following the Sino-Indian border war in 1962 and Beijing's nuclear test in 1964.[167] The nuclear explosive device that India tested in 1974 was built using plutonium extracted from the Canadian-supplied CIRUS reactor.

When engaging in nuclear cooperation with India, Canada was well aware of the proliferation risks. Chester Ronning, the Canadian high commissioner to New Delhi, worried about India's nuclear ambitions and was "disturbed" by his conversations with Indian leaders.[168] As Canada was negotiating with India over the sale of the RAPP-1 and RAPP-2 reactors in the mid-1960s, a briefing book prepared for Prime Minister Lester Pearson stated:

> The present and future development of India's civilian nuclear research and power programme will continue to ensure that a militarily significant scale in four to five years from now, could be initiated very quickly and without a prohibitive diversion of manpower resources. . . . It is known . . . that the operation of the Canada-India Reactor (CIR) at Trombay has been oriented towards optimizing the production of weapons-grade plutonium since the summer of 1963.[169]

Further, Paul Martin, the secretary of state for external affairs, claims to have realized that Homi Bhabha, who established the Atomic Energy Commission of India, coveted nuclear weapons.[170] Intelligence reports received by Canadian officials substantiated these assertions, although it is important to point out that Nehru did not share these sentiments—at least in the 1950s.[171] In 1965, Canadian defense minister Paul Hellyer received intelligence from his British counterpart suggesting that India was "making all necessary preparations for a test explosion sometime before the end of the year."[172] Indeed, Prime Minister Lal Bahadur Shastri had authorized a nuclear explosives program one year earlier (see chapter 7).

Canadian officials warned India that the development of nuclear weapons could lead to a termination of atomic cooperation. In 1971, Prime Minister Pierre Trudeau sent a letter telling Indian prime minister Indira Gandhi that the use of Canadian supplied technology or materials in the development of nuclear weapons would "inevitably call on our part for a reassessment of our nuclear cooperation arrangement with India."[173] But Ottawa continued to engage in civilian nuclear cooperation with India in spite of the overwhelming evidence that New Delhi was pursuing nuclear weapons. India's harsh rhetoric against the NPT, which was being negotiated in the mid-1960s, also did not dissuade Canada from moving forward with the plans to supply additional power reactors. It quickly became clear that India would not accept a treaty that treated the nuclear "haves" and the "have-nots" differently. Only a treaty that obligated *all states* to stop production of nuclear weapons would be acceptable to New Delhi.[174]

There was "nothing naïve" about Canada's nuclear cooperation with India, as I noted above.[175] Ottawa knew that the CIRUS reactor was an efficient plutonium producer and that India's commitment to nonproliferation was weak, but it went through with nuclear sales anyway. The dual-use dilemma partially explains this. Canadian officials were able to convince themselves that their assistance would not contribute to proliferation because there were legitimate peaceful uses of the technology it provided.[176] Moreover, they realized that the strategic benefits resulting from nuclear aid outweighed concerns about the spread of nuclear weapons.

ECONOMICS

There is scant evidence in favor of the financial profit argument. The initial transfers that Canada proposed were donations under the Colombo Plan, meaning that Ottawa did not generate a tremendous amount of hard currency from assisting India's peaceful nuclear program. When the Indians initially balked at the idea of receiving a Canadian reactor, Ottawa sweetened the pot. It gave an additional $7.5 million in Colombo Plan aid to India to cover additional costs associated with building the facility.[177] This made New Delhi more enthusiastic about the offer but resulted in greater financial losses for the Canadians. In the end, Canada was willing to forgo the chance to profit from the transfer because it had strategic incentives to provide India with nuclear assistance.

That Canada permanently terminated nuclear cooperation with India in 1974 further undermines the profit argument. Canada suffered immediate and future economic losses once it ended nuclear assistance to India. It lost $12 million worth of heavy-water sales and $1 million in spare parts for the RAPP program.[178] If economics were the driving force behind civilian nuclear cooperation, we would have expected Canada to resume sales after 1974 in pursuit of financial gains. That it did not lends further support to the argument that security considerations heavily influenced Indo-Canadian nuclear cooperation. Robert Bothwell summarizes this well: "It is clear that in its origins Canada's atomic connection to India was political; that it was aimed at influencing India's attitudes to and relations with the Western alliance, of which Canada was a self-conscious and active part."[179]

There is some evidence that Canada pursued atomic assistance to sustain its nuclear industry. Canadian officials perceived that by entering the nuclear market early, in 1955, they could "showcase Canadian nuclear technology and engineering talent abroad," which could "encourage the export of additional reactors."[180] Indeed, a cabinet memorandum concluded that Canadian business and the atomic industry could "gain competitive advantage in an emerging field and Ottawa could actively assist their position for constructing various types of atomic units in Canada or abroad in later years."[181] This was an especially salient consideration because the Canadian reactor used plutonium–heavy water–based technology, which was unique from the U.S.-made light water–enriched uranium technology.[182] Canadian officials recognized that it would be "most undesirable, especially from the longer-term commercial point of view, for us to lag behind" the Americans and the British.[183] These considerations continued to influence Canadian nuclear cooperation with India in the 1960s and 1970s. Ottawa recognized that providing nuclear power reactors to India (the RAPP-1 and RAPP-2) could establish Canada as a significant nuclear supplier. The Canadian minister of trade and commerce stated following the conclusion of the RAPP-2 agreement that "the sale confirmed Canada's position as one of the leading international suppliers of nuclear power stations."[184] At the same time, refraining from providing the reactors could have hurt the growth of the Canadian nuclear industry.[185] Although this evidence is significant, the termination of nuclear assistance to India in 1974 weakens the industry argument since these benefits were still attainable in the aftermath of the nuclear test.

American Nuclear Cooperation with India, 2001–2008

Much has been written about U.S. nuclear cooperation with India in large part because this policy is highly contentious.[186] After conducting its nuclear test in 1974, India began to assemble a nuclear arsenal that today contains around sixty to eighty warheads.[187] Nuclear cooperation with India therefore violates an international norm that states must make a legal commitment forswearing nuclear weapons by signing the NPT in order to receive nuclear technology for peaceful purposes. This norm was bolstered domestically in 1978, when the U.S. Congress passed the Nuclear Non-Proliferation Act (NNPA), which imposed severe restrictions on U.S. nuclear exports and effectively embargoed the sale of nuclear technology, materials, or knowledge to India.[188] The recent India deal reverses these policies and permits the sale of nuclear technology to New Delhi for the first time in decades.

The groundwork for the U.S.-India nuclear deal was laid in November 2001 when President Bush and Prime Minister Singh committed their countries to a strategic partnership. Further momentum toward the deal came in January 2004 when the two leaders established the Next Steps in Strategic Partnership initiative. Then, during Singh's visit to Washington, D.C., in July 2005, President Bush pledged to "achieve full civil nuclear energy cooperation with India."[189] After announcing its intention to cooperate with India in July 2005, the Bush administration needed to convince Congress to waive certain restrictions contained in the NNPA. After fairly heated deliberations, Congress passed legislation authorizing the necessary revisions, which President Bush signed into law in December 2006. The next hurdle was for Washington and New Delhi to agree to a formal nuclear cooperation agreement, which is required pursuant to the Atomic Energy Act. Following months of negotiations, in July 2007 the United States and India finalized the terms of the agreement. The finalized deal enabled "full civil nuclear energy cooperation" between the two countries.[190] In exchange for this cooperation, India agreed to separate its civilian and military nuclear facilities, declare its civilian facilities to the IAEA, voluntarily place its civilian facilities under IAEA safeguards, and sign an Additional Protocol for its civilian facilities.[191]

Additional steps were required before the deal could be implemented. On August 1, 2008, the IAEA board approved India's safeguards agreement, which broadens the agency's access to New Delhi's civilian nuclear facilities. Under the terms of the deal, the IAEA pledged to bring fourteen Indian reactors under safeguards by 2014.[192] Then, the Nuclear Suppliers Group passed an exception to its guidelines one month later, permitting nuclear commerce with India. Finally, the U.S. Congress officially approved the U.S.-India nuclear deal on October 1, 2008 and Secretary of State Condoleezza Rice and Indian minister of external affairs Pranab Mukherjee officially signed off on the agreement days later.

Strengthening India and the Indo-American Partnership

A review of statements made by key U.S. officials reveals that nuclear assistance is a key component of Washington's attempt to transform its bilateral relationship with India. President Bush suggested, for example, that civil nuclear cooperation would "deepen the ties of commerce and friendship between our two nations."[193] Similarly, a six-page memo sent by Secretary of State Condoleezza Rice to Bush in March 2005 stated that the time had come to make India one of the "two or three closest partners" of the United States and that nuclear cooperation was the way to help make this happen.[194] Rice later added in testimony before Congress in April 2006 that nuclear cooperation was an attempt to "seize this tremendous opportunity to solidify a key partnership that will advance American interests."[195] Nicholas Burns, the U.S. undersecretary of state for political affairs and a key negotiator of the 2007 agreement, echoed these sentiments. Burns said that "building a close U.S.-India partnership should be one of the United States' highest priorities for the future."[196] He went on to emphasize that the nuclear cooperation agreement is a crucial means to "deepen the strategic partnership."[197]

For New Delhi, the nuclear deal is tremendously symbolic. As Ashley Tellis observes, the agreement "becomes the vehicle by which the Indian people are reassured that the United States is a true friend and ally responsive to their deepest aspirations."[198] Thus, nuclear assistance is an instrument to facilitate future cooperation by signaling favorable intentions. Nuclear cooperation is especially valuable for this purpose because India desires U.S. aid as a means to help meet its future energy needs, a point that I will revisit below.

Part of the reason atomic assistance is necessary to improve relations in this case is that the United States had restricted nuclear sales to India since its first nuclear test. Peaceful nuclear cooperation, U.S. officials believed, would mend decades of mistrust and help transform the bilateral relationship. As Undersecretary Burns noted, "we sought [the nuclear deal] because India's nuclear weapons program and its status outside the nonproliferation regime has proven to be a longstanding stumbling block to enhanced U.S. India relations."[199] Burns also emphasized that atomic assistance to India was a crucial part of a confidence-building effort conducted at the highest levels of both governments to facilitate a closer relationship.[200] Likewise, Secretary Rice acknowledged that the relationship with India could not move forward "unless we deal with the problem before us, the impediments associated with civil nuclear cooperation, resolving them once and for all."[201]

U.S. nuclear assistance is also aimed at strengthening India economically and politically. India faces many challenges in the energy field due to its rapid growth. In order to emerge as a significant regional or global power, India requires foreign assistance in energy production.[202] As Ronen Sen, the Indian ambassador to the United States, articulated, "the biggest constraint on India's economic growth is energy. . . . We can't address issues of poverty, illiteracy, health care and the very basic fundamental needs of our people without energy."[203] Washington believes that peaceful nuclear assistance will help India meet its energy needs, in turn enhancing its economic growth and facilitating its transition to a major power in Asia.[204] According to Tellis, the nuclear deal "symbolizes . . . a renewed American commitment to assisting India [to] meet its enormous developmental goals and thereby take its place in the community of nations as a true great power."[205] In the words of another senior U.S. official, "We want India to emerge and to develop a worldview that gets them to the top. We want them at the high table [of international politics]."[206] Civil nuclear cooperation was one of the principal means to make this happen.

The preceding discussion indicates that the United States pursued atomic assistance to India to strengthen New Delhi and solidify Indo-American relations. But what explains U.S. interest in achieving these objectives? The available evidence indicates that peaceful nuclear assistance is intended to (1) strengthen relations with an enemy of an enemy, and (2) strengthen an existing democracy and relations with that democracy.

STRENGTHENING RELATIONS WITH ENEMIES OF ENEMIES

The United States and India are both threatened by China's rising influence in Asia. Bad blood has persisted between New Delhi and Beijing for decades. The two countries fought a war over a disputed border in October 1962 and India cited the threat stemming from China as one of the factors motivating its second round of nuclear tests in 1998.[207] Although Indian leaders are rarely openly hostile toward Beijing, China's rise makes New Delhi uneasy. In the words of Joseph Nye, there is a "strategic anxiety [that] lurks below the surface" in India.[208] The United States shares similar feelings about China. Washington ostensibly supports China's peaceful rise but many in the Pentagon view Beijing as a growing military threat. The 2007 Department of Defense Report on "Military Power of the People's Republic of China" indicates, for example, that "the pace and scope of China's military transformation has increased in recent years. . . . The expanding military capabilities of China's armed forces are a major factor in changing East Asian military balances; improvements in China's strategic capabilities have ramifications far beyond the Asia Pacific region."[209]

One of the reasons why Washington wants to strengthen its partnership with New Delhi is to counter the rising influence of China in Asia. This is a view shared by nonproliferation experts and supported by commentary from Bush administration officials. In a 2000 article in *Foreign Affairs*, Condoleezza Rice stated that the United States "should pay closer attention to India's role in the regional balance. There is a strong tendency conceptually to connect India with Pakistan . . . but India is an element in China's calculation, and it should be in America's too."[210] Rice expressed similar feelings after she became secretary of state in 2005. In April 2006, for instance, she argued that India is a strategically important country because it is "a rising global power that can be a pillar of stability in a rapidly changing Asia."[211] Robert Blackwill, the former U.S. ambassador to India, said China's rising influence in Asia provides incentives for the United States to closely engage India. Specifically, he noted that "India is properly attentive to the rise of Chinese power. . . . Indians understand that Asia is being fundamentally changed by the weight of PRC economic power and diplomatic skill. Which U.S. ally, except for Japan, thinks about China in such a prudent and strategic way?"[212]

Peaceful nuclear assistance is a key component of countering Chinese influence by strengthening India and the Indo-American partnership.[213] This is a view repeatedly expressed by senior U.S. government officials. For example, speaking in New Delhi in March 2006, President Bush emphasized that Indo-American cooperation—particularly the nuclear deal—was intended to "defeat our common enemies."[214] Undersecretary Burns suggested that the nuclear deal with India represents "a unique opportunity with real promise for the global balance of power" and added that "we see India as a stabilizing force in an often violent and unstable part of the world."[215] Additionally, Pentagon officials typically favored strict nonproliferation policies when it came to countries such as India, Iran, or North Korea but they generally supported the nuclear deal because they believed that New Delhi could serve as "a strategically important counterweight to China."[216] And Robert Blackwill, the former U.S. ambassador to India, pushed for atomic assistance to India as a means to counter China's regional dominance.[217]

Policymakers are sometimes reluctant to overplay the role that China plays in Indo-American relations. Neither the United States nor India wants to antagonize China by engaging in harsh rhetoric even though both countries are threatened by Beijing's accumulating influence. But even when the China factor is not explicitly highlighted, it weighs on the minds of key national security decision makers. As Ashton Carter notes, Washington views the nuclear deal as a means to "hedge against any downturn in relations with China," meaning that it would not hesitate to take advantage of its alliance with New Delhi if relations with Beijing deteriorate.[218]

STRENGTHENING A DEMOCRATIC INDIA

Democracy promotion played a key element in the George W. Bush administration's grand strategy. In interviews and speeches Bush routinely expressed a "deep desire to spread liberty around the world as a way to help secure [the United States] in the long-run" and noted that "in Europe, as in Asia, as in every region of the world, the advance of freedom leads to peace."[219] Part of Bush's democracy platform called for strengthening existing democracies and bilateral relationships with these countries.[220]

This strategy helps explain American interest in engaging India. When asked about his interest in India in 1999, Bush said, "a billion people in a functioning democracy. Isn't that something? Isn't that something?"[221] Bush and others in his administration believed that strengthening New Delhi and America's bilateral relationship with the world's largest democracy would serve U.S. strategic interests.[222] Undersecretary Burns and Secretary Rice made this clear almost every time they spoke publicly about Indo-American relations. For example, in testimony before Congress in 2005 Burns argued that propping up a democratic India would enhance its ability to work with the United States in achieving a variety of mutually beneficial objectives: "A strong, democratic India is an important partner for the United States. . . . India will play an increasingly important leadership role . . . working with us to promote democracy . . . and peace. . . . By cooperating with India now, we accelerate the arrival of the benefits that India's rise brings to the region and the world."[223]

Civilian nuclear assistance was an important cornerstone of the U.S. strategy to strengthen a democratic India. Secretary Rice said that peaceful nuclear aid would help incorporate India into the "tent of responsible democracies that wish to be anchors of stability and the promotion of democracy and peace in the world."[224] Another senior U.S. official, speaking on the condition of anonymity, indicated that the United States is assisting India's peaceful nuclear program to promote New Delhi's rise as "an independent, confident, and powerful state [because] our instincts and values are so close."[225] Peaceful nuclear assistance to India had the added benefit of making it more difficult for China, an authoritarian country, to emerge as the leader in Asia.[226] Washington hoped that a strong, democratic India could prevent authoritarian China from asserting regional dominance, which is an outcome that would adversely affect both U.S. and Indian interests.[227]

Why would nuclear energy assistance have these effects? First, aid of this nature would strengthen India politically in light of the prestige associated with nuclear technology. Indeed, in agreeing to share nuclear power plants and nuclear fuel with India, Washington sent a signal that it viewed New Delhi as an important player in a rapidly changing international system. Second, to the extent that energy production leads to growth, U.S. aid could strengthen India economically.

Alternative Explanations

NPT AND NONPROLIFERATION

Explanations rooted in norms and nonproliferation cannot explain nuclear cooperation between the United States and India. As I underscored above, India possesses a small nuclear arsenal[228] and it remains outside of the NPT. Scholars and policymakers recognized that providing aid to India under these circumstances violated the norms of the nuclear marketplace. For example, Jayantha Dhanapala and Daryl Kimball suggested that the deal "creates a dangerous distinction between 'good' proliferators and 'bad' proliferators [by] sending out misleading signals to the international community with regard to NPT norms."[229] Former secretary of state Colin Powell recognized the importance of engaging India but was also aware of the normative issues associated with atomic assistance. In 2003, Powell said that "we also have to protect certain red lines that we have with respect to proliferation."[230] Robert Einhorn, who was assistant secretary of state for nonproliferation during the second Clinton administration, offered a more direct criticism of the nuclear deal. He lambasted the agreement, arguing that it "sends the signal that bilateral relations and other strategic interests will trump nonproliferation . . . and . . . will reduce the perceived penalties associated with going nuclear."[231]

U.S. civilian nuclear assistance could also indirectly contribute to an increase in the size of the Indian nuclear arsenal. India relies on nuclear power plants to help ensure its energy security. It therefore must devote indigenous uranium resources to the production of fuel for these facilities, which leaves less uranium available for building nuclear weapons. If the United States provides fuel for India's civilian nuclear facilities—the argument goes—India would be able to use its scarce indigenous uranium resources for weapons production.[232] This argument rests on some fairly ambitious assumptions (such as India desiring a larger arsenal to implement its nuclear posture), but enabling India's arsenal expansion was a possible side effect associated with American exports of nuclear fuel.[233]

Proliferation concerns did not derail the India deal for two reasons. First, individuals who vocally opposed nuclear assistance were largely cut out of the decision-making process. The deal was pushed through quickly by a small group of senior officials with very little internal deliberation.[234] Only one U.S. official, NSC expert John Rood, represented nonproliferation interests during the negotiations with India in July 2005.[235] Second—and more importantly—senior officials believed that normative concerns associated with nuclear assistance were far outmatched by the strategic and political benefits the United States would obtain from offering atomic assistance.[236] This was certainly the case for President Bush and Secretary Rice.[237] Indeed, Bush had made up his mind that the United States should pursue nuclear cooperation with India after Rice's trip to New Delhi in early 2005. Bush said to Prime Minister Singh during his visit to Washington in July 2005: "You and I need to talk civilian nukes."[238] Even Robert Joseph, the undersecretary of state for arms control and international security who was generally a nonproliferation advocate, acknowledged that the strategic value of the India deal offset normative concerns. Joseph argued that "it is in our national security interest to establish a broad strategic partnership with India that encourages India's emergence as a positive force on the world scene. . . . India is a rising global power and an important democratic partner for the United States."[239]

To sell the deal to domestic and international audiences, U.S. officials claimed that it strengthened nonproliferation norms. For instance, Undersecretary Burns argued that the India nuclear deal brought New Delhi into the nuclear nonproliferation regime by requiring that some of its nuclear power reactors be placed under IAEA safeguards (although other plants would remain unsafeguarded and could therefore be used for military purposes with relative ease).[240] Undersecretary Joseph highlighted other nonproliferation concessions that India had made under the terms of the deal, including agreeing not to transfer sensitive enrichment and reprocessing technologies to states that do not already possess them and continuing a moratorium on nuclear testing.[241] But these considerations were afterthoughts. The United States moved forward with the deal because of the strategic benefits it offered, regardless of normative concerns, and then attempted to extract relatively minor nonproliferation concessions from New Delhi.

ECONOMICS

U.S. officials occasionally emphasized economic motives for assistance to India. They pointed out that the deal would generate revenue for U.S. companies such as General Electric.[242] Secretary Rice, for instance, suggested that nuclear assistance to India would "increase business opportunities" for the United States.[243] Other decision makers, including President Bush, emphasized the potential economic benefits of the deal much less frequently. Moreover, while the desire to make money was occasionally highlighted by Rice, there is no evidence that other economic considerations such as the need to jump-start the domestic nuclear industry or the lack of domestic demand for nuclear power were salient in explaining atomic assistance in this case.

Although generating hard currency is a significant benefit of atomic assistance to India, it is not a principal cause of the nuclear deal. From the very beginning, nuclear cooperation with New Delhi was pursued for strategic reasons, including strengthening relations with the world's largest democracy and countering the rising influence of China.[244] This is evident from Rice's statement that the economic benefits of the deal "must be viewed in the still larger, greater context of how this initiative elevates the U.S.-India relationship to a new strategic level."[245]

Chapter 5

A Thirst for Oil and Other Motives

Nine Puzzling Cases of Assistance

On average, suppliers use nuclear aid as a tool of economic statecraft to influence the behavior of their friends and adversaries. There are cases of nuclear cooperation, however, that are not successfully predicted by my theory. Out of all the cases in the dataset where nuclear cooperation agreements were signed about 20 percent do not appear to be influenced by the supplier state's political interests.[1] This chapter examines nine of these outlying cases to uncover the reasons for the onset of nuclear assistance. Why were the outliers not predicted by my theory and are they best explained by one of the alternative explanations or some other factor? What is the most important reason (or reasons) for nuclear cooperation in each of these cases based on the available evidence?

In some of the outliers, the results show that countries used peaceful nuclear cooperation to strengthen relationships with the recipient countries in ways that promoted their strategic interests. This is significant because it shows that support for my argument can be found even in cases that were not successfully predicted by the statistical model. The need to sustain domestic nuclear industries is salient in explaining three of the outlying cases, indicating that some cases not explained by my theory are accounted for by the economics argument. Interestingly, one additional alternative explanation emerges from this analysis. In three of the cases, suppliers provided nuclear technology to influence the recipient country on issues relating to oil supply. In other words, suppliers engaged in oil-for-nuclear technology swaps.

American Nuclear Cooperation with Indonesia, 1960–1965

U.S. nuclear cooperation with Indonesia began in June 1960 when the two countries signed a nuclear cooperation agreement. This deal, which was part of the U.S.-initiated Atoms for Peace program, authorized the United States to provide a $350,000 grant toward the cost of a research reactor and $141,000 toward the development of a nuclear research program.[2] The agreement also permitted the United States to provide enriched uranium to fuel the research reactor.[3] This assistance resulted in the construction of a small 250KW TRIGA-Mark II research reactor, which became the centerpiece of the Indonesian nuclear program during the 1960s.[4] In October 1964, the U.S.-supplied reactor produced Indonesia's first successful nuclear reaction.[5] Shortly after this event, in July 1965, Indonesian leader Sukarno proclaimed that his country would acquire nuclear weapons. He boasted: "God willing, Indonesia will shortly produce its own atom bomb."[6] Despite these nuclear aspirations, the United States continued its assistance to Indonesia's nuclear program. Washington renewed the 1960 agreement in September 1965, just as is was set to expire.[7]

The 1960 agreement with Indonesia was one of about forty bilateral agreements that were signed as part of the Atoms for Peace program.[8] Some have suggested that nuclear aid to Indonesia, like other cooperation initiated through this program, emerged in part because the United States wanted to "discourage the proliferation of nuclear weapons by shifting international attention from the development of weapons and toward the peaceful uses of nuclear energy."[9] In a fairly limited respect, one objective of U.S.-Indonesian cooperation might have been to decrease the likelihood that Jakarta would desire nuclear weapons by offering a "peaceful" alternative. The historical record offers a more compelling explanation, however.

Political considerations were on the minds of U.S. policymakers. Washington perceived that nuclear cooperation with Indonesia was an important means to limit the influence of the Soviet Union in Southeast Asia. This became especially salient after Moscow sought to ramp up its influence over Jakarta.[10] A declassified memo written by Richard Bissell, the Central Intelligence Agency's deputy director of plans, noted that "the new policy of the Soviet Union vis-à-vis Indonesia leaves us with only one practicable alternative," which is "to continue a moderate program of economic and military assistance sufficient to bolster the political position of our friends within Indonesia and to enable those who are willing to stand on principle to do so without being submerged by the overwhelming temptation of and pressures engendered by Soviet officers."[11] A National Security Council document published just prior to the signing of the first nuclear agreement echoed this sentiment: "the size and importance of Indonesia, together with its strategic position in relation to Australia and Free Asia, and the probable serious consequences of its loss to Communist control, dictate a vigorous U.S. effort to prevent these contingencies."[12] To counter the Soviet threat, Washington hoped to bring "new urgency and empathy to U.S. relations with the Third World" and felt that nuclear cooperation with Indonesia would be one policy that could help achieve this goal.[13] Peaceful nuclear assistance could improve bilateral relations between Washington and Jakarta while at the same time weakening Moscow's ability to woo Indonesia into the Soviet camp.[14]

U.S. enthusiasm about using nuclear cooperation with Indonesia as an instrument to constrain Soviet influence weakened when President Sukarno adopted an increasingly anti-American posture in the early 1960s. Relations between the two countries began to sour beginning in the summer of 1963 when Sukarno proclaimed that he was going to destroy neighboring Malaysia and began to develop a closer relationship with communist China.[15] Additionally, the United States began to worry about Sukarno's intent to acquire nuclear weapons, although Washington was skeptical of Jakarta's ability to actually acquire the bomb. According to a U.S. intelligence estimate written in 1964, "Indonesia will be unable to produce nuclear weapons from its own resources in the foreseeable future."[16]

Despite these concerns, Washington pledged to continue nuclear cooperation with Indonesia in 1965. When renewing the 1960 agreement, the United States added the condition that Indonesia must allow the International Atomic Energy Agency to inspect the American-supplied research reactor. Ultimately, cooperation continued because the United States did not want to lose complete control of the technology and materials it had supplied. If the agreement were terminated, Washington feared that Sukarno might refuse to return the low enriched uranium it had supplied to fuel the research reactor. Were this to happen, the credibility of the Atoms for Peace program would have been questioned.[17] In short, the United States was "persuaded to continue the cooperation basically by the argument that there was no way to end the program without the Indonesians taking full control of it."[18]

Brazilian Nuclear Cooperation with Iraq, 1980

In January 1980, Brazil and Iraq signed a nuclear cooperation agreement. This deal authorized Brazil to provide technology in uranium exploration and to train Iraqi scientists.[19] It also specified that Brazil would eventually supply unprocessed and enriched uranium and offer assistance in the construction of nuclear reactors.[20] An official statement by the Brazilian foreign ministry asserted that

> the bilateral cooperation will be effected in accord with the capacities and priorities of each side, in complete conformity with the agreements and international obligations of the respective countries. . . . The two parties reaffirm their support of the principle of the nonproliferation of nuclear arms and reaffirm their right to develop and use nuclear energy for peaceful purposes.[21]

Brazil initiated this nuclear cooperation with some reluctance. According to press reports, the Brazilian government feared that it could suffer diplomatically if it was too enthusiastic about nuclear aid to Iraq.[22] Yet this initial reluctance did not prevent Brazil from exporting nuclear materials beginning in 1980. Shortly after signing the nuclear cooperation agreement, two Iraqi planes traveled to Brazil and loaded eight tons of uranium oxide for Baghdad. Israeli agents reportedly forced the aircraft to the ground over Africa and unloaded the uranium, preventing it from reaching Iraq.[23] This incident created a fair amount of embarrassment for Brazil, which sought to keep its cooperation with Iraq quiet. It did not, however, deter Brasilia from offering additional aid to Baghdad. In September 1990, Brazilian president Fernando Collor de Mello admitted to President George H. W. Bush that Brazil had transferred "significant nuclear technologies" to Iraq but that he intended to immediately terminate this assistance.[24]

Brazil was not naïve about the proliferation risks associated with its peaceful nuclear assistance. It was well known that Iraq might use foreign aid to augment a military program. Iraqi leader Saddam Hussein proclaimed that "any country in the world that seeks peace and security . . . should assist the Arabs in one way or another to obtain the nuclear bomb in order to confront Israel's existing bombs. . . . No power can stop Iraq from acquiring technological and scientific know-how to serve its national objectives."[25] Brazil did not want Saddam Hussein to acquire nuclear weapons and it certainly did not want to support such an effort.[26] Brasilia was, nevertheless, willing to accept the risk that its aid would facilitate this outcome because it expected to receive important benefits that would offset the potential proliferation consequences. In particular, aiding the Iraqi "civilian" nuclear program could help Brazil secure a stable oil supply.

At the time the nuclear cooperation agreement was signed, the country depended on outside sources for 80 percent of its energy supply and Iraq provided 40 percent of this.[27] To illustrate the magnitude of this energy dependence, Saudi Arabia and Iran each provided less than 20 percent of the oil consumed in Brazil. It is easy to understand why one news source reported, "No other budding nuclear power is as dependent on Iraqi oil as Brazil" (France and Italy were also dependent on Iraqi oil and helped Iraq with atomic aid, as we will see below).[28] Brasilia simply could not afford to "ignore Iraq's control of a crucial part of Brazil's hydrocarbon energy supply."[29] It was also desperate to "negotiate payment in counter-trade to avoid exacerbating its grievous foreign debt load" that resulted in part from its dependence on Iraqi oil.[30] In the end, Brazil's thirst for oil made it difficult to say no to Iraqi requests for nuclear assistance.[31] Nuclear cooperation was part of an explicit quid pro quo to induce Baghdad to steadily increase the supply of oil, thereby enhancing Brazil's energy security.[32] Just two days after the signing of the nuclear agreement, Iraq agreed to supply an additional 20.7 million barrels (869 million gallons) of oil at current prices and to increase Brazil's daily allotment of oil by at least 160,000 barrels a day for the next thirteen years.[33]

Brasilia gambled that atomic assistance would result in greater access to Iraqi oil without contributing to nuclear proliferation. It is debatable whether the gamble paid off. On the one hand, after the onset of the Iran-Iraq War in September 1980, Iraqi oil exports to Brazil were drastically reduced to one hundred thousand barrels a day.[34] On the other hand, Baghdad did its best to meet its end of the bargain. As Narcisco Carvalho, an official from the Bank of Brazil, said in 1990, "While other suppliers were adding various requirements that raised the price of oil, Iraq always supplied Brazil regularly and even made shipments without having been paid for the last."[35]

British Nuclear Cooperation with South Korea, 1991

On November 27, 1991, the United Kingdom signed an agreement with South Korea authorizing cooperation in areas related to the nuclear fuel cycle. The deal called for the "transfer of nuclear material, equipment, and technology between the two countries" in the areas of "management, storage, and final disposal of irradiated fuel and radioactive waste, and transport of radioactive material.[36] In short, the arrangement authorized the United Kingdom to reprocess South Korean spent fuel in Britain and return separated plutonium and reprocessing wastes to South Korea. This agreement was signed at a time when there was considerable concern about potential nuclear weapons ambitions of both countries on the Korean Peninsula. South Korea appeared on the British government's "danger list" for British exports causing "strategic and proliferation concerns."[37] Media reports criticized British attempts to sell reprocessed fuel to South Korea, noting that such efforts increased the risk of nuclear proliferation in "one of the world's hottest troublespots."[38] North Korea—a country pursuing nuclear weapons at the time—chastised Britain for "criminal acts of encouraging" South Korean nuclear proliferation.[39] Further, the United States, which needed to approve this arrangement as the original supplier of the South Korean uranium, voiced concern about the proliferation significance of the deal.[40] These concerns and reorganization of the South Korean nuclear industry delayed the implementation of the 1991 agreement, but Britain remained interested in offering reprocessing services to Seoul throughout the 1990s.[41]

The historical record reveals that commercial incentives largely explain British attempts to initiate nuclear cooperation with South Korea in 1991. The British government recognized that South Korea was a "potential major customer" for reprocessing services due to its rapidly expanding nuclear industry.[42] Beginning in the early 1990s, South Korea increased civilian nuclear energy development; it planned to construct nine new nuclear power reactors in the 1990s and have a total of twenty-five in operation by 2015.[43] Britain perceived that this surge in nuclear power capacity would result in a large stockpile of spent fuel, creating opportunities for the British nuclear reprocessing industry. According to a statement issued by the British Department of Energy, "The [1991 nuclear cooperation] agreement is intended as a framework for the U.K. nuclear industry to pursue the commercial opportunities presented by South Korea's large and expanding civil nuclear power program."[44] John Wakeham, the British secretary of state for energy, further highlighted the commercial value of the agreement: "my hope is that companies . . . who have teamed up to develop business opportunities in Korea . . . will secure contracts on a commercial basis with their South Korean counterparts."[45] Ultimately, generating revenue and providing additional opportunities for the British nuclear industry were sufficiently powerful motivations to trump concerns about nuclear proliferation in East Asia. These commercial considerations were especially powerful because the nuclear agreement came at a time when some in the British government were questioning the future of British reactor sales and related fuel services.[46]

Canadian Nuclear Cooperation with Romania, 1977

Canada signed a nuclear cooperation agreement with Romania in 1977 after several years of negotiations. The terms of this deal authorized Ottawa to build a 600MW Canada Deuterium Uranium (CANDU) reactor at Romania's Cernavoda nuclear complex.[47] This arrangement would turn out to be "nothing less than a fiasco" for Canada as financial delays, safety concerns, and financial difficulties plagued the project.[48] Over a decade after construction began, the Canadian reactor was only 45 percent complete. In the early 1990s, Canada supplied Romania with $315 million in additional loans so that the project could be finished.[49]

Romania was an ally of the Soviet Union, a country that Canada sought to contain as part of its commitment to the North Atlantic Treaty Organization

(NATO). On the surface, it seems puzzling that a close U.S. ally would cooperate with a state in the Soviet camp during the cold war. A closer examination of the historical record, however, reveals that Canada's behavior was consistent with the Western strategy of limiting Moscow's influence. Relations between Romania and the Soviet Union were often rocky. In April 1964, for example, Romania's leadership issued a declaration expressing its dissatisfaction with the Warsaw Pact.[50] President Nicolae Ceausescu also established diplomatic relations with West Germany in 1967 and publicly criticized Soviet foreign interventions such as the invasion of Czechoslovakia.[51] Canada viewed these events opportunistically and employed atomic assistance to cultivate closer bilateral ties with Romania, drawing it further away from the Soviet bloc.[52] Officials in the Canadian Ministry of External Affairs believed that atomic aid could encourage Romania to further detach itself from the Soviet Union, which is conceivable for the reasons articulated in chapter 2.[53] The desired outcome was that the country would emerge as an independent Communist country, similar to Yugoslavia.[54] Canada also intended its nuclear aid to reduce East-West tensions. Beginning in 1972 the United States took a series of steps toward détente and sought to minimize tension with the Soviet Union and Ottawa felt compelled to assist in this effort.[55] Prime Minister Pierre Trudeau, who was elected in 1968, felt that Communist bloc countries should be "encouraged, enticed or cajoled into becoming full participants in the community of nations."[56] He believed that the way to make this happen "was to bind them to the world, not to cast them beyond the pale" by providing scientific and technical assistance.[57] In this context, atomic assistance to Romania could help Canada build a bridge between East and West.

Ottawa stood to gain economically by engaging in peaceful nuclear cooperation with Romania. CANDU exports can be quite lucrative, improving Canada's balance of payments by $1.6 billion per reactor. The Romanian sale also had the benefit of generating economies of scale at a time when Canada was searching for new foreign markets.[58] Economic considerations were not the driving force behind this arrangement, however.[59] It is unlikely that Canada would have moved forward with nuclear aid if it did not believe the political benefits outlined above were attainable. As Duane Bratt argues, "while Canada had obvious economic incentives to sell the [reactor], its political interests sealed the Cernavoda deal."[60]

Canadian-Romanian nuclear cooperation provides another illustration of how concerns about nuclear proliferation are often overlooked. Although Romania had signed the NPT prior to the onset of nuclear cooperation, Ottawa feared that it might retransfer Canadian-supplied technology to states such as Pakistan or the Soviet Union.[61] This did not stop Canada from ultimately supplying nuclear technology, which is remarkable because just three years had passed since India used a Canadian reactor to produce plutonium for its first nuclear explosive device. At this time, perhaps more than any other, Ottawa should have been especially wary of proliferation risks. It discounted these dangers in part because it believed that Moscow would constrain Romania's ability to acquire nuclear weapons. Canadian external affairs minister Don Jamieson conveyed this when he indicated, "Romania is clearly within the Soviet orbit [and therefore] it is not very likely [that] there would be any development of nuclear weapons in a country such as Romania."[62] This argument, whether true or not, was likely crafted to downplay the proliferation perils of nuclear cooperation so that Canada could justify reaping the aforementioned political and economic benefits.

French Nuclear Cooperation with Iraq, 1975–1981

France signed a nuclear cooperation agreement with Iraq on November 18, 1975, during a visit to Baghdad by the newly named prime minister of France, Jacques Chirac.[63] This deal authorized France to supply Baghdad with a 40 MW research reactor, a zero-power reactor, and a radioactive waste treatment station.[64] The research reactor, which came to be known as Osiraq, was significant in part because it operated with 93 percent enriched uranium, which is suitable for use in a nuclear bomb.[65] The nature of Franco-Iraqi nuclear cooperation raised concerns around the globe that Baghdad might acquire nuclear weapons. These fears were perpetuated by provocative rhetoric from Iraqi leaders, such as the statement from Saddam Hussein cited above when discussing Brazil's cooperation with Iraq.[66] This did not prevent France from continuing nuclear cooperation. In July 1980, France shipped 13kg of highly enriched uranium to Iraq for use in the Osiraq reactor.[67] According to those who were involved in nuclear transactions with Iraq, "the question of nuclear proliferation really was not that important. The French concerns were far more prosaic, and reflected the kind of practical economic considerations that so often lie behind great political decisions."[68]

Many countries, most notably Israel, grew increasingly intolerant of Franco-Iraqi cooperation as it became clear that Paris had no intention of ending its assistance to Baghdad. On June 7, 1981, eight Israeli bombers entered Iraqi airspace and destroyed the Osiraq reactor.[69] The Israeli government proclaimed the attack a success and argued that it was necessary because "Iraq was preparing to produce atomic bombs" and "the target of those bombs was Israel."[70] Although the Israelis proclaimed the attack a success, the Osiraq raid was an indictment of IAEA safeguards. As I will discuss in chapter 9, Iraq had declared the facility to the agency and the reactor was inspected it in accordance with the IAEA's standard procedures. The strike brought an end to French nuclear cooperation with Iraq.[71]

Why did France overlook the clear proliferation risks associated with selling Iraq nuclear technology? The answer is strikingly similar to what we

saw in the Brazil-Iraq case. The empirical evidence reveals that the French
nuclear sales were part of a broader strategy to secure a reliable supply of oil
from Iraq. France was highly dependent on energy imports in the 1970s. Indeed, more than 75 percent of the country's energy needs were met through
foreign sources.[72] To put this in perspective, only 20 percent on the countries
in the world were more energy dependent during this period.[73] Following the
Yom Kippur War and the subsequent Arab oil embargo, enhancing energy
security became a top priority for Paris.[74] As one scholar observed, "the real
danger for France [in the late 1970s was] the oil crisis, not the SS-20."[75]

France developed a multifaceted strategy to reduce its dependence on foreign energy suppliers. Part of this strategy involved a massive expansion of
domestic nuclear power. In 1973, nuclear power provided only 8 percent of
all electricity production in France. This figure spiked to 24 percent by 1980
and by 2005 nearly 80 percent of electricity produced in the country came
from nuclear power plants.[76] Another component of France's energy security
strategy was to seek assurances from key petroleum exporters. France was
"madly scrambling to secure oil supplies" after the embargo because oil accounted for two-thirds of all energy consumption.[77] Twenty percent of all
its oil imports came from Iraq, which made assurances on oil supply from
Baghdad especially valuable.[78]

Nuclear cooperation was an attempt to receive such assurances. French
officials did not expect that this policy would be a panacea to the problem
of energy security but they believed it could be a meaningful part of the
solution.[79] Paris explicitly linked oil supply to the transfer of nuclear technology. The nuclear cooperation agreement indicated that Iraq would provide "permanent and reliable" oil supplies in exchange for nuclear exports.[80]
Policymakers in France rarely mentioned nuclear assistance to Iraq without referring to oil. For instance, when Prime Minister Raymond Barre visited Baghdad in 1979 he focused on two things: (1) reaffirming that France
was committed to supplying Iraq with a nuclear reactor; and (2) ensuring
that Baghdad continued to export oil at favorable prices.[81] Iraqi officials
recognized that French aid constituted a quid pro quo. Vice-President Saddam Hussein (soon to be president) assured Barre that Paris could count on
"the continuity and security" of oil supplies from Iraq.[82] This not-so-subtle
trade-off was widely recognized around the world. For instance, Winston
Churchill, a British member of parliament and the grandson of the former
prime minister, noted that French "lust for oil" caused it to export a nuclear
reactor to Iraq despite the proliferation risks.[83] A report on the transfer in the
Christian Science Monitor plainly stated: "France agreed to sell the Baghdad
regime [a nuclear plant] in exchange for plentiful supplies of Iraqi crude."[84]

Iraq followed through on its end of the bargain—at least in the short term.
French nuclear sales led to an increase in Iraqi oil shipments from twenty
million tons per year in 1978 to thirty million tons in 1980, an increase of

50 percent in just two years.[85] This was a nontrivial gain given the importance of imports from Iraq in meeting France's overall energy needs. There was, however, a limit to the benefits France reaped. Just as in the case of Brazil, Iraq's oil exports to France steeply dropped following the initiation of the Iran-Iraq War. Thus, the rewards that motivated Paris to engage in nuclear cooperation with Iraq did not materialize to the extent that it expected.[86]

German Nuclear Cooperation with Brazil, 1975

On June 27, 1975, West Germany and Brazil signed one of the most controversial nuclear cooperation agreements to date. The German-Brazilian deal was monumental because it was the first to include a self-sufficient "fuel cycle package sale," meaning that it offered Brazil the capability to sustain its nuclear program indigenously.[87] The arrangement included five elements: (1) uranium exploration and mining; (2) uranium enrichment; (3) fuel fabrication; (4) reprocessing of spent fuel; and (5) nuclear power plants.[88] From a financial perspective, the "centerpiece of the deal" was the sale of between two to eight nuclear reactors. This transaction was valued at between $2 billion and $8 billion.[89] The United States strenuously objected to this arrangement:

> The political context for the trade is not at all reassuring. Although relatively stable at the moment, Brazil has suffered recurring upheavals of government, internal repression, terrorism, and saber-rattling toward neighbors. Brazil has steadfastly refused to subscribe to the Nonproliferation Treaty, and it has never been more than a reluctant participant in the discussion of nuclear restraint.[90]

The deal was further criticized in the U.S. press as "a reckless move that could set off a nuclear arms race in Latin America, trigger the nuclear arming of a half-dozen nations elsewhere and endanger the security of the United States and the world as a whole."[91]

The German-Brazilian pact, once dubbed the "deal of the century," did not materialize to the degree that either country expected. The project "ran into trouble from the very start."[92] This was in part because of technical problems, economic hardship, and a lack of funding from the Brazilian government. Germany constructed a 1,300MW power reactor at Angra dos Reis (this reactor is known as Angra II) and transferred a limited amount of fuel-cycle knowledge. Construction on a second German reactor, Angra III, was halted after the 1986 Chernobyl disaster.[93] In 2004, Germany and Brazil replaced their 1975 NCA with a more general agreement on renewable energy sources that did not focus on nuclear cooperation.[94] Brazil authorized the resumption of construction on Angra III in 2007 and the plant is expected to come online in 2015.[95] Germany decided to get out of the nuclear export business in the wake of the March 2011 accident at Japan's Fukushima nuclear power station, so the Angra III project is unlikely to be finished with German aid.[96]

The empirical evidence reveals that Germany wanted to export nuclear technology to Brazil in large part to receive assurances of uranium supplies.[97] In the 1970s, West Germany decided to increase its investment in nuclear power and construct forty nuclear power reactors to meet its growing energy needs. The need for uranium to fuel these plants "created both a reason and an opportunity for Brazil and West Germany to act together to implement separate strategic aims."[98] Bonn hoped to secure uranium from Brazil in exchange for enrichment technology and other nuclear items. One of the "main hopes" was that West German cooperation with Brazil would lead to the discovery of additional uranium reserves, offering Bonn an alternate source of nuclear fuel.[99] This was especially important in light of the 1973 oil crisis and West Germany's growing dependence on imports to meet its energy needs.

Other economic considerations also affected the deal. In the 1970s, Germany's nuclear energy program experienced financial difficulties because of the lack of sufficient domestic demand for a robust nuclear industry. Bonn may have sought to export nuclear reactors to Brazil as a means to revive the industry at a time when it planned to increase its national reliance on nuclear energy.[100] Additionally, Germany may have viewed cooperation with Brazil as a means to develop future ties with the developing world. After signing the pact with Brazil, Bonn proclaimed that "the day of the industrial nations' hegemony is over—perhaps particularly that of the United States. The developing countries cannot and should not be denied; they will no longer tolerate, no do they have to tolerate, the kind of discrimination they have had to suffer in the past."[101] Klaus Scharmer, head of the international section of the Julich nuclear research center, echoed these sentiments: "We must combat the 'development gap' that tends to grow between countries that are more and less developed. We must try to hasten the advance of the underdeveloped."[102]

Concerns about proliferation were not salient in explaining the signing of the German-Brazilian agreement. Despite U.S. warnings, West Germany was undeterred by Brazil's hostile attitude toward the NPT and suspicions that it coveted nuclear weapons. Bonn even went so far as to claim that: "Germany is simply following the provisions of the Nonproliferation Treaty's Article 4, which binds subscribers to facilitate "the fullest possible exchange of equipment, materials and scientific and technological information for the peaceful uses of nuclear energy."[103] This statement fails to mention, however, that Brazil was not entitled to the Article IV privilege because it had not ratified the NPT.

Indian Nuclear Cooperation with Vietnam, 1999

On January 19, 1999, India and Vietnam initiated nuclear assistance by signing an agreement on cooperation in the peaceful uses of nuclear energy.[104]
This deal authorized India to construct a "Nuclear Training Center" in
Dalat, Vietnam, and provide training to Vietnamese scientists at Indian nuclear facilities.[105] After the conclusion of this agreement, Indian prime minister Atal Bihari Vajpayee asserted that "science and technology . . . are the
backbone of a modernizing society and India is pleased to assist Vietnam in
this direction. I am glad cooperation in the area of peaceful uses of nuclear
energy is progressing well."[106] Nuclear cooperation became a subject of interest during bilateral meetings subsequent to 1999. For example, a joint statement issued on May 2, 2003, called for additional cooperation in the area of
nuclear power.[107] To date, however, these sentiments have yet to yield further
tangible nuclear cooperation.

Nuclear cooperation with Vietnam should be viewed in the broader context of India's "Look East" policy. Indian prime minister Narasimha Rao
initiated this strategy in the 1990s to develop stronger economic and political ties to the countries in Southeast Asia. The purpose of this policy was to
(1) institutionalize linkages between India and the Association of Southeast
Asian Nations (ASEAN); (2) improve bilateral relations with each of the
ASEAN countries; and (3) prevent China from exerting too much influence
in the region.[108] These goals generated a lot of hype but progress was stymied until Prime Minister Vajpayee revived the policy following the 1997–98
Asian economic crisis and began to closely engage several Southeast Asian
countries.

Vietnam was, in the words of Vajpayee, a "critical element" of India's
Look East policy.[109] India provided atomic assistance to support the guid-
ing principles of this strategy, especially the strengthening of its bilateral
relationship with Vietnam. The technical assistance India pledged would not
singlehandedly forge an Indo-Vietnamese partnership, but it could aid in
the pursuit of this objective. Vuong Huu Tan, the chairman of the Vietnam
Atomic Energy Commission, recently stated that nuclear power development
is important because it "is a solution for the country to achieve its energy
security goal."[110] Vietnam's demand for power continues to grow, raising
questions about whether future energy needs will be met. As Tan revealed,
"the power shortage will be severe from 2015 as demand outpaces capacity
so finding new sources of electricity is an urgent task."[111] At the same time,
Vietnam depends on foreign assistance in order to benefit from nuclear energy since it lacks the indigenous capacity to build the necessary plants. Under
these conditions, atomic aid can be an effective means of statecraft (see chapter 2). Indeed, Vietnamese officials said that civilian nuclear cooperation,

and the Look East Policy more generally, had brought Hanoi and New Delhi closer together. Prime Minister Nguyen Tan Dung said in 2006, for example, that "relations between the two countries have increased considerably in the last five years" because of India's engagement.[112] He added that enhanced cooperation had taken "Vietnam-India relations to a new height" and emphasized the importance of Indian nuclear aid in fostering closer relations.[113]

Why does New Delhi care about improving its bilateral relationship with Hanoi? The principal reason has to do with limiting the influence of China in Southeast Asia. The region is the historical battleground of influence between New Delhi and Beijing.[114] Vietnam's "geostrategic location" makes it an especially important country for the purposes of constraining Chinese influence.[115] Moreover, India and Vietnam are both threatened by Beijing's growing capabilities, having fought wars with China in 1962 and 1979, respectively. The salience of this shared threat increased in the late 1990s following China's emerging presence in the Indian Ocean.[116] In light of these circumstances, Vietnam is an important partner for India.[117] If bilateral ties between India and Vietnam were stronger, the two countries would have better results in working together to limit China's influence in the region.

It is telling that the 1999 nuclear agreement was signed just as both countries' concerns about China's growing presence in the region were escalating.[118] India offered nuclear aid and other forms of assistance to Vietnam because it needed Vietnam to serve "as a spear in the Chinese underbelly to counter the threatening Beijing-Islamabad-Rangoon entente now taking shape against New Delhi."[119] Nuclear cooperation could help accomplish this by strengthening an Indo-Vietnamese partnership, thus making it harder for Beijing to cultivate a relationship with Hanoi—not by enabling Vietnam's ability to build nuclear weapons.[120] Using nuclear energy assistance to achieve this objective was beneficial from India's perspective because it did not overly agitate China. New Delhi is apprehensive about Beijing's future intentions but it hesitates to engage in provocative acts because it does not want to inflame a complicated bilateral relationship. In short, atomic assistance allows New Delhi to have its cake (i.e., strengthen ties with a key state in Southeast Asia) and eat it too (i.e., avoid exacerbating a potentially explosive relationship with Beijing).[121]

Italian Nuclear Cooperation with Iraq, 1976–1981

Italy signed a nuclear cooperation agreement with Iraq on January 15, 1976. The terms of this deal enabled Italy to supply Iraq with a "radiochemistry lab" that would include lead-shielded hot cells and glove-boxes for the remote manipulation of radioactive material.[122] This lab's "peaceful" use was the safe handling of highly radioactive substances, but it could also be used to extract plutonium from spent nuclear fuel. In other words, it would provide Iraq with a reprocessing capability. Italy provided this facility for a mere $1.67 million—$600,000 less than Italy's initial offer.[123] The 1976 agreement also provided the basis for Italy to build four additional laboratories in Iraq, train Iraqi nuclear scientists, and supply 1,767kg of uranium enriched to 2.6 percent. These additional transactions generated slightly more than $50 million for the Italian firm SNIA Technit.[124] Italy completed construction on the radiochemistry lab in 1978 and ended its other nuclear assistance to Iraq following the Israeli raid of Osiraq.

Italian nuclear assistance to Iraq between 1976 and 1981 had implications for Iraqi nuclear weapons development. As one U.S. State Department official frighteningly observed, it was "of sufficient magnitude to permit Baghdad to obtain enough plutonium to produce a nuclear weapon in about a year's time."[125] Understanding Italian motivations in selling nuclear technology to Iraq is particularly important given the historical significance of this case.

A familiar story emerges from the available historical evidence. Italy's nuclear cooperation with Iraq was largely an attempt to secure a stable oil supply.[126] The Italians were "hit hard by the energy crisis, and were even hungrier than the French to export high technology, especially to the Arab oil producers on whom they depended for a large part of their oil supply."[127] At the time the nuclear cooperation agreement was signed, Italy imported roughly one third of its oil from Iraq.[128] Key Italian decision makers such as Umberto Colombo, president of the National Committee for Nuclear Energy (CNEN), saw Iraq as a "great commercial opportunity" because it provided Italy with a means to trade nuclear technology for oil. Other Italian officials agreed. There is little doubt that the persuasive power of oil motivated Rome to provide Iraq with peaceful nuclear assistance.[129] These motives were known outside Italy as well. For instance, Carter administration officials in the United States "believe[d] Italy established its close nuclear ties with Iraq in order to secure long-term access to Iraqi oil."[130]

A second consideration may have also influenced the transactions. The Italian nuclear industry was in its infancy and it was struggling due in part to the lack of domestic demand. Colombo noted that by engaging in nuclear cooperation with Iraq "we are also trying to give our nuclear industry work, and it needs to work if we are to develop it for the future."[131] Italy might have perceived that nuclear cooperation with Iraq could kick-start its nuclear industry, but this motivation was not the driving force behind the sales. The nuclear transactions were clearly about oil and not about making money. Italy had to work at near cost and just barely broke even, having agreed to build the nuclear labs for a very modest sum.[132]

Italian policy does not make atomic assistance conditional on having a pristine nonproliferation record. As Carlo Mancini and Giuseppe Maria Borga point out, Italy "does not exclude cooperation and commercial relations with countries remaining outside the circle of the NPT parties or otherwise unwilling to accept full scope safeguards."[133] This does not mean that Rome shells out technology indiscriminately to any state that might be seeking to build nuclear weapons. It does, however, reveal that Italy is willing to accept certain proliferation risks if it can obtain certain benefits. In the Iraqi case, the prospect of enhanced energy security helped tip the balance in favor of providing nuclear assistance despite the danger for nuclear weapons proliferation. Steve Weissman and Herbert Krosney succinctly make this point when they argue that because of the oil considerations "the Italians were in even less of a position than the French to turn down a deal that could possibly lead to a bomb for Iraq."[134]

Italian officials knew exactly what they were doing. They recognized that their aid could enhance Saddam Hussein's ability to build nuclear weapons, although they hoped that this would not happen. Elites sought to downplay the proliferation hazards and shift the blame to other suppliers, but this was mostly an attempt to justify the policy to domestic and international audiences. Colombo explained, for example, that "everything is dangerous for proliferation, even buying a book, if you have the intention to proliferate." He went on to suggest that "Italy is not giving something that is not available elsewhere. Should we withdraw, a lot of people will take our place. . . . The real problem in nuclear sales comes from the French."[135] This type of fatalistic attitude is sometimes expressed by suppliers when they recognize the trade-offs they are making—but it is not necessarily true. As we saw in chapter 3, nuclear exporters provide aid for specific politico-strategic reasons. It is not as though every country with civilian nuclear energy ambitions will have droves of suppliers knocking down their doors.

Soviet Nuclear Cooperation with Yugoslavia, 1956–1967

The Soviet Union initiated assistance to Yugoslavia in January 1956, when the two countries signed their first nuclear cooperation agreement. This arrangement permitted Moscow to construct a 6.5MW heavy water moderated reactor at Vinča. As part of the terms of this agreement, the Soviet Union also supplied heavy water and uranium fuel for the reactor.[136] The fuel was not subjected to IAEA safeguards and Yugoslavia assumed ownership of it upon delivery.[137] The Soviets eventually supplied 48.2kg of weapons-grade uranium—nearly enough to manufacture two nuclear weapons—to Yugoslavia.[138] Using this Soviet-supplied fuel, the Vinča reactor began operation in December 1959. Following this initial transfer, the Soviet Union and Yugoslavia signed additional nuclear agreements in 1963 and 1975 but Belgrade began to rely increasingly on other suppliers including France and the United States.

During Soviet-Yugoslav cooperation in the 1950s, the government of Josip Broz Tito was exploring the development of nuclear weapons.[139] The Tito regime proclaimed in January 1950: "We must have the atomic bomb. We must build it even if it costs us one-half of our income for years."[140] These ambitions were well known by intelligence agencies around the world by January 1954, before the onset of Soviet assistance.[141] Yugoslavia suspended its nuclear weapons program in the early 1960s, but Soviet-Yugoslav nuclear cooperation provides another illustration of how concerns about proliferation can be subordinated to other political objectives.[142] Consistent with Soviet policy at the time, Moscow perceived that it could maintain political control over Yugoslavia's nuclear weapons ambitions to minimize the risk of proliferation.[143]

The evidence reveals that Moscow initiated nuclear cooperation with Yugoslavia as part of its policy of rapprochement to improve relations with Belgrade, which had deteriorated during the tenure of Josef Stalin.[144] After World War II, Tito launched an ambitious industrialization plan and worked toward a Balkan alliance to balance against Soviet dominance. These actions alienated Moscow and led to the Soviet Union's disassociation from Yugoslavia in 1948.[145] Desperate for assistance from other countries, Yugoslavia forged closer relations with the West. Beginning in 1950, it "slowly but steadily built up its economic, military, and diplomatic ties with the Western Powers, capitalizing skillfully on Western interest in keeping it free of Soviet domination."[146]

After Stalin's death in March 1953, Moscow sought to normalize relations with Yugoslavia. In May 1955, Soviet leader Nikita Khrushchev traveled to Belgrade to improve relations with Yugoslavia.[147] A joint declaration issued during Khrushchev's trip stated that "the two Governments have agreed to assist and facilitate cooperation among the social organizations of the two countries through the establishing of contacts, the exchange of Socialist experiences and a free exchange of opinions."[148] Shortly thereafter, Moscow ceased issuing propaganda critical of Tito's regime, reopened communications from Eastern Europe to Yugoslavia, and lifted a blockade on Yugoslav–Soviet Bloc trade.[149] An important part of normalizing relations was the assistance offered by Moscow in the development of nuclear energy in Yugoslavia.[150]

By using nuclear aid as an instrument to improve relations with Yugoslavia, Moscow hoped to reintegrate Belgrade into the Soviet Bloc.[151] Short of reintegration, Moscow hoped to weaken Yugoslavia's developing relationship with the West. Soviet nuclear assistance—and foreign aid more generally—to Yugoslavia was "aimed at the destruction or weakening of the existing defense agreements or military alliances which [recipient] countries may have with Western powers."[152] This strategy proved to be at least partially successful. As a result of "wooing" Belgrade, the Soviet Union "arrested" Yugoslavia's ties with the Western powers and enhanced its opportunities for influence over the Tito regime.[153]

Summary of Suppliers' Motives

The analysis of nine outliers yields some interesting results. As table 5.1 shows, four considerations are especially salient in explaining nuclear cooperation in outlying cases. These factors include (1) a desire to jump-start the domestic nuclear industry; (2) the perceived need to strengthen bilateral relations; (3) an attempt to constrain the influence of a threatening state; and (4) securing a stable petroleum supply from oil-producing states. Other motivations influence nuclear aid in only one case. In one instance (U.S.-Indonesia) nuclear cooperation was partially aimed at convincing the recipient country to abandon nuclear weapons and in another (Germany-Brazil) it was intended to secure a continuous flow of uranium to fuel nuclear power plants. These findings are significant for several reasons.

To begin, they reveal that support for my argument can be found even in cases that are not successfully predicted by my statistical model. Two of the four significant motivations to emerge from the analysis of

Table 5.1 Determinants of nuclear cooperation in outlying cases

Motivation for nuclear cooperation	Cases
Jump-start domestic nuclear industry	Germany-Brazil Italy-Iraq UK-South Korea
Strengthen bilateral relations	Canada-Romania India-Vietnam U.S.-Indonesia USSR-Yugoslavia
Limit the influence of a threatening state	India-Vietnam U.S.-Indonesia USSR Yugoslavia
Secure a stable oil supply	Brazil-Iraq France-Iraq Italy-Iraq
Discourage proliferation	U.S.-Indonesia
Reserve assurances of nuclear fuel supply	Germany-Brazil

outliers—strengthening bilateral relations and constraining the influence of threatening states—are part of my theory of atomic assistance. The outlying cases of nuclear cooperation are those that I least expected to offer support for my theory. That I find at least some evidence in favor of my argument in three outliers suggests that my theory applies to more cases than the statistical analysis leads one to believe. This should inspire further confidence that my argument is salient in explaining peaceful nuclear cooperation. It may also suggest that my key concepts are not perfectly measured, which is a common problem in quantitative analysis. For example, based on my coding, at the time nuclear cooperation occurred between India and Vietnam they were not military allies, they did not share an enemy, and Vietnam was not a superpower enemy. But an examination of this case reveals Vietnam and India were both threatened by the same state—China—and the presence of a common enemy helped inspire New Delhi's nuclear aid to Hanoi.

Although the cases analyzed in this chapter offer some support for my theory of nuclear cooperation, they also illuminate its limitations. The U.S.-Indonesia and USSR-Yugoslavia cases provide useful illustrations. U.S. nuclear aid to Indonesia in the 1960s was intended in part to prevent the Soviet Union from developing close ties with the country. Similarly, Soviet assistance to Yugoslavia was intended to transform its relationship with Belgrade after diplomatic contact was severed in 1948. Moscow was growing worried about increased Yugoslav ties to the West and it used nuclear assistance as a means to improve its relationship with Belgrade while preventing the West from exerting further influence against it. These motivations are consistent with my argument that states use nuclear cooperation as a means to improve bilateral relations, with the overall objective of constraining the influence of a third party. However, part of my argument is that supplier states constrain the influence of an enemy by providing nuclear technology to states that share the same adversary. This was not the case in either of these instances. U.S. aid to Indonesia was intended to constrain the power of the Soviet Union and limit Communist influence in the country more generally, but Indonesia was not necessarily threatened by Moscow. Soviet assistance to Yugoslavia was intended to prevent the United States from influencing that country, but Belgrade was directly adversarial toward Washington. Indeed, Washington was also using civilian nuclear assistance to compete for influence in Yugoslavia.[154] This suggests that suppliers can use nuclear assistance as a means to constrain the power of a state they are threatened by even if the recipient state does not share its threat perception. If an exporting state and one of its enemies are competing for influence against the same state, it may offer atomic assistance as a means to stymie the enemy's efforts. Although partially consistent with my argument, this logic is not fully captured by it.

The analysis of outliers further sheds light on the limits of nonproliferation-related explanations of civilian nuclear trade. In seven of the nine cases, concerns existed regarding the recipient state's commitment to nonproliferation (recall that similar concerns existed in the cases examined in chapter 4). Suppliers were aware that their assistance *might* be used to build nuclear weapons although this was never their hope or intent. In some instances, suppliers voiced reservations about these proliferation risks. They went through with nuclear cooperation despite these qualms because the perceived economic and/or strategic gains of assistance outweighed fears of contributing to proliferation. This provides further evidence in favor of my argument that nuclear suppliers will provide nuclear aid to countries that are pursuing nuclear weapons or that have not signed the NPT if doing so is otherwise in their political or economic interests. Some of the cases analyzed in this chapter underscore that recipient states can be successful in obtaining atomic assistance to bolster a nuclear weapons program. As we saw in chapter 3, however, there is not a clear relationship between receiving nuclear aid and pursuing nuclear weapons. The important point is that suppliers do not systematically withhold assistance to states that are pursuing the bomb as the nonproliferation argument predicts.

Some of the outliers are explained by the economics argument. A common motivation for engaging in atomic assistance was to sustain domestic nuclear industries. For example, part of the reason Italy provided aid to Iraq was because its nuclear industry was relatively small and it felt a need to compete with more established suppliers such as the United States. This evidence suggests that some of the cases not explained by my theory can be accounted for by one of the economic arguments. Economic factors, therefore, cannot be dismissed as easily as nonproliferation in explaining atomic assistance.

One of the primary purposes of this chapter is to determine whether there is a variable that is important in explaining nuclear cooperation that I omitted from my initial statistical analysis. The examination of outliers reveals a novel explanation for atomic assistance. In three of the cases analyzed in this chapter—Brazil-Iraq, France-Iraq, and Italy-Iraq—the supplier provided nuclear assistance to secure a stable supply of oil from the recipient country. These examples suggest an alternative hypothesis for nuclear cooperation: oil-producing countries are more likely to receive peaceful nuclear assistance than non-oil-producing states. In the next chapter, I evaluate this proposition.

Chapter 6

Oil for Peaceful Nuclear Assistance?

Oil is a critical resource in contemporary international politics. It is essential for economic growth and energy security—particularly for countries that use oil to generate electricity.[1] States therefore often pursue foreign policies that ensure a stable supply of oil.[2] For example, countries may export strategic commodities such as arms to states that are oil producers in order to receive oil imports on favorable terms.[3]

Do supplier countries swap nuclear assistance for oil? Civilian nuclear assistance provided to Iraq in the 1970s and early 1980s suggests that there may be a relationship between oil and atoms for peace. When Brazil, France, and Italy assisted the Iraqi nuclear program they were heavily dependent on imports to meet their energy needs, and they were desperately searching for ways to enhance their energy security. The 1973 Yom Kippur War and the subsequent Arab oil embargo provided uncertainty about the future sources of petroleum. All three suppliers provided nuclear aid to Iraq with the explicit expectation that doing so would increase their oil imports from Baghdad at a time when reliable energy imports were especially coveted.

Is exchanging oil for nuclear technology a compelling motivation for peaceful nuclear assistance across a broader set of cases or does it explain only a handful of outliers? In general, countries do not trade reactors or other related technology for oil. But when oil prices rise substantially, nuclear suppliers provide technology to petroleum-producing countries to improve their energy security. This explains, in part, why oil-producing countries received so many pledges of atomic assistance in 2007 and 2008 when oil prices escalated to record highs.

The Historical Record

Statistical Tests

If countries swap nuclear assistance for oil, oil-producing states should receive a disproportionate amount of atomic assistance. Yet we would not expect oil producers to be more likely to sign all types of nuclear cooperation agreements. In the case studies presented in Chapters 4 and 5 suppliers occasionally exchanged nuclear technology, materials, or know-how for oil, but in no instance did states use safety NCAs as a means to influence petroleum producers on oil price or supply. Thus, in this chapter I explore whether nonsafety agreements are correlated with a variable measuring whether the recipient country is a significant oil exporter.

To begin exploring this issue, table 6.1 displays a simple cross-tabulation of nonsafety NCAs against oil-producing countries for the period from 1950 to 2000. This analysis reveals a negative and statistically significant relationship between being an oil producer and receiving nonsafety assistance in developing a civil nuclear program. About 1.2 percent of observations in the dataset that include a non-oil-producing recipient experience peaceful nuclear cooperation. A smaller percentage of oil-producing observations, 0.9 percent, experienced atomic aid. Based on these figures, oil-producing countries were 25 percent less likely than countries that do not produce oil to receive nuclear aid between 1950 and 2000. Petroleum producers also received only a small fraction (about 10 percent) of all nuclear aid. These findings contradict the oil-for-nuclear-assistance logic. This negative relationship might emerge because oil producers can often meet their energy needs through their domestic petroleum reserves and are less likely to need nuclear energy or atomic assistance.

This analysis does not imply, however, that oil producers are always less likely to receive peaceful nuclear assistance. All of the oil-for-nuclear swaps analyzed in the previous chapter occurred at a time when oil prices were rising substantially in the wake of the 1973 Arab-Israeli War. In the 1970s, the nuclear supplier countries were extremely worried about a stable oil supply

Table 6.1 Cross-tabulation of oil producers and civilian nuclear assistance, 1950–2000

| Civilian nuclear assistance | Oil producer | | |
	No	Yes	Total
No	116,742 (98.80%)	18,173 (99.10%)	134,915 (98.84%)
Yes	1,421 (1.20%)	165 (0.90%)	1,586 (1.16%)
	118,163 (100%)	18,338 (100%)	136,501 (100%)

$\chi^2 = 12.67\ p < .001$

and they desperately raced to meet their energy needs. This suggests that oil and nuclear technology are especially likely to be exchanged when oil prices spike.

Figure 6.1 plots the percentage change in oil prices between 1950 and 2000. This figure reveals that oil prices were relatively stable prior to 1973 but rose quickly in 1974 and again around 1980. In the mid-1980s, prices dropped but began to spike again by 2000. If the logic advanced above is correct, then we should see oil for nuclear exchanges during the mid- and late 1970s and around 2000.

To evaluate whether this is true, I conduct a cross-tabulation analysis of peaceful nuclear assistance against oil-producing countries only for years where oil prices spiked substantially. The results, which are displayed in table 6.2, reveal a positive and statistically significant relationship between signing nonsafety NCAs and oil production on the part of the recipient state. About 0.8 percent of the observations in the dataset that do not include a petroleum-producing recipient experience peaceful nuclear cooperation. On the other hand, a higher percentage of observations including an oil producer—about 1.4 percent—experience nonsafety NCAs. This means that oil producers are about 80 percent more likely to receive nuclear assistance. And when oil prices spike, petroleum producers sign about 23 percent of all nonsafety NCAs. This is a highly disproportionate percentage given that oil-producing states such as Iraq, Iran, Libya, and Saudi Arabia should be

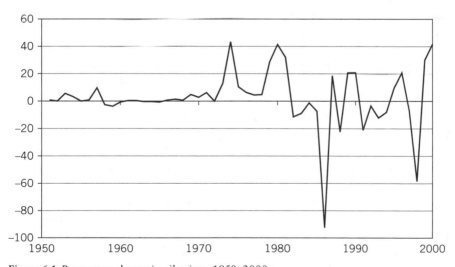

Figure 6.1 Percentage change in oil prices, 1950–2000
Source: Susan Carter, Scott Sigmund Gartner, Michael Haines, Alan Olmstead, Richard Sutch, and Gavin Wright, "Table Db56–59 Crude Petroleum—Average Value, Foreign Trade, and Proved Reserves: 1859–2001," *Historical Statistics of the United States* (Cambridge: Cambridge University Press, 2006).

Table 6.2 Cross-tabulation of oil producers and civilian nuclear assistance, years following large increases in oil price

Civilian nuclear assistance	Oil producer		
	No	*Yes*	*Total*
No	8,256 (99.22%)	1,343 (98.60%)	9,599 (99.13%)
Yes	65 (0.78%)	19 (1.40%)	84 (0.87%)
	8,321 (100%)	1,362 (100%)	9,683 (100%)

$\chi^2 = 5.13 \; p = 0.024$

less likely to turn to nuclear energy because they can meet their energy needs through other sources. Collectively, these results indicate that suppliers swap nuclear technology for oil when oil prices increase.

Do these same results hold when we account for the other factors that could affect peaceful nuclear assistance? To determine whether this is the case, I replicate the statistical models presented in chapter 3 while adding a variable measuring whether the recipient country is an oil producer. The results show that the statistical significance of the oil producer variable washes away when I analyze the entire period from 1950 to 2000. This indicates that once we account for confounding variables, the relationship between nonsafety NCAs and oil production in the recipient state is indistinguishable from zero. The other findings are similar to those I presented in Chapter 3. Most notably, all five variables relevant to my argument remain statistically significant in the expected direction. This is significant because it indicates that accounting for oil-for-nuclear-assistance exchanges does not change the fact suppliers use atomic assistance to strengthen their bilateral relationships with recipients in ways that promote their strategic interests.

The findings of the multivariate statistical analysis reaffirm the results of the cross-tabulation analysis presented in table 6.2. Petroleum producers are more likely to receive nuclear technology when oil prices spike, even when accounting for the relevant control variables. Interestingly, the factors that are generally important in explaining atomic assistance are not correlated with nonsafety NCAs during periods when oil prices rise suddenly. Other than the oil producer measure, only four variables achieve conventional levels of statistical significance: NPT, supplier's GDP, recipient's GDP, and distance (see appendix 6.1 for further details).

Figure 6.2 illustrates the substantive impact that each of these variables has on the likelihood of nuclear cooperation when oil prices rise. The figure shows that oil exporters are 87 percent more likely than nonoil producers to sign nonsafety NCAs. The other variables generally produce more modest changes in the probability of nuclear cooperation. Note, however, that

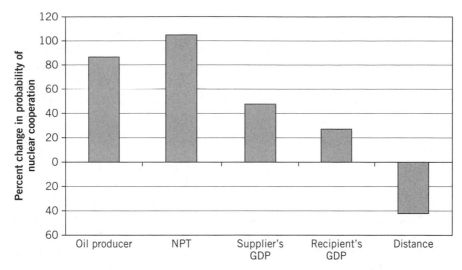

Figure 6.2 Percentage change in probability of peaceful nuclear cooperation resulting from statistically significant independent variables, years following large increases in oil prices
Note: Calculations are based on model 2.

NPT members are 104 percent more likely than non-NPT members to sign nonsafety NCAs. This is somewhat unexpected because in chapter 3 I consistently found that the NPT variable was unrelated to nuclear cooperation or correlated with assistance in the negative direction. Overall, these results suggest that countries behave slightly differently when they are desperate to enhance their energy security. It is important to note, however, that such spikes occur infrequently. Indeed, we see large increases in oil prices in fewer than 7 percent of the observations in the dataset.

This analysis adds to our understanding of atomic assistance. In the vast majority of the cases, countries provide atomic assistance to manage their bilateral relationships. And generally speaking, oil producers get less help developing civil nuclear programs because they have a lower demand for nuclear energy. But under specific and predictable conditions, supplier countries behave differently because they feel the need to enhance their energy security.

Additional Qualitative Evidence

The cases I analyzed in chapter 5 illustrated the causal mechanisms linking oil production and peaceful nuclear assistance. In the remainder of this section I present some additional qualitative evidence in favor of the oil argument.

REVISITING THE U.S.-IRAN CASE

I showed in chapter 4 that the United States provided peaceful nuclear assistance to Iran in large part to enhance its politico-strategic interests by cultivating closer ties with Tehran.[4] This was important, I argued, because of the military alliance the two countries shared and because of America's desire to limit the influence of the Soviet Union. What about the role of oil?

Primary documents reveal that Washington perceived a linkage between civilian nuclear assistance and oil pricing.[5] The United States perceived that by providing Iran with nuclear technology, it could convince the shah to lower the price of oil, which in turn would benefit American economic and security interests. On March 11, 1974, a declassified telegram from Henry Kissinger to Richard Helms, the American ambassador to Iran, clearly illustrated the oil motive: "We will want to [bring] Iran into intensive dialogue on how our ties can be further strengthened [focusing on how] technology and help with industrialization [can be] traded for guaranteed oil supply at specified prices."[6] Kissinger went on to say that "we want to establish a framework of consultation within which we can pursue a constructive dialogue bilaterally with respect to oil supply."[7] Nuclear technology was perceived to be an important part of this dialogue. As Joseph Sisco indicated in a secret government report, "discussions on oil are bound to be sensitive . . . however, the detailed estimates of the costs of alternative energy sources we will be developing should be of major interest to the Iranians and could gradually help persuade them of the necessity of modifying their present stance on oil prices."[8] Nuclear power was among the alternative energy sources that Sisco had in mind when he made this statement.[9]

This case nicely illustrates the statistical findings because, like the cases presented in chapter 5, the attempts to link oil and nuclear assistance occurred following large spikes in petroleum prices during the 1970s. Below, I will discuss how spikes in oil prices over the last decade have influenced the nuclear marketplace.

THE OIL MOTIVE SINCE 2000

The quantitative analysis I conducted above only covers the period from 1950 to 2000. But since 2000 the world has witnessed another spike in oil prices. Figure 6.2 plots the price of oil by month between January 2001 and December 2008. As the graph shows, oil prices began to rise in mid-2004 and increased steadily until September 2006. After a short-lived and modest decline, prices increased dramatically beginning in December 2006. In 2008, oil reached $100 a barrel for the first time in history and prices hovered around that for several months. In October 2008, prices began to decline and approach 2004–05 levels. Despite this decline, countries remain deeply concerned about the price of oil and the possibility that petrol supplies could subside in the future. President-elect Barack Obama stated in

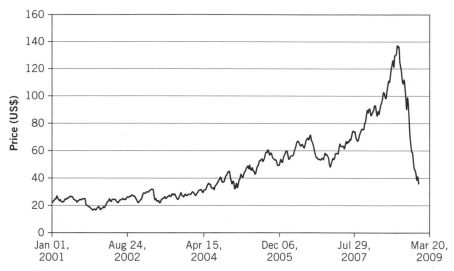

Figure 6.3 Oil prices by week, January 2001–December 2008
Source: United States Department of Energy, "Weekly All Countries Spot Price FOB Weighted by Estimated Export Volume" (Washington, D.C.: Energy Information Administration, January 14, 2009).

December 2008, for instance, that oil prices were an urgent concern and a threat to American interests.[10]

Since oil prices increased in 2006 many petroleum producers have received pledges of support for their civil nuclear programs, just as the oil-for-nukes argument predicted. India, Japan, and the United Kingdom agreed to provide atomic assistance to Kazakhstan.[11] The United States and France each signed agreements with Saudi Arabia and the United Arab Emirates (UAE).[12] Paris also inked nuclear deals with Libya and Algeria.[13] And Russia pledged to support the Bahraini and Venezuelan civil nuclear programs.[14] Newer suppliers have also lined up to assist oil producers. For example, South Korea signed a nuclear cooperation agreement with Indonesia, Iran agreed to give atomic assistance to Nigeria, and Argentina offered aid to Algeria.[15] Some of these countries also received nuclear aid prior to 2006. Venezuela, for instance, received a modest amount of assistance from the United States, Argentina, and Brazil in the early days of the atomic age. Others had a hard time building their civil nuclear infrastructures prior to the recent spike in oil prices. Qatar and Bahrain did not sign a single nuclear cooperation agreement before 2006. And Saudi Arabia and the UAE each only signed one atomic pact (it is no coincidence that these two deals were concluded during the oil crisis of the 1970s).

There are a few reasons why oil-producing countries received so much nuclear aid after 2006. Consistent with the argument I presented in chapter 2, in many cases nuclear suppliers provided atomic assistance to strengthen their bilateral relationship with the recipient country. Russia, for example, used atomic assistance to strengthen its relations with Venezuela and counter U.S. influence, as I underscored in chapter 2.[16] The United States likewise viewed peaceful nuclear cooperation with the UAE as an important means to strengthen its partnership with a strategically important Arab country.[17] But oil also played a key role in the deals described above. There is substantial evidence indicating that nuclear suppliers provided aid to enhance their energy security by swapping technology for oil. Japan, for example, gave pledges of atomic assistance to Kazakhstan in part because of rising petroleum prices and its dependence on foreign oil. After concluding a nuclear cooperation agreement with the Kazakhs, Japanese prime minister Junichiro Koizumi asserted that the deal provided "a basis for even broader cooperation between our countries," which is important because Japan imports almost all of its oil and Kazakhstan possesses vast oil deposits.[18] Similar motivations drove Russian nuclear cooperation with Bahrain—even though Moscow is also an oil producer. The two countries signed agreements dealing with oil joint ventures and supply at the same time that they concluded the nuclear cooperation agreement.[19] Oil likewise played a key role in American-Saudi civil nuclear cooperation. According to a White House statement, the United States agreed to assist Riyadh in developing nuclear reactors because "our global economy depends greatly on Saudi Arabian energy."[20] Washington hoped that helping Saudi Arabia in the nuclear field would persuade it to increase oil supplies, which would, in turn, lower petroleum prices. To help ensure that this could happen, the United States offered assistance in protecting Saudi oil infrastructure at the same time the nuclear deal was concluded.[21]

Oil played an especially powerful role in French nuclear cooperation with Arab states. Paris believed that nuclear cooperation with these countries would improve bilateral relations. Indeed, French president Nicolas Sarkozy referred to atomic assistance as "one of the foundations of a pact of trust," implying that it is the centerpiece of a strategy to forge closer partnerships with these countries.[22] Improved relations are viewed as a means to keep oil supplies flowing.[23] This is important for France because it remains dependent on foreign oil to meet its energy needs.[24] Consequently, when oil prices spiked, France worried about its energy security. By offering Saudi Arabia assistance in developing a nuclear power program, President Sarkozy believed that he would have more leverage in persuading Riyadh to lower crude oil prices. Sarkozy put this thinking to the test when he pushed the Saudi government on this issue during a trip to Riyadh in January 2008.[25] The Libyan deal emerged for similar reasons. French foreign minister Bernard Kouchner indicated that the civil nuclear agreement would symbolize Libya's "return to the international community" and improve French-Libyan relations.[26] According to Sarkozy, this would, in turn, help France secure oil-related contracts in Libya.[27] Prior to the French pledges of atomic assistance,

Paris believed that it was missing out on the chance to get involved in Libya's oil sector and improve its energy security.[28] Although Libya is rich in petroleum, the French share of the Libyan oil market was quite modest. In 2005, France received only 6 percent of Libya's oil exports, while fellow European Union countries Italy and Germany received 21 percent and 11 percent, respectively.[29] Paris hoped to increase this percentage by helping Libya develop its peaceful nuclear program.

A critic of the oil-for-nuclear assistance logic might argue that Iran's nuclear ambitions were the real driver of increased atomic assistance after 2006. Oil-producing states—especially those in the Middle East—are threatened by the prospect of an Iranian bomb and they may be developing civil nuclear infrastructures to hedge against Tehran's actions. According to this argument, other countries want to have the option to build the bomb if Iran does so and the best way to accomplish this is to develop nuclear energy programs. This argument is insightful, but it is incomplete for two reasons. First, nuclear suppliers provide atomic assistance for very specific reasons, as I discussed in chapter 2. They do not merely shell out nuclear technology to any country that demands it. There is no reason to assume, therefore, that oil producers would be able to secure atomic aid just because they are threatened by Iran's nuclear program. Second, Iran's nuclear weapons program began in the mid-1980s.[30] Tehran may have kept its program secret for a few years, but by the early 1990s it was well known that Iran coveted the bomb and was pursuing the technology to develop it.[31] The alternative explanation cannot account for why most of the oil producers did not develop civil nuclear infrastructures prior to 2006.

Appendix 6.1: Data Analysis and Empirical Findings

Dataset, Variables, and Method

I generally employ the same procedures used to estimate model 3 in chapter 3 (see table 3.7 and appendix 3.1). I make three modifications to test the oil argument.[32] First, as noted above, I employ a dependent variable that excludes safety agreements because the case studies presented in the previous two chapters did not provide any evidence that states signed this type of agreement to influence the recipient state on oil pricing or supply.

Second, I add an additional independent variable to the statistical model. I create a dichotomous measure and code it 1 if the importing state is an oil producer and 0 otherwise. To determine whether a state is an oil producer, I consult data compiled by James Fearon and David Laitin.[33] About 13 percent of all observations in the dataset include an oil-producing state as the recipient.

Third, I modify the sample size to test the argument that suppliers are especially likely to sign NCAs with petroleum producers following large increases in oil prices. Some models will be estimated in a sample that only includes observations when there is a large increase in oil prices from year t-2 to year t-1 and 0 otherwise. Data on oil prices come from the *Historical Statistics of the United States*.[34] I define a large increase in prices as percentage changes in price greater than one standard deviation above the mean (33.4 percent).

Results

Table 6.3 displays the results of a multivariate statistical analysis designed to test the oil arguments articulated above. Model 1 includes all of the control variables discussed in chapter 3 as well as the oil producer variable. Model 2 adjusts the sample size so that only years following large increases in oil prices are included. Model 3 is a trimmed model that includes the oil producer variable and the five variables testing my argument. This model is also estimated using the limited sample.

Table 6.3 Oil producers and peaceful nuclear cooperation, 1950–2000

	(1)	(2)	(3)
Oil producer	−0.083	0.628**	0.625**
	(0.104)	(0.285)	(0.265)
Strategic factors			
Alliance	1.107***	0.098	0.504
	(0.099)	(0.409)	(0.450)
Conflict	−1.126**		
	(0.508)		
Shared enemy	0.203*	−0.044	0.275
	(0.110)	(0.493)	(0.468)
Superpower enemy	0.925***	0.289	0.580*
	(0.085)	(0.412)	(0.349)
Joint democracy	0.698***	0.311	0.411
	(0.083)	(0.310)	(0.314)

(*Continued*)

Table 6.3—*cont.*

Nonproliferation			
Nuclear weapons	0.410***	0.401	
	(0.154)	(0.659)	
NPT	−0.187***	0.711*	
	(0.072)	(0.383)	

Generating revenue			
Supplier's GDP	0.000***	0.000***	
	(0.000)	(0.000)	
Importer's GDP	0.000***	0.000*	
	(0.000)	(0.000)	
Distance	−0.000***	−0.000***	
	(0.000)	(0.000)	
Exports	−0.000***	−0.000	
	(0.000)	(0.000)	

Sustaining industry			
Domestic demand	−4.122***	−1.583	
	(1.301)	(5.273)	

Temporal controls			
No NCA Years	−0.092***	0.174**	0.247***
	(0.019)	(0.080)	(0.070)
Spline 1	−0.000	0.004***	0.005***
	(0.000)	(0.001)	(0.002)
Spline 2	0.000	−0.003	−0.004*
	(0.000)	(0.002)	(0.002)
Spline 3	−0.000	0.001	0.001
	(0.000)	(0.001)	(0.001)
Constant	−3.877***	−4.404***	−5.093***
	(0.123)	(0.271)	(0.178)
Observations	134745	9491	9660

Notes: Robust standard errors in parentheses; * significant at 10%; ** significant at 5%; *** significant at 1%. Conflict is dropped from models 2 and 3 because it predicts failure perfectly.

As noted above, there is not a statistically significant relationship between oil producers and civilian nuclear assistance when I examine the period from 1950 to 2000. Adding the oil variable does not undermine support for my theory of peaceful nuclear cooperation and the controls behave similar to the way they did in chapter 3, with one exception.[35] Based on the results from models 2 and 3, there are important differences in patterns of atomic assistance during periods when oil prices rise substantially. The coefficient on the variable measuring whether the recipient state is an oil producer is positive and statistically significant in both models, indicating that suppliers are more likely to swap technology for oil when they are concerned about their energy security. Notice that superpower enemy is statistically significant in model 3 but none of the other variables testing my theory attain conventional levels of significance. There are also some differences with respect to the control variables. The NPT variable is positively correlated with nuclear assistance while the statistical significance of nuclear weapons pursuit washes away. Exports and domestic demand are likewise statistically insignificant.

PART II

ATOMS FOR WAR

Chapter 7

Spreading Temptation

Why Nuclear Export Strategies Backfire

Nuclear suppliers transfer nuclear technology, materials, and know-how to enhance their politico-strategic influence in international politics. In particular, countries offer aid to strengthen allies and alliances; forge closer relationships with enemies of enemies; strengthen ties with other democracies; and enhance their energy security by trading technology, materials, and know-how for oil—but only when petroleum prices increase sharply.

Does atomic assistance inadvertently raise the likelihood of nuclear weapons proliferation? This chapter introduces a theory of proliferation, which suggests that civilian nuclear assistance raises the likelihood that countries will begin nuclear weapons programs, particularly if they experience an international crisis after receiving aid.

Statistical tests support this argument. There is a correlation between nuclear cooperation agreements and bomb program initiation that holds up even when accounting for other potential causes of nuclear proliferation. Consistent with my theoretical expectations, security threats condition the relationship between atomic assistance and nuclear weapons program initiation; the relationship between civilian nuclear assistance and nuclear weapons pursuit becomes stronger as countries experience a greater number of militarized disputes with other states. The type of assistance matters when it comes to the proliferation potential of peaceful nuclear cooperation. Yet NCAs raise the likelihood of proliferation even when they explicitly prohibit the transfer of "restricted" technology (i.e., enrichment facilities, heavy water reactors, and reprocessing centers).

I do not find support for the argument that states without weapons programs engage in "nuclear hedging." Indeed, countries that are not already pursuing the bomb do not appear to seek peaceful aid because they have an interest in developing nuclear weapons. Recipients generally have peaceful intentions when assistance commences but circumstances evolve over time. Once states initiate a bomb program they may seek help under false pretenses but they do not always get it.

This chapter analyzes two cases using qualitative analysis to test the causal logic of my argument more directly. An examination of South African nuclear decision making shows that atomic assistance created conditions in the country that contributed to the initiation of a weapons program in 1974. This case demonstrates that nuclear cooperation can raise the likelihood of weapons pursuit in the absence of protracted security threats. An analysis of nuclear decision making in India reveals how atomic assistance and security threats can interact to produce proliferation decisions.

A Theory of Nuclear Weapons Program Initiation

There are trade-offs associated with initiating a nuclear weapons program. The bomb can offer those countries that possess it some strategic advantages.[1] Countries that possess nuclear weapons, for instance, may be less likely to be attacked by their adversaries.[2] States that live in dangerous neighborhoods and expect to be involved in conflict in the future, therefore, can enhance their security by going nuclear.[3]

On the other hand, significant political and economic costs usually accompany a decision to pursue nuclear weapons. Weapons programs are extraordinarily expensive.[4] Given the costs of going nuclear, countries that go down the weapons path usually have to sacrifice other national priorities. Oftentimes, it is economic development programs that must suffer in order to build nuclear weapons. This is aptly illustrated by Zulfikar Ali Bhutto's proclamation in the 1960s that acquiring nuclear weapons might require the Pakistani people to go hungry or "eat grass." There are also indirect financial costs that accompany bomb programs. Countries that initiate weapons programs, for example, can face economic sanctions and reductions in foreign direct investment.[5] All of these burdens can deter countries from going for the bomb. As Michael Horowitz argues, "The enormous level of financial intensity necessary to acquire nuclear weapons has always functioned as a significant constraint on the diffusion of nuclear weapons."[6]

A host of diplomatic and strategic costs can likewise result from going for the bomb. Given that most countries find the spread of nuclear weapons threatening, they usually respond negatively to states that attempt to get the bomb. Other countries might roll back their diplomatic relationships with the proliferating state and label it as a "rogue" entity. Even allies of the state pursuing the bomb often express concern about its activities, which can decrease the strength of the military alliance. For instance, South Korea's nuclear weapons program strained Seoul's alliance with the United States. In an extreme case, states going for the bomb might experience a military attack against their nuclear facilities.[7] In September 2007, for example, Israel bombed a Syrian nuclear reactor that could have been used to produce plutonium for nuclear weapons.

Building the bomb is also technically difficult. Successful weapons production requires three steps. The first is the production of plutonium-239 or weapons-grade uranium containing a high percentage (generally at least 90 percent) of U-235. These materials are produced using sophisticated dual-use technology and equipment. The second is weapon assembly. To convert the enriched uranium or plutonium into a weapon several components are needed including electronics to trigger chemical explosives and a neutron generator to start the nuclear detonation at the appropriate time. The third is integrating the weapon with a delivery system such as a ballistic missile or other military vehicle.[8] Of these stages the most difficult, by far, is the acquisition of nuclear materials.[9]

With these costs in mind, figure 7.1 illustrates how peaceful nuclear cooperation influences nuclear weapons program initiation. I argue that atomic assistance typically begins innocuously as a means to help states develop or expand their civilian nuclear programs. Yet over time the accumulation of nuclear aid creates conditions in the recipient country that raise the likelihood that a leader will attempt to proliferate.

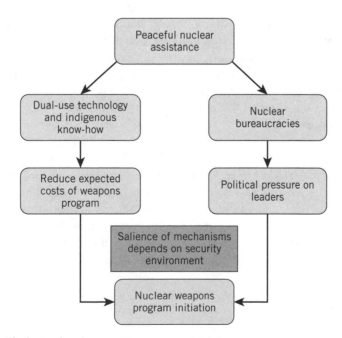

Figure 7.1 The logic of nuclear weapons program initiation

Civilian nuclear assistance creates two potential pathways to nuclear weapons programs. First, it reduces the expected costs of pursuing the bomb even if countries have no interest in proliferating at the time that they receive outside help.[10] Nuclear cooperation provides countries with technology that can be used to build nuclear weapons since all nuclear equipment and materials are dual use in nature. As discussed elsewhere in the book, nuclear reactors can be used for research or electricity production but they can also produce plutonium for nuclear weapons. Uranium enrichment and plutonium reprocessing facilities can be used to produce fuel for power reactors or fissile material for nuclear weapons.[11] Civilian nuclear cooperation also lowers the hurdles to weaponization by building up a knowledge base in nuclear engineering and other related fields. When countries provide atomic assistance they often train scientists and technicians from the recipient country. The United States, for instance, provided training to foreign scientists at the Argonne Laboratory School of Nuclear Science and Engineering located just outside of Chicago in the 1950s and 1960s. It is common for suppliers to provide training in the handling of radioactive materials; processes for fuel fabrication and materials having chemical or nuclear properties; and the operation and function of reactors and electronic control systems. States also routinely share experiences in other crucial fields such as metallurgy and neutronics.[12] This training enables countries to operate facilities that are provided via peaceful nuclear cooperation. More important, intangible transfers help establish "a technology base upon which a nuclear weapon program could draw."[13] Scientists that receive training in metallurgy abroad could return home and assist in the construction of a covert uranium enrichment facility that was dedicated to a weapons program. Likewise, personnel that helped construct or operate a reactor built primarily with foreign assistance could establish a new reactor designed principally to produce plutonium for a bomb.

The diffusion of technology and know-how as a result of atomic assistance influences how leaders perceive nuclear weapons programs. If leaders are going to take on the political and economic costs discussed above, they generally want confidence that the bomb can be successfully produced in a relatively short period of time.[14] Otherwise, leaders fear that they will subject their countries to unnecessary hardship. Receiving peaceful nuclear assistance provides greater assurances to leaders that bomb programs can pay off without dragging on for decades. This, in turn, makes pursuing the bomb relatively more attractive. As Stephen Meyer argues, "When the financial and resource demands of [beginning a weapons program] become less burdensome, states might opt to proceed . . . under a balance of incentives and disincentives that traditionally might have been perceived as insufficient for a proliferation decision."[15]

The right-hand side of figure 7.1 illustrates a second mechanism linking peaceful nuclear assistance and nuclear weapons programs. Drawing on the literature on bureaucratic politics, I argue that atomic aid establishes or strengthens atomic energy commissions or other nuclear-related bureaucracies.[16] States require foreign assistance to get their nuclear programs off the ground, even when they are technologically developed. As Itty Abraham argues, "No atomic program anywhere in the world has ever been purely indigenous."[17] In many cases, atomic energy commissions would not exist—except on paper—in the absence of nuclear cooperation. Atomic assistance empowers bureaucracies once they are established by enabling technological advances. Scientists use these breakthroughs for political purposes to demonstrate that they can successfully harness the peaceful uses of nuclear energy.[18] This, in turn, enhances the prestige of the scientific community. Consequently, scientists earn greater access to key decision makers and are able to secure additional resources for nuclear programs.

The empowerment of scientists has unintended consequences for proliferation. As they become more influential, members of atomic energy commissions sometimes lobby leaders to begin a military program on the grounds that bomb production is technologically possible and can be done with relatively limited costs.[19] Jim Walsh captured this logic well when he argued that "the biggest proliferation impact of [civil nuclear cooperation] may not have been the spread of little reactors but the spread of little Atomic Energy Commissions, many of which became bomb advocates."[20] Leaders are often persuaded by this lobbying when it occurs because they are keenly aware that the quicker the bomb can be developed, the less that other national priorities need to suffer. Thus, this second mechanism relates to the first mechanism articulated earlier.

The notion that nuclear bureaucracies can become bomb advocates is not new. Yet existing arguments linking bureaucracies and proliferation are underspecified. As Scott Sagan notes, "There is no well-developed domestic political theory of nuclear weapons proliferation that identifies the conditions under which . . . coalitions are formed and become powerful enough to produce their preferred outcomes."[21] My argument is that peaceful nuclear cooperation helps influence whether scientists and atomic energy commissions will facilitate the development of a nuclear program for military purposes. Atomic assistance can enhance the power and influence of scientists and nuclear energy advocates, potentially leading to nuclearization.

The argument advanced above suggests that, on average, countries receiving atomic assistance are more likely to initiate nuclear weapons programs. As figure 7.1 illustrates, there is an additional element to this basic proposition. The salience of the two mechanisms discussed above depends on a country's security environment. In the absence of security threats emanating from militarized conflict with neighboring countries, countries receiving atomic aid should still be more likely than states receiving no atomic aid to pursue the bomb. Yet prior nuclear assistance is substantially more likely to influence the onset of a weapons program if a state suffers a defeat in a war with a rival or feels threatened for another reason. Under such conditions, the strategic advantages that the bomb offers would be especially attractive *and* states would possess dual-use technology and trained scientists that collectively shorten the expected time to weaponization. Thus, nuclear aid followed by militarized conflict is an especially strong prescription for the onset of nuclear weapons programs. If countries have not received prior civilian assistance, the perceived costs of beginning a weapons program might be prohibitively high—even for states that face frequent militarized conflict.

The argument advanced above suggests two testable implications of nuclear weapons program onset:

Hypothesis 7.1: Countries receiving higher levels of peaceful nuclear assistance are more likely to begin nuclear weapons programs than states receiving lower levels of assistance.

Hypothesis 7.2: Countries receiving higher levels of peaceful nuclear assistance are more likely than states receiving lower levels of assistance to begin nuclear weapons programs when security threats arise.

Statistical Analysis

Challenges Associated with Statistics and Nuclear Proliferation

I test the hypotheses on nuclear weapons pursuit using quantitative and qualitative methods. Using large-N methods to study nuclear proliferation presents researchers with some challenges. Scholars must identify when countries began pursuing nuclear weapons and when their efforts yielded success. This is an inherently difficult task because nuclear weapons programs are shrouded in secrecy. Proliferators naturally want to conceal their programs to avoid preventive strikes against their nuclear facilities or other punitive measures. Once they acquire nuclear weapons, states do not always test them in public. Given that information is incomplete, researchers may disagree about whether certain countries pursued nuclear weapons. For example, there is some evidence that Egypt pursued the bomb during the 1960s but some datasets do not classify it as a proliferator.[22] Even if there is consensus that a country had a bomb program there may not be agreement regarding the years that the program was active. It is widely known that Israel possesses nuclear weapons, for instance, but it is less clear when it acquired the bomb.

A second challenge arises because nuclear proliferation is a rare event. Most countries in the world never seriously considered pursuing nuclear weapons and only ten states have ever acquired them. Methodologists have developed statistical tools that allow researchers to study rare events using statistical analysis.[23] Although the estimates may be technically correct, one must bear in mind that the findings could be driven by a small number of cases. Moreover, statistical results may change when researchers take into account new information. This is always the case when it comes to statistics but this is especially true when studying an event that occurs infrequently such as nuclear proliferation. The substantive effects of variables may change even if their statistical significance does not.[24]

These problems do not imply that scholars should dismiss statistics as a tool for understanding proliferation. With appropriate caution, quantitative analysis can tell us things about the spread of nuclear weapons that would be difficult to ascertain solely on the basis of qualitative tests.[25] I address the issues identified above in part by basing my analysis on multiple data sources. I make changes to the coding of the dependent variable—in this case, nuclear proliferation—to account for disagreements about who was proliferating and when. If the findings remain consistent regardless of how I measure nuclear proliferation it would be easier to conclude that the historical record supports my argument. I also aim to provide readers with a nuanced understanding of how peaceful nuclear assistance shapes the probability of nuclear proliferation, both in absolute and relative terms. Moreover, I use qualitative historical analysis to test my hypotheses on nuclear proliferation so that my conclusions do not rest entirely on evidence from large-N tests.

Dataset

I construct a nuclear proliferation dataset similar to the one produced by Sonali Singh and Christopher Way.[26] It includes information on all states in the international system from 1945 to 2000. The unit of analysis is the country-year, meaning that countries have distinct observations for every year over this time period. Poland, for instance, is included in the dataset fifty-six times. Countries exit the dataset once they begin a weapons program. For example, Pakistan exits the sample in 1973 since it launched a bomb program in 1972. It is important to structure the dataset in this way because I want to empirically evaluate whether the nuclear assistance a state has received in the *past* has any effect on how it behaves in the *present*.[27] Given these parameters, the dataset includes around fifty-five hundred country-year observations.

Preliminary Analysis

Table 7.1 sorts each observation in the dataset based on whether a state participated in at least one NCA and whether it began a weapons program. This simple cross-tabulation is illuminating. To begin, it underscores that nuclear weapons programs are relatively rare. Decisions to pursue weapons occur in seventeen of the observations in the sample (0.24 percent). Although bomb programs are initiated infrequently, this cross-tabulation analysis shows that nuclear cooperation increases the likelihood that countries will go down the nuclear path. Of the non-NCA observations in the dataset, nuclear weapons programs were initiated in 0.10 percent of the cases. Among the NCA observations in the dataset, 0.43 percent of the cases experienced the onset of a bomb program. This difference between these two figures may seem small. But, as I will discuss below, the difference is magnified once we account for the likelihood of proliferation over a period of years (rather than the probability that a state will decide to build nuclear weapons in a single year).

This analysis emphasizes that these relationships are *not* deterministic. The data show that 13 percent (12/94) of all states receiving peaceful nuclear assistance eventually begin weapons programs. This means that while countries receiving peaceful assistance are more likely to begin weapons programs, the majority of countries that benefit from such aid do not proliferate. Also note that 80 percent of the countries that began programs did so after receiving civilian aid. This means that peaceful nuclear cooperation is not a necessary condition for beginning a bomb campaign. It is important to realize, however, that almost all of the countries that initiated nuclear weapons programs without first receiving atomic assistance did so in the 1940s and early 1950s when peaceful nuclear cooperation did not occur. Since 1954, only Libya has begun a nuclear weapons program without first signing an NCA.

What about Other Variables?

The statistics presented in table 7.1 support my theory of proliferation, but they do not take into account other factors that could influence the onset of weapons programs. There are several possible confounding variables:

- Military assistance. Recall from chapter 1 that there were eight cases of military assistance between 1945 and 2000. These transfers could

Table 7.1 Nuclear cooperation and nuclear weapons program onset, 1945–2000

Nuclear weapon program onset	Civilian nuclear cooperation		
	No	*Yes*	*Total*
No	4,015 (99.90%)	3,022 (99.57%)	7,037 (99.76%)
Yes	4 (0.10%)	13 (0.43%)	17 (0.24%)
Total	4,019 (100%)	3,035 (100%)	7,054 (100%)

$\chi^2 = 7.78$ p = 0.005

influence nuclear weapons program onset because their principal purpose is to facilitate the recipient state's ability to produce the bomb.

- Militarized interstate disputes and rivalry. As I discussed above, the conventional wisdom is that states pursue nuclear weapons when they face security threats. We might expect, therefore, that involvement in militarized interstate disputes and the existence of a rivalry with another country would increase states' incentives to build nuclear weapons.

- Nuclear allies. Countries may be less likely to pursue nuclear weapons when they have a security guarantee from a state that possesses the bomb.[28] States with this type of guarantee believe that their ally's arsenal will help deter third party attacks and prevail in the conflicts they fight. Simply put, having a nuclear-armed ally can be a substitute for actually possessing nuclear weapons.[29] As we will see in chapter 9, this contributed to nuclear restraint in the case of Japan.

- NPT membership. International institutions could affect proliferation aims, as I will discuss in greater detail in chapter 9.[30] In 1970 the nuclear Nonproliferation Treaty entered into force. The NPT established a norm against proliferation by obligating all signatories to forgo the bomb. NPT members might refrain from pursuing nuclear weapons because they view the treaty as a moral and legal barrier to bomb-related activities.[31] Countries that proliferate within the NPT might be labeled as "rogue states" and face economic or political sanctions.[32]

- Democracy and democratization. A state's regime type could also influence whether it initiates a nuclear weapons program. Drawing on the logic of the democratic peace theory, democracies may be less likely to pursue the bomb because their domestic institutions have pacifying effects that reduce both the opportunity and willingness to proliferate.[33] On the other hand, democracies may be more prone to proliferation because leaders face temptations to pander to nationalist constituencies.[34] Pressures to proliferate are especially strong in countries that are democratizing because competing elites face demands to shore up their domestic support.

- Economic openness and liberalization. Economic interests may affect whether states go nuclear. In particular, leaders who seek to integrate with the global economy have strong incentives to avoid nuclearization because they depend on foreign investment and trade.[35] Liberalizing coalitions might trade away the opportunity to develop the bomb for the opportunity to make money.[36] Conversely, states that are relatively insulated from the international economy have less to lose from going down the weapons path.

• Economic development. A state's general level of economic development could influence its decision to begin a nuclear weapons program.
States that lack sufficient levels of development may be hesitant to
build the bomb due to technical shortcomings. Consequently, a country's gross domestic product and its general industrial capacity might
play an important role in explaining nuclear proliferation.[37]

I conducted multivariate statistical analysis to test my argument while
controlling for these confounding variables. NCAs remain positively correlated with nuclear weapons program onset, indicating that there is evidence
in favor of my argument even after I account for the other factors that are
thought to influence nuclear proliferation.

How substantively important is peaceful nuclear assistance in shaping
the likelihood that states will initiate nuclear weapons programs? Because
nuclear proliferation occurs infrequently the probability that any one country will initiate a nuclear weapons program in a given year is exceedingly
low. The probability of nuclear weapons program onset when a country has
not signed any NCAs and is "average" in every other respect is 0.00025.[38]
Consider this number to be the "baseline probability" of nuclear weapons
program initiation in a given year. Figure 7.2 illustrates how the baseline
probability changes as countries accumulate greater levels of peaceful nuclear assistance. As the figure shows, increasing the number of NCAs from
zero to five raises the probability of weapons pursuit by 48 percent. It is fairly
common to participate in five additional NCAs; countries such as Argentina, Brazil, Canada, and Sweden did so during the 1960s. Larger increases
in nuclear assistance produce more substantial changes in the probability
of nuclear proliferation. When the number of NCAs increases from zero
to twenty-five, for example, the likelihood of bomb pursuit goes up by
400 percent. More than two dozen countries participated in at least twenty-
five NCAs during the period of study. An extreme increase in agreements
(from zero to fifty) raises the probability that a country will initiate a nuclear
weapons program in a given year by 2,380 percent. About twenty countries
signed at least fifty NCAs from 1945 to 2000.

The changes reported above are rather large but readers should note that
the overall probability of proliferation remains low even after increases in
peaceful nuclear assistance. The probability of weapons pursuit when a country has signed fifty agreements and all other factors are set at their average
values is around 0.006, for instance. Recall that this probability is based on
the risk of nuclear proliferation in a single year. Yet what policymakers care
about is whether aid may lead to the spread of nuclear weapons programs
over time. If the United States was considering signing an NCA it would
probably want to know how nuclear exports would influence the risk of proliferation decades into the future. If we assume that all variables were fixed

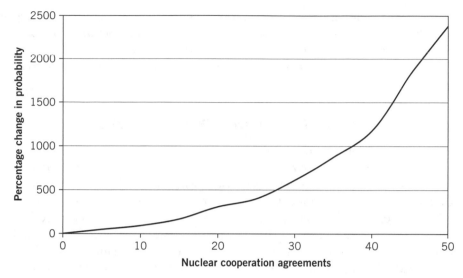

Figure 7.2 Effect of peaceful nuclear assistance on predicted probability of nuclear weapons program initiation
Note: Calculations are based on model 2 (see appendix 7.1).

over time, the previously cited probability of proliferation (0.006) would rise to 0.14 over a twenty-five-year period and 0.27 over a fifty-year period. As these numbers reveal, peaceful nuclear assistance has a cumulative effect on nuclear weapons programs as time passes. I will show later in this chapter that the probability of weapons program initiation also rises steeply when peaceful nuclear assistance and interstate conflict are both present.

Civilian nuclear assistance is not the only factor that explains weapons program onset. Indeed, my analysis lends support to the notion that nuclear proliferation is a multicausal phenomenon that cannot be explained by a single variable. A state's security environment is particularly important in explaining nuclear weapons program initiation. Rivalries with other countries and militarized conflict are positively correlated with this stage of proliferation (see appendix 7.1 for further details).

Other factors are unrelated to nuclear weapons pursuit. Perhaps most surprising is the insignificance of NPT membership. Note, however, that this variable is statistically significant in some of the tests reported in appendix 7.1. Military assistance is also unrelated to nuclear weapons program initiation. The reason for this finding is that countries tend to receive it *after* they have initiated a bomb program. To be sure, it would be strange if a state provided military assistance to a country that was not attempting to proliferate.

Industrial capacity, gross domestic product, nuclear protection (i.e., having an alliance with a nuclear armed state), democracy, democratization, economic openness, and liberalization are all statistically insignificant.[39] Many of these results are consistent with other research on the spread of nuclear weapons.[40]

Which Comes First: Atomic Assistance or Weapons Programs?

The results presented above reveal a relationship between atomic assistance and nuclear weapons program initiation. We see evidence of this relationship, according to my argument, because assistance increases the likelihood that states will make political decisions to pursue the bomb. Some readers might wonder whether the causal arrow really goes in the other direction. Recall, however, that states are not always successful in securing nuclear aid once they begin a weapons program. Nuclear weapons pursuit was positively correlated with receiving atomic aid in some tests but statistically insignificant in others (see chapter 3). The critical issue for the purposes of this chapter is whether the onset of peaceful nuclear assistance precedes the initiation of a nuclear weapons program—or a serious interest in such a program.

One way to address this issue is to compare when states first received nuclear assistance with when they initiated a weapons program. This information is displayed in table 7.2. We know from the cross-tabulation analysis conducted above that four countries signed their first NCA after beginning a weapons program. And table 7.2 reveals that China concluded its first agreement in the same year that it pursued the bomb. Yet in most cases, there is a large time lag between the signing of an agreement and weapons program onset. Argentina, Brazil, Iran, Iraq, and North Korea all signed NCAs more than twenty years prior to pursuing nuclear weapons. Four additional countries—India, Pakistan, South Africa, and South Korea—signed their first NCA more than ten years before they initiated a bomb program. Collectively, these nine countries account for nearly 70 percent of the weapons programs between 1953 and 2000. In the French and Israeli cases, atomic assistance preceded weapons pursuit, but only by three and four years, respectively.

On the whole, the information presented in table 7.2 suggests that these countries sought out nuclear assistance with peaceful intentions but that over time they were motivated to draw on civil technology and know-how for military purposes. Many of these states continued to seek peaceful nuclear aid after they pursued the bomb and some were successful, but this evidence shows that a significant amount of assistance occurred prior to the weapons program.

Even if nuclear cooperation precedes political decisions to proliferate in the overwhelming majority of the cases, it is possible that states seek out aid because they have an interest in developing nuclear weapons.[41] Countries

Table 7.2 Dates of first nuclear cooperation agreement and nuclear weapons programs for nonnuclear weapon states, 1945–2000

Country	Year of first NCA (recipient of assistance)	First year of weapons program initiation	Years before weapons program
Argentina	1955	1978	+23
Brazil	1955	1978	+23
China	1955	1955	0
France	1951	1954	+3
India	1951	1964	+13
Iran	1957	1985	+28
Iraq	1959	1982	+23
Israel	1954	1958	+4
Korea, North	1956	1980	+24
Korea, South	1956	1970	+14
Libya	1971	1970	−1
Pakistan	1955	1972	+17
South Africa	1957	1974	+17
Soviet Union	1956	1945	−11
United Kingdom	1955	1947	−8
United States	1954	1945	−9

Source: Singh and Way, "Correlates of Nuclear Proliferation."

that expect their security environments to deteriorate in the future, for instance, could pursue nuclear assistance because they believe that developing the bomb might be necessary at a later date. The basic idea is that states want to shorten the time it would take to build nuclear weapons in the event that future security threats necessitate bomb development. This strategy is similar to what Ariel Levite and other scholars have termed "nuclear hedging."[42] Given that civil nuclear technology and know-how can serve military purposes, this is a plausible story. If this alternate account were true, atomic assistance would not cause proliferation; it would simply enable outcomes that states already preferred. Statistical tests designed to account for the possibility of joint causality show that this alternative logic is not supported by the historical record (see appendix 7.1 for details).

Does the Type of Assistance Matter?

The preceding analysis tells us a great deal about the relationship between peaceful nuclear assistance and the spread of nuclear weapons. The analysis to this point, however, aggregates all types of nuclear cooperation. There are five types of agreements authorizing atomic assistance: safety NCAs, intangible NCAs, nuclear material NCAs, comprehensive research NCAs, and comprehensive power NCAs. Readers might wonder whether some of these NCAs are more likely than others to contribute to weapons programs.

I replicate the multivariate statistical model using disaggregated NCA variables. Some interesting findings emerge from this analysis. Intangible NCAs and comprehensive power NCAs are statistically correlated with weapons program initiation. On the other hand, nuclear material NCAs and comprehensive research NCAs do not achieve conventional levels of statistical significance. It may be tempting to conclude that these two treaty types are innocuous from a proliferation standpoint. We cannot yet definitively reach this conclusion for two reasons. First, this assistance might influence weapons program initiation if states experience frequent militarized conflict (I explore this possibility below). Second, these treaties could be correlated with nuclear weapons acquisition (see chapter 8).

I probe the relationship between power NCAs and weapons program onset a little further. Scholars and policymakers routinely draw a red line between transfers of certain fuel cycle facilities—especially uranium enrichment centers, plutonium reprocessing facilities, and heavy water production plants—and other types of atomic assistance. The conventional view is that these fuel cycle facilities are the most sensitive from a proliferation standpoint. For example, Matthew Kroenig argues that "the diffusion of civilian nuclear technologies does not present a significant proliferation risk, it is only the spread of sensitive nuclear technologies that directly contributes to the international spread of nuclear weapons."[43]

Readers may recall that comprehensive power NCAs sometimes authorize the transfer of "sensitive" fuel cycle facilities and wonder whether these exports may be driving the results presented above. To explore this issue, I conduct further research on the 330 power NCAs signed between 1945 and 2000. I identify two different types of power agreements. The first type explicitly prohibits the transfer of "restricted technology," which is defined as reprocessing facilities, enrichment centers, and heavy water production plants. These agreements, which I label "limited power NCAs," make up 63 percent (207/330) of all power deals. The second type of power NCA does not prohibit these transfers. I label these agreements "unlimited power NCAs" and they comprise 37 percent (123/330) of all power NCAs. If the conventional wisdom is correct, we would expect that limited power NCAs would be statistically unrelated to nuclear weapons program onset.

I replicate the statistical analysis using these alternate measures of assistance. The findings show that both limited power NCAs and unlimited power NCAs are statistically related to weapons programs in a positive direction. These results are important because they show that transfers of "restricted technology" are not solely responsible for the relationship between power NCAs and weapons program onset. The connection between atomic assistance and proliferation is broader than scholars and policymakers often imply.

Does Interstate Conflict Magnify the Peaceful-Military Link?

Hypothesis 7.2 suggests that the relationship between atomic assistance and weapons program onset is conditional on militarized interstate conflict. As a state's security environment worsens, the probability that peaceful nuclear cooperation contributes to bomb program initiation should increase. Figure 7.3 illustrates that the historical record supports this notion. The figure plots the marginal effect of nuclear cooperation agreements on weapons program onset as militarized interstate disputes increase and all other factors are held constant.[44] Peaceful cooperation has a greater effect on proliferation as the security environment deteriorates. The probability of proliferation at low levels of conflict remains quite small even after increases in atomic aid.[45] As countries experience greater levels of conflict, atomic aid substantially increases the likelihood that they will initiate weapons campaigns. States in extremely dangerous neighborhoods are very likely to go for the bomb if they receive nuclear aid.[46]

I replicate figure 7.3 using the disaggregated independent variables to explore whether we observe similar patterns. Comprehensive power NCAs are

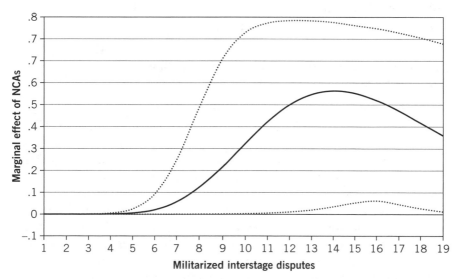

Figure 7.3 The marginal effect of nuclear cooperation on nuclear weapons program onset as militarized interstate disputes increase
Note: Dotted lines represent 95 percent confidence interval.

the most proliferation prone in combination with frequent militarized conflict. Intangible agreements are statistically related to nuclear weapons pursuit at all levels of conflict but the marginal effect increases at a slower rate compared to power treaties. The results for comprehensive research NCAs are more nuanced. At high levels of conflict research treaties produce large changes in the likelihood of proliferation in both relative and absolute terms. Research agreements are statistically unrelated to weapons program initiation in the absence of conflict, however. Safety NCAs and nuclear material NCAs are statistically insignificant across the entire range of the militarized interstate dispute variable.

Case Studies

This section explores two historical cases of nuclear weapons program initiation. It is meant to provide a rich description of my argument and illustrate how the causal processes operate in actual cases of nuclear weapons program onset. The case studies will underscore four aspects of my argument. First, both states began their nuclear programs with peaceful intentions. Second, the accumulation of peaceful nuclear assistance over time contributed to proliferation decisions by reducing the expected costs of a bomb program. Third, atomic aid empowered scientists and members of atomic energy commissions, who pressured elites to authorize a weapons project on the grounds of technological feasibility. Finally, the Indian case will show how militarized conflict can enhance the likelihood that nuclear cooperation leads to a weapons program.

South Africa: Peaceful Nuclear Assistance and Momentum Toward the Bomb

South Africa began its nuclear program with peaceful intentions.[47] The country's large uranium deposits motivated political leaders to exploit this resource to "maintain industrial development."[48] Tapping indigenous uranium reserves would promote development because materials could be exported for a sizable profit.[49] More important, uranium could be used in a domestic program to produce electricity. South African elites "saw nuclear power as a panacea for its energy problems."[50] Further, the development of a civilian nuclear program would help South Africa overcome its stigma as a "technological colony" of the West.[51] With these motives in mind, President Jan Smuts established the South African Atomic Energy Board (AEB) in March 1949 to promote nuclear activities for peaceful purposes.[52]

The U.S. Atoms for Peace initiative resulted in critical nuclear assistance during the infancy of Pretoria's nuclear program. Washington inked an NCA with South Africa in 1957 that led to the training of eighty-eight nuclear scientists.[53] J. Wynand de Villiers, who later directed the country's nuclear explosives project, was among the scientists trained as a result of this peaceful nuclear cooperation. The United States also transferred a nuclear research reactor named Safari-1, which became operational in 1965.[54] The Safari reactor utilized highly enriched uranium, which was also supplied by the United States.[55] Washington provided enriched uranium and heavy water for a second indigenously constructed reactor that went critical in 1967.[56] A. J. "Ampie" Roux, the president of the AEB, later underscored the importance of this early American cooperation: "We can ascribe our degree of advancement today in large measure to the training and assistance so willingly provided by the United States of America during the early years of our nuclear program."[57]

British and German nuclear assistance likewise played a significant role in the development of South Africa's civil nuclear program. On the basis of a 1957 agreement, the United Kingdom provided training to South African professionals. London allowed South African engineers and scientists to observe work on the peaceful uses of the atom at British facilities and shared technical information.[58] The British also aided South Africa's uranium mining efforts.[59] Germany provided substantial intangible assistance, frequently inviting scientists to visit nuclear installations beginning in the 1950s.[60] Ampie Roux visited Germany frequently, as did Waldo Stumpf, who became the chief executive officer of the board in the 1990s.[61]

The peaceful nuclear cooperation discussed above was, according to David Albright, "undoubtedly important in [South Africa's] efforts to obtain nuclear weapons."[62] Consistent with one of the key mechanisms of my theory, foreign nuclear assistance empowered scientists and members of the AEB. One expert on South Africa's nuclear history illustrated this well when he noted that the burgeoning nuclear infrastructure became a "formidable instrument" for "the growing body of scientists South Africa was accumulating."[63] These scientists wielded tremendous political influence because they were well respected and information on the civilian nuclear program was tightly guarded, just as we will see in the Indian case below. Laurence Alberts, the vice president of the AEB from 1971 to 1977, conveyed this in an interview when asked about the influence of scientists: "[The] priest in the white coat had more impact than the priest in the black coat."[64] AEB chairman Roux became especially influential. Roux reportedly quipped, "I can ask this government for anything I want and I'll get it."[65] This statement represented more than simply self-aggrandizement; he had the ears of the political leadership.

Roux lobbied Prime Minister H. F. Verwoerd to begin work on uranium enrichment in the early 1960s. The prime minister supported Roux's efforts. By 1967 Pretoria produced enriched uranium on a laboratory scale using an unusual vortex-tube enrichment method.[66] In light of this development, Roux pushed new Prime Minister John Vorster to fund construction of a small uranium enrichment plant. The political leadership rejected this proposal, most likely because of concerns about costs and technological feasibility.[67] But Roux persisted and ultimately Vorster authorized the building of the facility, known as the Y-plant or Valindaba. The implication of this reversal is that scientists persuaded Vorster that an enrichment plant was both cost-effective and could be built relatively easily given South Africa's existing civil nuclear infrastructure.[68] Ultimately, these arguments turned out to be correct. The prime minister publicly announced in 1970 that South Africa had mastered a technique for enriching uranium.[69] Roux subsequently proclaimed that South Africa was now "in a position to make its own nuclear weapons."[70] Nevertheless, the nuclear program remained peaceful—at least for the time being.

Scientists were also a driving force behind the program to develop nuclear explosives for peaceful purposes (e.g., digging harbors). This program was *not* the first indication of South Africa's intent to develop nuclear weapons, as some scholars have conjectured. Rather, it was a "by-product of the AEB's organizational incentives and influence."[71] One indication of this is that the prime minister did not authorize this program. A lower level official instead approved it in 1971. The AEB aggressively pushed for research on explosives primarily to enhance the organization's budget and influence, according to Andre Buys, who served as a former official in the government-owned arms corporation.[72] As a result of the AEB's lobbying, scientists were able to conduct successful tests in May 1974 demonstrating the feasibility of a nuclear explosive device.[73]

These technological advances contributed to a shift in the aim of the South African nuclear program. Indeed, the country's "technological 'can-do' mentality" motivated the pursuit of nuclear weapons.[74] Scientists, especially Roux, began making persuasive arguments that South Africa should develop nuclear bombs because doing so was technologically feasible.[75] These pleas resonated with Vorster, who authorized a nuclear weapons program and the preparation of a test site in the Kalahari Desert shortly after the May 1974 experiment.[76] He made this decision in large part because he recognized that South Africa's civil nuclear infrastructure would permit the quick and successful development of these weapons.[77] Vorster's plan prior to 1974 was to "let the program develop at its own pace."[78] Now that Pretoria had sufficient fissile material for at least a few nuclear weapons and competency in nuclear explosives, the prime minister's "wait and see" approach yielded a political decision to proliferate.[79] As Mitchell Reiss notes, "with the [civil nuclear] capability already in place, the subsequent decision to build nuclear weapons was made that much easier."[80] The only thing left to do was assemble the weapons, which South Africa did just a few years after the initiation of the program (see chapter 8).

I argued above that external security threats enhance the likelihood that atomic assistance will contribute to nuclear weapons program onset. The South African case shows that external threats are not necessary for nuclear cooperation to raise the risk of bomb pursuit. Some analysts contend that threats stemming from the Soviet Union motivated South Africa to pursue nuclear weapons.[81] This contention emerged in large part because President F. W. de Klerk publicly stated in the early 1990s that a nuclear weapons capability was necessary to deter "a Soviet expansionist threat to Southern Africa."[82] A closer look at the historical record, however, reveals that external threats cannot account for the onset of the bomb program. For starters, the timing of events does not add up. The "threat" did not become salient until Cuban forces intervened in Angola in October 1975.[83] But at the time of this intervention, Vorster had already made a political decision to pursue nuclear weapons, as noted above. Moreover, recently declassified documents show that security motivations—particularly the need for a deterrent against a Soviet-supported attack from Angola or Mozambique—had little role in influencing the onset of the weapons program.[84]

India: Nuclear Aid and the China Factor

India established the Atomic Energy Commission (AEC) in 1948, marking the beginning of the country's civilian nuclear program. The motives of the program were peaceful in nature but Indian elites—like leaders in other countries—understood the dual-use dilemma. During a legislative debate in April 1948, for instance, Prime Minister Jawaharlal Nehru said: "If we are to remain abreast in the world as a nation which keeps ahead of things, we must develop this atomic energy quite apart from war—indeed I think we must develop it for . . . peaceful purposes. It is in that hope that we should develop [nuclear technology]."[85] Nehru believed that a civilian nuclear program would enhance India's prestige and he genuinely believed that it would "maintain a purely peaceful mission."[86] When India began receiving civilian nuclear assistance in the 1950s it did not yet have a strong interest in nuclear explosives.[87]

In 1955 India built its first research reactor using British-supplied designs. This facility, known as the Apsara research reactor, became operational in 1956 using enriched uranium fuel also supplied by the United Kingdom. As I discussed in chapter 4, Canada agreed to supply India with a 40-megawatt reactor in April 1956 known as the Canada-India-Reactor-United States.[88] It was intended to help the Indians build up their knowledge in nuclear engineering.[89] The United States provided heavy water to moderate the CIRUS reactor, enabling it to begin operating in 1960. In addition, beginning in 1955, it invited 1,104 Indian nuclear scientists to train at the Argonne Laboratory School of Nuclear Science and Engineering, among other facilities.[90]

U.S. and Canadian assistance continued in the 1960s. In April 1961 India began construction of a reprocessing plant designed to extract plutonium from spent nuclear fuel. This facility, named Phoenix, was designed in part by an American firm, Vitro International, and based on declassified U.S. plans for reprocessing using the PUREX method.[91] In 1964 Canada agreed to assist India in developing its first power reactor, known as the Rajasthan Atomic Power Plant (RAPP-1), and supply one-half of the initial uranium fuel charge. This assistance enabled India to obtain "detailed design data, including plans and working drawings regarding the design and construction of nuclear power stations of the heavy water type."[92] In December 1966 it agreed to offer assistance in the design and construction of a second nuclear power reactor at Rajasthan (RAPP-2). At the same time, the United States agreed to supply plutonium to India for research purposes.[93]

All of this help was critical for India's civilian nuclear development. Indeed, the country could not have developed a nuclear program at the pace it desired in the absence of foreign aid.[94] The assistance also had the unintended consequence of spurring India's decision to begin a nuclear weapons program.

In early 1964, Prime Minister Nehru wrote a private note stating that "apart from building power stations and developing electricity, there is always a built-in advantage of defense use if the need should arise."[95] According to Ashok Kapur, this message, which was written in the wake of the war with China in 1962, represented "a major shift in Nehru's thinking and policy."[96] This should not be interpreted as the beginning of a nuclear weapons program. Yet it does reveal that the Indian leader was entertaining the idea of pursuing the bomb because nuclear cooperation reduced the expected costs of a nuclear weapons program. Nehru did not come to this realization alone, however.

Consistent with my argument, atomic assistance empowered members of the scientific community—especially AEC Chairman Homi Bhabha, who subsequently lobbied the prime minister to pursue nuclear weapons on the grounds of technological feasibility. Unlike Nehru, Bhabha "never lost sight of the military uses of atomic energy" and had limited utility for peaceful applications.[97] Bhabha reportedly told an English friend during an exchange in 1958 that he hoped to develop nuclear weapons and others observed in the 1960s that the head of India's nuclear program "appeared to be in favor of making bombs."[98] He repeatedly conveyed to Nehru that India could develop nuclear weapons quickly if necessary because of advances in the civilian nuclear program, which were made possible by nuclear cooperation. In the late 1950s, for instance, Bhabha appears to have told the prime minister that India could develop the bomb within three or four years from the time of a political decision to proliferate.[99] These types of assertions were excessively optimistic but, over time, they had a persuasive impact on Nehru's thinking about the nuclear program.

On May 27, 1964, Nehru died. The new prime minister, Lal Bahadur Shastri, initially opposed the pursuit of the bomb. After taking office, Shastri repeatedly stated that India would develop nuclear energy exclusively for peaceful purposes.[100] Yet as time progressed the same factors that motivated Nehru to consider the military applications of nuclear energy influenced Shastri's decision making. Ultimately, on November 27, 1964, Shastri officially endorsed a nuclear explosives program during a speech in the Lok Sabha, the Indian parliament. This represented a major shift in Indian nuclear policy and it is generally regarded as the beginning of the country's weapons program.[101] Indeed, Shastri's comments marked the first time that a prime minister publicly supported an explosives program.[102]

Raja Ramanna, who directed an Indian study on nuclear explosives, remarked in his memoirs that "getting the Prime Minister to agree to [the explosives program] must have required great persuasion, as Shastri was opposed to the idea of atomic explosions of any kind."[103] He was highly sensitive to the expected costs of a bomb program ($42–$84 million) because India faced economic hardship and massive food shortages during his tenure.[104] Initiating a weapons program would force New Delhi to abandon plans for economic development and divert substantial resources away from other domestic programs.[105] This antibomb argument was captured in an editorial published in the *Statesman*: "Both bomb production and effective delivery could be secured if the price is paid for it in terms of economic deprivation. But no responsible person has suggested that the object is worth that price."[106] A former secretary general of the Ministry of External Affairs agreed: "The enormous diversion of resources [required for a nuclear weapons program] will retard India's economic and social development programs indefinitely and . . . not only weaken India internally but eliminate it as a political factor in Asia and Africa."[107]

Bhabha was the driving force behind Shastri's political decision to begin work on nuclear explosives. Ramanna underscored this when he said in an interview that the weapons program "was really Bhabha's crusade."[108] The AEC chairman convinced the prime minister to authorize this program on the grounds that it was technologically feasible and could be done without sacrificing India's economic development. Atomic assistance factored in to this argument because it facilitated critical advances in the nuclear program. In June 1964 the first spent fuel from the Canadian-supplied CIRUS reactor entered the reprocessing plant at Trombay. This meant that India would soon separate plutonium for the first time. Bhabha seized this opportunity to underscore the relative ease with which the bomb could be produced.

In October 1964 he proclaimed that India could acquire a nuclear weapon within eighteen months of a political decision to develop it and that a ten-kiloton blast would cost only $350,000.[109] India would not acquire weapons-usable plutonium until 1965 (even though the spent fuel was loaded into the reprocessing facility in June 1964), however, and it lacked a reliable bomb design.[110] These challenges were overlooked, in part, because Bhabha had an extraordinary amount of power, and information relevant to the nuclear program was so tightly guarded that others did not have a chance to question his assertions. Eventually, Bhabha's arguments persuaded Shastri to go down the nuclear weapons path. The prime minister authorized the explosives program just after meeting with Bhabha on November 27, 1964.[111]

External threats from China also contributed to the proliferation decision.[112] Indeed, the Indian case shows how militarized conflict with neighboring states can magnify the connection between atomic assistance and nuclear weapons program onset. India's defeat in a 1962 border war with China raised concern about the prospect of future Chinese aggression. Beijing's first nuclear test in October 1964 left many Indian elites feeling especially threatened and led to some calls for an Indian nuclear deterrent. This view was exemplified by Mushtaq Ahmed, the president of the Delhi Pradesh Congress Committee, who boldly stated in the aftermath of the Chinese test that "the only course for India is to produce her own atom bomb to defend herself."[113] One hundred members of Shastri's own Congress Party similarly urged India to acquire "an independent nuclear deterrent to protect herself against any possible threat from China."[114]

The Chinese threat did not independently cause Indian proliferation, as a standard realist argument might imply.[115] It did, however, increase the extent to which arguments advanced by Bhabha resonated with Indian elites. Not only could nuclear weapons be built relatively easy because of nuclear cooperation, but there was also a strategic rationale for possessing the bomb. Bhabha made this clear when trying to convince the government to authorize work on nuclear explosives. He said, for instance, that "the explosion of a nuclear device by China is a signal that there is no time to be lost."[116] Bhabha further pointed to the strategic utility of bomb possession in a historic radio address following the Chinese test: "Atomic weapons give a State possessing them in adequate numbers a deterrent power against attack from a much stronger State."[117] Against this backdrop, key cabinet level officials and Prime Minister Shastri were more persuaded by bomb lobbying on the grounds of technological feasibility.[118]

With the caveat that counterfactual reasoning is difficult, the available historical evidence indicates that Shastri and other key officials might have been unconvinced by the feasibility argument in the absence of external threats from China.[119] Equally important, concerns about China would have most likely been insufficient to produce Shastri's decision on nuclear explosives in the absence of peaceful nuclear assistance. The issues of cost and technological capacity weighed heavily on Shastri, as the above discussion makes clear. Moreover, if Bhabha had not been empowered as a result of atomic assistance, the most vocal bomb advocate would have been removed from the scene.

Shastri's authorization of the nuclear explosives project in 1964 is the most important event for the purposes of this chapter because it marked the onset of India's bomb program. It is worth noting, however, that my argument helps us better understand another important decision in India's nuclear history: Prime Minister Indira Gandhi's authorization of a nuclear test in 1972. To recognize why this is the case we must pick up the story in 1966, which was a tumultuous year in Indian nuclear history.

In January of that year Shastri and Bhabha both died unexpectedly, the former of a heart attack and the latter in a plane crash. This left the nuclear program in the hands of new Prime Minister Gandhi and AEC chairman Vikram Sarabhai. In a matter of weeks after taking the reins of the AEC, Sarabhai ordered a stop to the nuclear explosives project. He did not agree with Bhabha's estimates regarding the expected costs of building the bomb. He believed that producing an explosive device would be costly and would require the diversion of substantial resources from economic development programs. Sarabhai maintained that "the real problem in this whole question relates to the utilization of national resources for productive and social welfare against the burden of defence expenditure."[120] For Sarabhai, the civil nuclear program had not progressed far enough by 1966 to sufficiently reduce the costs of building a nuclear explosive device. Indira Gandhi appears to have shared this view. She argued in May 1966 that India should build up its atomic infrastructure and increase know-how and competence in nuclear matters but not develop the bomb.[121] This evidence shows that perceptions about the extent to which nuclear cooperation reduces the expected costs of a weapons program can vary among individual decision makers. As I will discuss below, once Gandhi perceived that nuclear cooperation had sufficiently reduced the expected costs of a bomb program, she authorized a nuclear test.

Events that occurred during and after 1966 reinforce my argument that peaceful nuclear cooperation can increase the risk of weapons program onset by empowering scientists and members of atomic energy commissions. Remarkably, research on the nuclear explosives project continued even though Sarabhai had terminated the program in June 1966. A small group of key scientists including Homi Sethna, Raja Ramanna, P. K. Iyengar, and Rajagopala Chidambaram held a tremendous amount of power and pushed ahead with research on nuclear weapons. Ramanna boldly asserted that "Sarabhai could not keep scientists from doing their work. He couldn't look over our shoulders."[122] By 1968, scientists were doing serious studies on nuclear weapons. Two years later they began building the Purnima research reactor. Purnima was intended to facilitate the design of nuclear explosives by providing experience in the properties of plutonium.[123] The reactor was fueled by 20kg of plutonium that was most likely extracted from spent fuel produced by the CIRUS reactor, which was supplied by Canada and moderated by U.S.-supplied heavy water.[124] The Purnima project unfolded because of an initiative taken by scientists without authorization from political leaders. Sethna, who was directly involved in the explosives research, claims that the Purnima project did not appear on AEC's books and did not have the explicit approval from either Sarabhai or Gandhi.[125]

As technological momentum toward the bomb ensured behind closed doors, Indian politicians openly debated the merits of nuclear weapons. Just as had been the case in the 1960s, the policy discourse centered on the economic costs of initiating a bomb effort. India continued to suffer from food shortages and high inflation, fueled in part by a severe drought. In 1972, more than 40 percent of Indians lived below the poverty line.[126] Many Indians favored beginning a nuclear weapons program, but not at the expense of development and the economy. A poll conducted by the Indian Institute of Public Opinion taken in September 1971 found that 63 percent of Indians believed that New Delhi should go for the bomb but only 38 percent supported such a policy if it meant that cuts would be made in areas relevant to economic development.[127]

This made it especially significant that India had accumulated an indigenous base of technology and know-how as a result of foreign assistance to its civil nuclear program. India had enough expertise to construct new reactors strictly dedicated to plutonium production for nuclear bombs.[128] It also had the opportunity to divert plutonium directly from the Canadian-supplied CIRUS and RAPS-I reactors, which were supplied for peaceful purposes. By 1972, significant amounts of plutonium had been separated from spent fuel produced by these reactors at the Trombay reprocessing facility. India believed that it could use this plutonium in a nuclear explosives device with few repercussions. For instance, Homi Sethna indicated that he and other decision makers believed that Washington and Ottawa would continue to supply India's civil nuclear program even if plutonium was used from CIRUS to conduct a nuclear test.[129] Since India was comfortable using plutonium from the CIRUS reactor, it believed that the costs of bomb production would be relatively cheap. This was clearly articulated by Madhu Dadwate, a member of parliament, who argued in March 1972 that building nuclear weapons would not be too costly because much of the expense had already been incurred in the development of nuclear energy. He argued that "the first five or six processes in the production of atomic energy and production of the bomb were the same," making the costs of bomb production relatively low for India since it had developed a peaceful nuclear program over the last two decades.[130]

Prime Minister Gandhi gave her approval to test a nuclear device in September 1972. There is little information in the available historical record that tells us exactly what she was thinking when she made this decision. Referring to the meeting where the decision to develop the bomb was made, Ramanna said "[Gandhi] didn't speak . . . it's difficult to say what was on her mind. She said, 'let's have it.' "[131] Existing evidence does suggest that the prime minister was inclined to build the bomb at least in part because it could be done relatively cheaply, due to advances in the civilian sector, and it would not divert substantial resources away from programs for economic development.[132] It is also clear that Gandhi was heavily influenced by key scientists, just as Homi Bhabha persuaded Shastri to begin an explosives program in 1964. Particularly influential was Ramanna, who had a close relationship with the prime minister. Scientists lobbied the prime minister to authorize a nuclear test, partially on the grounds that it was now technically feasible to do so, and she ultimately acquiesced.

Appendix 7.1: Data and Empirical Findings

This appendix describes the methodology employed in chapter 7 in greater detail. It also presents the complete regression tables so that interested readers can take a closer look at the empirical findings.

Dependent Variable

Following procedures taken by Singh and Way, I create a dichotomous variable and code it 1 if a cabinet-level official makes "an explicit political decision demonstrating a strong willingness to acquire nuclear weapons" in year t and 0 otherwise.[133] Table 7.2 listed the states that pursued nuclear weapons between 1945 and 2000 along with the year of program initiation.[134]

Independent Variables

To test my argument I code three independent variables. I operationalize civilian nuclear assistance based on the dataset I described in chapter 1. I create an independent variable that measures the aggregate number of NCAs that a state has signed between 1945 and year t-1 entitling it to nuclear technology, materials, or knowledge from another country.[135] If a state signs an NCA but is only supplying—and not receiving—nuclear assistance as part of the terms of the deal, then this would not be captured by this variable.[136] The disaggregated NCA variables are based on these same criteria (see chapter 1 for further details).

To operationalize security threats, I create a variable measuring the five-year moving average of the number of militarized interstate disputes a country experiences between year t-6 and year t-1. This variable is based on Version 3.0 of the Correlates of War's MID dataset.[137] I code a third variable that interacts these two measures to test for the conditional effect that nuclear cooperation has on nuclear weapons program initiation.[138]

Control Variables

Below I describe the coding procedures and data sources for the control variables. Unless otherwise noted, all of these variables are taken from the Singh and Way dataset.[139]

- Military assistance. I define military nuclear assistance as transfers of complete nuclear warheads, bomb designs, or dual-use technology provided explicitly to help the recipient country build nuclear weapons. I create a variable measuring the number of agreements a state has signed entitling it to proliferation assistance between 1945 and year t-1. I self-code this variable by consulting the historical literature on nuclear proliferation (see chapter 1 for additional details).

- Economic development. To operationalize economic capacity, I include a variable measuring a country's GDP per capita and a squared term of this measure to allow for the possible curvilinear relationship between development and nuclear weapons pursuit.

- Industrial capacity. To measure a state's industrial capacity, I code a dichotomous variable 1 if it produces steel domestically and has an electricity generating capacity greater than 5,000MW and 0 if it does not.

- Rivalry. I measure rivalries based on D. Scott Bennett's coding of enduring rivalries. I create a dichotomous variable and code it 1 if a state is involved in at least one interstate rivalry in year t-1 and 0 if not.

- Nuclear allies. To operationalize nuclear protection, I code a dichotomous variable 1 if a state shares a formal defense pact with a major power that possesses nuclear weapons and 0 if it does not.

- Democracy. I use the Polity IV composite indicator of regime type as a proxy for democracy. The polity variable that I include in the model measures the relative openness of political institutions based on a twenty-one-point scale.[140]

- Democratization. To measure whether a state is democratizing, I calculate movement toward democracy over a five-year time span by subtracting a state's polity score in year t-6 from its polity score in year t-1.

- Economic openness. To account for a state's exposure to the global economy, I include a variable measuring the ratio of exports plus imports as a share of GDP.[141]

- Liberalization. I operationalize liberalization by calculating movement in the economic openness variable over time. Specifically, I subtract a state's level of openness in year t-6 from its openness in year t-1.

- NPT membership. I measure NPT membership by coding a dichotomous variable 1 if a state has ratified the treaty in year t-1 and 0 if it has not done so.[142]

In addition to the variables outlined above, I include the standard controls for temporal dependence recommended by Nathaniel Beck, Jonathan Katz, and Richard Tucker.[143] Specifically, I include the linear term counting the number of years that pass without a country pursuing nuclear weapons in all statistical models, along with three cubic splines.

Method

Unless otherwise noted, the models of nuclear weapons program acquisition are estimated using rare events logit.[144] This estimator is appropriate because the dependent variable is dichotomous and decisions to initiate a bomb program occur relatively infrequently.[145] I use clustering over states to control for heteroskedastic error variance. All of the independent variables are lagged one year behind the dependent variable to control for possible simultaneity bias. In other words, I only measure assistance prior to year t to empirically assess whether prior nuclear cooperation is correlated with the initiation of a weapons program.

Results

This section presents the regression tables from the multivariate statistical analysis and discusses a series of robustness checks.

FIRST-CUT RESULTS

Table 7.3 presents the findings of the first-cut statistical analysis that were illustrated above in Figure 7.2.[146] Models 1 and 2 are identical except that the former model excludes military assistance. The findings discussed earlier in this chapter were based on model 2.[147] Note that the results are similar in both models.[148]

Table 7.3 Determinants of nuclear weapons program initiation, 1945–2000

	(1)	(2)
Nuclear cooperation agreements	0.065**	0.064**
	(0.028)	(0.027)
Military assistance		0.968
		(1.277)
Militarized interstate disputes	0.312***	0.307***
	(0.095)	(0.093)
GDP per capita	0.001	0.001
	(0.000)	(0.000)
GDP per capita squared	−0.000**	−0.000*
	(0.000)	(0.000)
Industrial capacity	0.864	0.918
	(0.820)	(0.810)
Rivalry	2.026*	1.980*
	(1.065)	(1.055)
Nuclear allies	−0.308	−0.218
	(0.861)	(0.913)
Polity	0.062	0.061
	(0.054)	(0.055)
Democratization	−0.059	−0.061
	(0.057)	(0.056)
Openness	−0.004	−0.004
	(0.016)	(0.016)
Liberalization	−0.008	−0.007
	(0.018)	(0.017)
NPT	−1.092	−1.076
	(0.808)	(0.808)
Years without nuclear weapons program initiation	0.005	−0.015
	(0.253)	(0.248)
Spline 1	0.001	0.000
	(0.002)	(0.002)
Spline 2	−0.001	−0.001
	(0.002)	(0.002)
Spline 3	0.001	0.001
	(0.001)	(0.001)
Constant	−8.445***	−8.246***
	(1.896)	(1.794)
Observations	5519	5519

Notes: *** $p<0.01$, ** $p<0.05$, * $p<0.1$; robust standard errors in parentheses.

ACCOUNTING FOR ENDOGENEITY

I account for the possibility that peaceful nuclear cooperation is endogenous to nuclear weapons programs by devising an instrument for civilian nuclear assistance.[149] This instrument is produced by estimating a state's expected number of NCAs in a dataset of country-years between 1945 and 2000. Good instruments should be uncorrelated with the error term but correlated with the explanatory variable of interest (i.e., nuclear cooperation).[150] In international relations it is often difficult for scholars to meet this assumption when using simultaneous equations models. I instrument for nuclear cooperation using three variables that are described in appendix 3.1: "Neighbor NCA, Energy Demand, and Oil Price." These variables are associated with signing NCAs (albeit to varying degrees) and there is little theoretical reason to expect that they would be directly associated with nuclear proliferation.

I estimate the endogenous model using a technique originally developed by G. S. Maddala and practically implemented by Omar Keshk.[151] The two-stage estimation technique generates the instrument for nuclear cooperation and then substitutes it in a structural equation that is estimated using probit. Note that the dependent variable for this analysis is slightly different than the one I used for the initial empirical tests because I am modeling the simultaneous relationship between nuclear cooperation and weapons programs. I create a dichotomous variable and code it 1 if a nonnuclear weapons state is pursuing the bomb and 0 otherwise.[152] This variable is coded as missing once a state acquires nuclear weapons, meaning that nuclear weapons powers (i.e., France after 1960) are not included in the estimation sample.[153]

Table 7.4 displays the results of the endogenous model designed to address the simultaneity issue. Model 3 only includes the instrument for nuclear cooperation and controls for temporal dependence. Model 4 adds the same control variables that were included in model 1. Model 5 is a full model that also includes the military assistance variable.

These results are generally consistent with the findings presented above. Most important, the instrument for nuclear cooperation is positively and statistically correlated with nuclear weapons pursuit. This finding, which is robust to alternate model specifications, indicates that the effect of nuclear assistance on weapons program onset is *not* conditioned by the causes of nuclear cooperation.[154] The correlation between peaceful nuclear assistance and weapons program onset displayed in table 7.3 is attributable to the aid itself. Proliferators certainly seek out aid after they initiate a weapons program, but they do not always get it (see chapter 3).

ANALYSIS OF DISAGGREGATED NCAS

Table 7.5 displays the findings of this analysis using disaggregated measures of peaceful nuclear assistance. Model 6 operationalizes nuclear cooperation by measuring intangible NCAs. Model 7 uses material NCAs as the key

Table 7.4 Endogenous models of nuclear weapons program initiation, 1945–2000

	(3)	(4)	(5)
Nuclear cooperation agreements	0.118***	0.164***	0.162***
	(0.017)	(0.039)	(0.038)
Military assistance			0.592
			(0.452)
Militarized interstate disputes		0.245***	0.231***
		(0.054)	(0.053)
GDP per capita		0.000**	0.000**
		(0.000)	(0.000)
GDP per capita squared		−0.000***	−0.000***
		(0.000)	(0.000)
Industrial capacity		−0.404	−0.402
		(0.364)	(0.359)
Rivalry		0.881***	0.862***
		(0.215)	(0.215)
Nuclear allies		−0.470*	−0.454*
		(0.265)	(0.263)
Polity		−0.042***	−0.041**
		(0.014)	(0.014)
Democratization		0.018	0.016
		(0.021)	(0.021)
Openness		0.006*	0.006*
		(0.003)	(0.003)
Liberalization		−0.003	−0.003
		(0.006)	(0.006)
NPT		−0.376	−0.379
		(0.240)	(0.236)
Years without nuclear weapons program initiation	−0.657***	−0.609***	−0.599***
	(0.065)	(0.078)	(0.078)
Spline 1	−0.0047***	−0.004***	−0.004***
	(0.001)	(.000)	(0.001)
Spline 2	0.003***	0.003***	0.003***
	(0.000)	(.000)	(0.001)

(*Continued*)

Table 7.4—*cont.*

	(3)	(4)	(5)
Spline 3	−0.000*	−0.000*	−0.000*
	(0.000)	(0.000)	(0.000)
Constant	−1.020	−2.04***	−2.06***
	(0.209)	(0.365)	(0.376)
Observations	7802	5807	5807

Notes: *** $p<0.01$, ** $p<0.05$, * $p<0.1$; robust standard errors in parentheses.

independent variable, while model 8 employs comprehensive research agreements. Model 9 uses comprehensive power NCAs as a proxy for peaceful nuclear cooperation. Models 10–11 disaggregated these deals further to explore whether limits on "restricted" technology make a difference. Model 10 codes limited power NCAs while model 11 includes a measure of unlimited power treaties.

As previously discussed, these findings indicate that intangible NCAs and comprehensive power NCAs raise the risk of nuclear weapons pursuit but material agreements and research treaties do not. The results from models 10 and 11 underscore that the initial connection between comprehensive power NCAs and proliferation is not driven by transfers of enrichment centers, reprocessing facilities, or heavy-water reactors.

Turning to the control variables, the findings are similar to those presented in table 7.3, but there are a few differences worth mentioning. Rivalry is statistically significant in models 6–10 but insignificant in model 11, indicating that this finding is sensitive to model specification. GDP per capita squared is significant in model 6 and models 9–10 but insignificant in models 7–8. Finally, the NPT is statistically significant and negative in models 9 and 10; this variable does not achieve statistical significance in the other models.

MILITARIZED CONFLICT, NUCLEAR ASSISTANCE, AND WEAPONS PROGRAMS

Table 7.6 displays the findings relevant to hypothesis 7.2. To test this proposition, it is necessary to include an interaction term between NCAs and militarized interstate disputes in the statistical models, along with the two constituent parts.[155] Models 12–13 use the aggregated measure of nuclear assistance. Both models include all of the variables from model 2 along with the relevant interaction term. Model 10 is estimated with rare events logit (like models 1–9) while model 11 is estimated using probit to assess whether the findings are sensitive to the choice of estimator. Models 14–17 explore the conditional effect of disaggregated nuclear cooperation agree-

Table 7.5 Disaggregated nuclear cooperation agreements and nuclear weapons program onset, 1945–2000

	(6) Intangible NCAs	(7) Material NCAs	(8) Research NCAs	(9) Power NCAs	(10) Limited power NCAs	(11) Unlimited power NCAs
Nuclear cooperation agreements	0.108*	0.140	0.192	0.318**	0.551**	0.381**
	(0.055)	(0.255)	(0.290)	(0.137)	(0.251)	(0.175)
Military assistance	0.839	1.251	1.318	1.075	1.282	0.942
	(1.369)	(1.306)	(1.306)	(1.235)	(1.181)	(1.305)
Militarized interstate disputes	0.294***	0.274**	0.283***	0.348***	0.321***	0.317***
	(0.096)	(0.108)	(0.097)	(0.097)	(0.080)	(0.102)
GDP per capita	0.001	0.000	0.000	0.001	0.001*	0.000
	(0.000)	(0.001)	(0.001)	(0.000)	(0.000)	(0.000)
GDP per capita squared	−0.000**	−0.000	−0.000	−0.000***	−0.000**	−0.000*
	(0.000)	(0.000)	(0.000)	(0.000)	(0.000)	(0.000)
Industrial capacity	1.043	1.191	1.112	0.597	0.855	0.840
	(0.807)	(0.868)	(0.808)	(0.879)	(0.756)	(0.883)
Rivalry	1.737**	1.671**	1.644**	2.461*	1.918**	2.365
	(0.847)	(0.851)	(0.818)	(1.348)	(0.897)	(1.483)
Nuclear allies	−0.182	−0.110	−0.019	−0.208	−0.101	−0.260
	(0.916)	(0.911)	(0.833)	(0.829)	(0.812)	(0.925)
Polity	0.051	0.056	0.058	0.076	0.069	0.077
	(0.051)	(0.053)	(0.052)	(0.055)	(0.056)	(0.058)

Democratization	−0.054	−0.056	−0.059	−0.077	−0.086	−0.068
	(0.053)	(0.053)	(0.055)	(0.063)	(0.066)	(0.058)
Openness	−0.006	−0.010	−0.006	0.002	−0.001	0.000
	(0.016)	(0.018)	(0.015)	(0.017)	(0.019)	(0.014)
Liberalization	−0.007	−0.004	−0.006	−0.005	−0.008	−0.006
	(0.018)	(0.018)	(0.017)	(0.016)	(0.017)	(0.018)
NPT	−1.136	−1.043	−1.069	−1.311*	−1.630*	−1.073
	(0.833)	(0.887)	(0.858)	(0.778)	(0.896)	(0.790)
Years without nuclear weapons program initiation	−0.025	−0.067	−0.086	0.011	−0.029	0.019
	(0.247)	(0.229)	(0.218)	(0.299)	(0.277)	(0.278)
Spline 1	0.001	0.000	0.000	0.000	0.000	0.001
	(0.002)	(0.002)	(0.002)	(0.002)	(0.002)	(0.002)
Spline 2	−0.001	−0.001	−0.001	−0.001	−0.001	−0.001
	(0.002)	(0.002)	(0.002)	(0.002)	(0.002)	(0.002)
Spline 3	0.001	0.001	0.001	0.001	0.001	0.001
	(0.001)	(0.001)	(0.001)	(0.001)	(0.001)	(0.001)
Constant	−7.887***	−7.288***	−7.233***	−9.292***	−8.453***	−8.754***
	(1.704)	(1.917)	(1.871)	(2.168)	(2.014)	(2.195)
Observations	5519	5519	5519	5519	5519	5519

Notes: Robust standard errors in parentheses; *** p<0.01, ** p<0.05, * p<0.1.

Table 7.6 The conditional effect of nuclear cooperation on nuclear weapons program initiation, 1945–2000

	(12) Aggregated NCAs	(13) Aggregated NCAs	(14) Intangible NCAs	(15) Material NCAs	(16) Research NCAs	(17) Power NCAs
Nuclear cooperation agreements	0.050**	0.018**	0.031*	−0.014	−0.060	0.106***
	(0.022)	(0.007)	(0.016)	(0.123)	(0.148)	(0.033)
Militarized interstate disputes	0.281***	0.149***	0.160***	0.147***	0.128***	0.165***
	(0.080)	(0.030)	(0.035)	(0.038)	(0.031)	(0.032)
NCAs * MIDs	0.050**	0.024**	0.034	0.013	0.079*	0.069**
	(0.023)	(0.011)	(0.028)	(0.045)	(0.042)	(0.031)
Military assistance	1.061	0.249	0.227	0.331	0.342	0.376
	(1.158)	(0.545)	(0.601)	(0.597)	(0.556)	(0.565)
GDP per capita	0.001	0.000***	0.000**	0.000	0.000	0.000***
	(0.000)	(0.000)	(0.000)	(0.000)	(0.000)	(0.000)
GDP per capita squared	−0.000**	−0.000***	−0.000***	−0.000	−0.000	−0.000***
	(0.000)	(0.000)	(0.000)	(0.000)	(0.000)	(0.000)
Industrial capacity	0.931	0.280	0.374	0.481*	0.492*	0.137
	(0.871)	(0.263)	(0.256)	(0.269)	(0.265)	(0.284)
Rivalry	1.554*	0.565**	0.573**	0.549**	0.494**	0.728***
	(0.910)	(0.248)	(0.232)	(0.234)	(0.241)	(0.269)
Nuclear allies	−0.192	0.007	0.007	0.065	0.092	−0.029
	(0.928)	(0.303)	(0.305)	(0.301)	(0.275)	(0.299)

Polity	0.053	0.019	0.017	0.020	0.017	0.023
	(0.052)	(0.018)	(0.017)	(0.018)	(0.016)	(0.019)
Democratization	−0.070	−0.027	−0.024	−0.024	−0.025	−0.033
	(0.058)	(0.021)	(0.020)	(0.020)	(0.020)	(0.022)
Openness	−0.007	−0.002	−0.003	−0.004	−0.003	0.000
	(0.019)	(0.006)	(0.006)	(0.006)	(0.005)	(0.005)
Liberalization	−0.003	−0.000	−0.003	−0.002	−0.000	0.003
	(0.020)	(0.006)	(0.006)	(0.006)	(0.006)	(0.006)
NPT	−1.007	−0.678**	−0.613**	−0.623**	−0.641**	−0.923***
	(0.787)	(0.282)	(0.281)	(0.288)	(0.302)	(0.309)
Years without nuclear weapons program initiation	−0.009	0.006	0.010	−0.016	−0.033	0.027
	(0.267)	(0.095)	(0.094)	(0.083)	(0.077)	(0.113)
Spline 1	0.001	0.000	0.000	0.000	0.000	0.000
	(0.002)	(0.001)	(0.001)	(0.001)	(0.001)	(0.001)
Spline 2	−0.002	−0.001	−0.001	−0.001	−0.001	−0.001
	(0.002)	(0.001)	(0.001)	(0.001)	(0.001)	(0.001)
Spline 3	0.001	0.001	0.001	0.001	0.001	0.001
	(0.001)	(0.000)	(0.000)	(0.000)	(0.000)	(0.001)
Constant	−7.822***	−4.117***	−4.141***	−3.759***	−3.606***	−4.477***
	(1.774)	(0.588)	(0.584)	(0.590)	(0.571)	(0.683)
Observations	5519	5519	5519	5519	5519	5519

Notes: Robust standard errors in parentheses; *** $p<0.01$, ** $p<0.05$, * $p<0.1$

ments on nuclear weapons pursuit. Each of these models includes one of the disaggregated NCA variables along with the appropriate interaction term.

All of the models displayed in table 7.6 include an interaction term with a constituent part that is continuous, which means that we need additional information in order to adequately interpret the findings. The appropriate way to interpret these results is to plot the marginal effect of atomic assistance and the corresponding confidence intervals across the full range of the militarized interstate dispute variable, as I did in figure 7.3.[156] The statistical significance of the NCA variable in models 12 and 13 tells us that agreements are correlated with proliferation even when the number of disputes is zero. Additional conflict only magnified the effect of atomic assistance, as the previous discussion illustrated (see figure 7.3).

The coefficient on the variable measuring intangible NCAs is positive and statistically significant in model 14, indicating that intangible aid raises the risk of weapons program initiation in the absence of militarized disputes. The interaction term is insignificant in the model but this does not necessarily imply that conflict does not condition the relationship between intangible NCAs and proliferation. On the contrary, a replication of figure 7.3 with the intangible NCAs substituted for the aggregated NCA variable shows that the marginal effect of these treaties rises as the number of disputes increases, consistent with the prediction of hypothesis 7.2. Neither the NCA variable nor the interaction term is significant in model 15, making it unlikely that the effect of nuclear material NCAs on proliferation is conditional on militarized conflict. The interaction term is statistically significant and positive in model 16 but the variable measuring the total number of research NCAs is insignificant. This tells us that when the disputes variable equals zero research NCAs are uncorrelated with weapons program initiation but these agreements are related to proliferation as the number of disputes increases. Finally, the coefficient on the variable measuring the number of comprehensive power NCAs is statistically significant and positive in model 17, indicating that these agreements are correlated with the onset of bomb programs even when the number of disputes is set to zero. The interaction term is also positive and significant, suggesting that the marginal effect of power NCAs only increases as the threat environment worsens (a replication of figure 7.3 confirms this).

The findings with respect to the controls are mostly consistent with what we have seen up to this stage. The key difference relates to the NPT. The variable measuring whether a state has ratified this treaty is negative and statistically significant in models 13–17, indicating that the NPT matters once we account for the interactive relationship between atomic assistance and militarized conflict. Although there is some support for the argument that treaty membership reduces the risk of proliferation, evidence in favor of this proposition is tenuous.

ADDITIONAL ROBUSTNESS CHECKS

To explore whether my results are sensitive to proliferation codings, I code several alternate dependent variables. I begin by consulting an alternate set of proliferators and dates compiled by Dong-Joon Jo and Erik Gartzke.[157] Jo and Gartzke measure whether states have active nuclear weapons programs from 1939 to 1992.[158] For states that are not recognized as nuclear powers under the NPT, the authors "adopt the year in which a suspect state's nuclear activities are seen to increase noticeably as the start of the weapons program."[159] I also make three changes to the Singh and Way data on weapons pursuit. First, I add a few proliferators that they excluded but that might have initiated weapons programs based on their criteria.[160] Second, I code India as pursuing nuclear weapons on one occasion (1964), not two (1964 and 1980). Third, I alter the coding for the United States and the Soviet Union.[161]

I replicate model 2 using these alternate dependent variables (findings not reported).[162] The NCA variable remains positive and statistically significant, indicating that support for my argument is not sensitive to my initial coding of the dependent variable. The results for the control variables are also similar to those presented in table 7.3.

Chapter 8

Who Builds Bombs?

How Peaceful Nuclear Cooperation Facilitates the Spread of Nuclear Weapons

Peaceful nuclear assistance raises the risk that countries will pursue nuclear weapons, especially if security threats later arise. Does nuclear cooperation also increase the likelihood that states will successfully build the bomb? If so, how?

Statistical tests reveal that there is a correlation between nuclear cooperation agreements and nuclear weapons production even when accounting for military assistance and the other factors that are thought to influence the spread of the bomb. Analysis using disaggregated independent variables shows that comprehensive power NCAs are strongly associated with weapons production but other types of agreements are not. Further tests indicate that militarized conflict amplifies the relationship between comprehensive power treaties and proliferation.

Using qualitative historical analysis, I examine ten cases of acquisition and another six cases where states did not build the bomb despite having an interest in developing nuclear weapons. The findings show that most of the states that built nuclear weapons after 1953, in the era of Atoms for Peace, benefited from atomic assistance, albeit to varying extents. Some countries used aid they received for peaceful purposes directly to enable bomb production. Others benefited from nuclear cooperation in a more nuanced way, drawing on the knowledge base established as a result of assistance to build military facilities. Consistent with my argument, the historical analysis also shows that four states were unable to build nuclear weapons largely because they did not benefit from sufficient levels of atomic aid. In two cases, ample assistance did not lead to bomb acquisition in part because the relative absence of militarized conflict reduced the strategic incentives to deploy nuclear weapons. These cases underscore that receiving large amounts of peaceful nuclear assistance does not guarantee that a state will acquire the bomb.

The Connection between Atomic Assistance and Bomb Acquisition

The argument I made in chapter 7 also has implications for nuclear weapons acquisition. The benefits that countries expect to accrue from peaceful nuclear aid when they begin bomb programs should materialize. Peaceful nuclear cooperation could enable bomb production in two respects. First, dual-use technology and materials can be used directly to produce fissile material for bombs (recall that the production of fissile material is the most difficult step in building the bomb).[1] A country could, for instance, use a reactor provided for civil purposes to produce plutonium for nuclear weapons. Second, states could draw on the technical knowledge base established through civil nuclear cooperation to augment a nuclear weapons program.[2] Individuals trained via civil nuclear cooperation sometimes head their country's nuclear weapons programs. More generally, atomic assistance establishes a cadre of trained scientists and technicians that can help countries overcome the technological challenges associated with constructing reactors, enrichment centers, or reprocessing plants.[3] This is important because states often attempt to build covert facilities dedicated to a weapons program from the beginning; states that can draw on a knowledge base from a peaceful nuclear program have a greater chance of successfully building these facilities in a short amount of time. On the other hand, countries that do not benefit from civilian assistance are likely to have difficulty overcoming the technical challenges associated with bomb production.

Militarized conflict magnifies the connection between assistance and bomb acquisition, just as it strengthens the relationship between aid and weapons pursuit. External threats are not necessary for nuclear cooperation to facilitate weapons acquisition since states sometimes pursue the bomb during periods of relative peace. Yet security threats enhance a state's resolve to deploy nuclear weapons expeditiously by underscoring the strategic need for the bomb. Therefore, developed civilian nuclear infrastructures are likely to produce proliferation quickly when states have a high demand for nuclear weapons compared to a similar scenario where states do not face external threats. Although militarized conflict is an important part of this story, it is not the driving force behind my argument. States that do not benefit from atomic assistance may be unable to build nuclear weapons even if they are determined to become nuclear powers, a point illustrated by some of the case studies presented later in this chapter.

Hypothesis 8.1: Countries receiving higher levels of peaceful nuclear assistance are more likely to acquire nuclear weapons than states receiving lower levels of assistance.

Hypothesis 8.2: Countries receiving higher levels of peaceful nuclear assistance are more likely than states receiving lower levels of assistance to acquire nuclear weapons when security threats arise.

Below, I employ quantitative and qualitative analysis to evaluate these hypotheses. I begin with statistical analysis and then analyze historical cases to evaluate the causal processes of my argument. This multimethod approach should inspire greater confidence in the robustness of my findings, especially given that nuclear weapons acquisition is a rare event.

Statistical Analysis

Limitations

The limitations of statistical analysis I discussed in chapter 7 are especially acute for the analysis conducted in this chapter because nuclear weapons acquisition is even rarer than nuclear weapons program initiation. Although large-N analysis remains useful in this context, it is important to be cautious about the inferences that are drawn. I employ the same strategies for coping with the rare events problem that I discussed in the previous chapter. Yet I rely even more heavily on qualitative historical analysis here. I examine every case where countries successfully built nuclear weapons—as well as cases where states failed to produce the bomb—to take more of the burden off of the quantitative analysis.

Dataset

I use a dataset similar to the one employed in chapter 7 to analyze whether there is a correlation between NCAs and nuclear weapons acquisition (interested readers should see appendices 7.1 and 8.1 for further details). One main difference is that countries exit the dataset once they acquire the bomb—not once they initiate a weapons program. France, for example, exits the sample after 1960 because this is the year it first assembled a nuclear bomb.

Preliminary Analysis

I code a dependent variable that measures the first year in which a country assembled a nuclear weapon. The key independent variable is based on the NCA dataset discussed in chapter 1 and employed throughout the book. The variable employed for the initial analysis measures the total number of non-safety NCAs a state signed between 1945 and the present year entitling it to receive nuclear technology, materials, or know-how for peaceful purposes. I exclude safety NCAs because there is little theoretical reason to expect that they are associated with bomb acquisition.[4]

A simple bivariate regression reveals that nonsafety NCAs are correlated with nuclear weapons acquisition. This result provides preliminary support for my argument but it does not account for the confounding variables discussed in chapter 7, which could also influence weapons acquisition. The next step is to use multivariate statistical analysis to evaluate whether this relationship holds up once we account for other factors that could influence proliferation. As expected, nonsafety NCAs remain statistically related to bomb acquisition in the positive direction. This is true whether I include all countries in the dataset or limit the sample to states that have in interest in developing nuclear weapons.

Figure 8.1 reveals that increases in peaceful nuclear assistance produce large substantive increases in the likelihood of proliferation. The probability of nuclear acquisition when a state has not signed any NCAs and all other variables are set at their sample means (i.e., the baseline probability) is 0.00018. Raising the number of nonsafety NCAs from zero to five while holding all other factors constant raises the likelihood of bomb production by nearly 70 percent. It is relatively common for states to experience this type of increase in atomic assistance. For example, France's total number of nonsafety NCAs increased from zero to five from 1950 to 1957. Signing ten

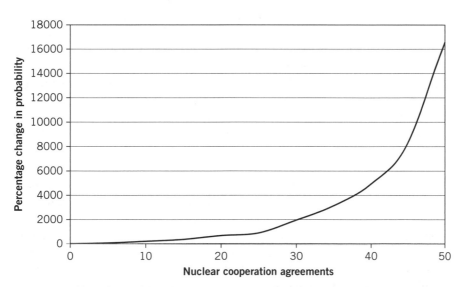

Figure 8.1 Effect of peaceful nuclear assistance on predicted probability of nuclear weapons acquisition
   Note: Calculations are based on model 2 (see appendix 8.1).

agreements raises the probability of acquisition by about 200 percent, relative to the baseline probability. This increase in NCAs roughly corresponds to the change in assistance experienced by India from 1973 to 1987. The risk that states will build nuclear weapons continues to rise as states accumulate greater levels of aid. For instance, increasing the NCA variable from zero to twenty-five and holding all other factors constant raises this probability by roughly 900 percent. As figure 8.1 shows, even more extreme increases in assistance result in huge spikes in the probability of nuclear weapons acquisition. A state that has participated in fifty NCAs, for example, is around 17,000 percent more likely than a similar state that has not signed any treaties to build the bomb. Note, however, that only about 2 percent of the observations in the dataset reach this threshold for civilian nuclear assistance.

The overall probability of nuclear weapons acquisition remains fairly small even after increases in peaceful nuclear assistance. To cite one example, when the number of nonsafety NCAs equals twenty-five and all other factors are set at their average values the probability of acquisition in a given year is roughly 0.002. This probability increases over time, of course, just as the probability of nuclear weapons program initiation spikes as time passes. If a state participates in twenty-five nonsafety NCAs, the probability that it will acquire the bomb at some point over a twenty-five-year period is 0.056. That same probability over a fifty-year period is 0.11. Taken together, these findings provide strong evidence in favor of hypothesis 8.1.

Two other variables achieve conventional levels of statistical significance. Military assistance intended to help states build the bomb is statistically related to nuclear weapons acquisition in the positive direction. But note that this finding is sensitive to model specification; in some tests, military assistance is statistically insignificant (see appendix 8.1). That this finding is not robust is at least mildly surprising. A closer look at the data explains, in part, why military assistance is less closely associated with nuclear proliferation than many assume. Some states (China, Israel, Pakistan, and North Korea) build nuclear weapons after receiving military aid. Yet Iran and Libya did not acquire the bomb despite receiving such aid because they lacked the indigenous knowledge base that typically emerges as a result of peaceful nuclear cooperation, a point that the case studies presented below will illuminate. And like peaceful aid, military assistance is not a necessary condition for bomb acquisition. Six countries built nuclear weapons in the absence of proliferation aid—France, India, Russia, South Africa, the United Kingdom, and the United States. As we will see below, three of these countries (India, South Africa, and France) benefited from peaceful nuclear cooperation prior to building nuclear weapons.

Economic liberalization is statistically significant but the sign on the variable is positive, contrary to theoretical expectations. The other control variables are statistically insignificant.[5] Although the variables measuring a state's security environment are not correlated with proliferation, I will show below that militarized conflict conditions the relationship between atomic assistance and nuclear weapons acquisition. Security threats emanating from interstate conflict and peaceful nuclear assistance combine to produce very large increases in the probability of nuclear weapons acquisition.

Does the Type of Assistance Matter?

In chapter 7 we saw that the type of assistance mattered when it came to nuclear weapons pursuit. Comprehensive power NCAs and intangible NCAs were strongly related to the onset of weapons programs while nuclear material NCAs and comprehensive research NCAs were not. It is reasonable to wonder whether we see a similar pattern when examining nuclear weapons acquisition. I disaggregate the nonsafety NCA variable based on procedures discussed in the previous chapter and replicate the statistical tests using these alternate independent variables.

The results indicate that comprehensive power agreements are positive and highly statistically significant. Disaggregating these treaties further, I find that they are correlated with nuclear weapons acquisition even when they explicitly restrict transfers of enrichment facilities, reprocessing plants, and heavy water reactors. Thus, the connection between atomic assistance and bomb production is not driven by transfers of technology that is often deemed to be the most sensitive.[6] On the other hand, intangible NCAs, nuclear material NCAs, and comprehensive research NCAs are not statistically significant. These agreements can facilitate the spread of nuclear weapons, as I will demonstrate below. But, on average, states that sign these deals are no more likely than those that do not to acquire the bomb.

Does Interstate Conflict Magnify the Peaceful-Military Link?

The analysis up to this point indicates that comprehensive power NCAs, on average, raise the risk of nuclear weapons acquisition. Does militarized conflict further strengthen this connection? To address this question, I explore how the relationship between comprehensive power NCAs and bomb production changes at varying levels of interstate conflict.[7] To illustrate the findings, figure 8.2 plots the marginal effect of comprehensive power NCAs on nuclear weapons acquisition as the number of militarized interstate disputes increase. I calculate the marginal effect by taking the difference in the probability of bomb acquisition arising from a one standard deviation increase in the NCA variable.

The figure illuminates an interesting pattern. The 95 percent confidence interval does not include zero at any point, indicating that the marginal effect is statistically significant across the entire range of the conflict variable. In terms

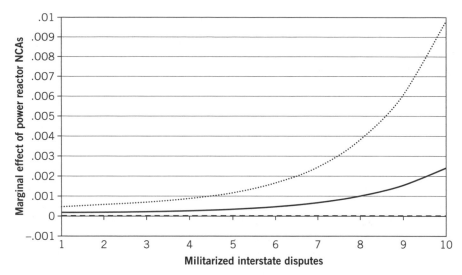

Figure 8.2 Marginal effect of comprehensive nuclear power cooperation agreements on the probability of nuclear weapons acquisition as militarized interstate disputes increase, 1945–2000

Note: Dotted lines indicate 95 percent confidence interval. MID variable extends from 0 to 9 for ease of presentation (the maximum value of the variable is 18).

of substantive significance, notice that the marginal effect increases as the security environment worsens. These results are supportive of hypothesis 8.2.

Case Study Analysis

The previous section demonstrated that there is a strong correlation between comprehensive power NCAs and nuclear weapons acquisition. This section explores particular cases to determine how and why atomic assistance made a difference. I analyze all ten states that built nuclear weapons from 1945 to 2006. I also examine six cases of nonacquisition: Argentina, Brazil, Iran, Iraq, Libya, and South Korea. These particular states are selected because they attempted to develop nuclear weapons but did not produce a single operational warhead. We can learn more about the connection between atomic assistance and bomb acquisition by analyzing these cases than by analyzing those where states expressed little interest in proliferating (e.g., Canada).

Table 8.1 categorizes these cases based on (1) whether states ultimately built the bomb; and (2) whether peaceful nuclear cooperation contributed to the success or failure of the program. This framework is useful for qualitatively assessing the two hypotheses. States listed on the left-hand side of table 8.1 are outliers for my argument while those listed in the right-hand side

Table 8.1 Weapons acquisition and peaceful nuclear cooperation

Built nuclear weapons?	Did NCAs influence outcome?	
	No	Yes
No	Argentina	Iran
	Brazil	Iraq
		Libya
		South Korea
Yes	*Russia*	China
	USA	France
	UK	India
		Israel
		North Korea
		Pakistan
		South Africa

Note: Italics denote cases where nuclear weapons acquisition occurred before peaceful nuclear cooperation was conceived by nuclear suppliers.

are successfully explained, albeit to varying extents. I begin by analyzing the outliers and then turn my attention to the successfully predicted cases, focusing on how atomic assistance enabled nuclear weapons acquisition or the ways in which a lack of aid constrained bomb development.

Outliers

States in the lower-left quadrant of table 8.1 are outliers for my argument because they built nuclear weapons without benefiting from atomic assistance. Countries in this category, Russia, the United Kingdom, and the United States, demonstrate that peaceful nuclear cooperation is not a necessary condition for bomb acquisition. These three states built nuclear weapons in an era where civilian nuclear assistance did not yet occur. Since President Eisenhower's Atoms for Peace address in December 1953, every state that acquired the bomb first signed at least one NCA.

Argentina and Brazil, the states in the upper-left quadrant of table 8.1, did not build nuclear weapons despite accumulating a substantial amount of nuclear aid. These cases are outliers for the most basic version of my bomb acquisition argument because an inability to overcome technological hurdles cannot account for the nonacquisition of nuclear weapons. When Argentina terminated its weapons program in 1990 it had signed numerous NCAs entitling it to nuclear technology, materials, or know-how. Agreements with Canada and Germany led to the transfer of heavy water reactors that many Western analysts feared would be used to produce plutonium for nuclear weapons.[8] The knowledge base created as a result of atomic assistance also helped Buenos Aires indigenously build a uranium enrichment facility near Pilcaniyeu in the early 1980s; China supplied the initial uranium hexafluoride (UF_6) for the plant.[9] This plant could have produced enough weapons-grade uranium for roughly five nuclear weapons per year.[10] Brazil likewise concluded NCAs that supported its covert nuclear weapons program. For instance, West German equipment and plans transferred for peaceful purposes were being used in Brazil's parallel military program.[11] Peaceful nuclear assistance ultimately helped Brazil build laboratory-scale enrichment and reprocessing facilities during the 1980s.[12] Based on technical considerations alone, Brasilia could have produced a small nuclear arsenal prior to 1990.[13]

Why did these two states refrain from building nuclear weapons despite possessing adequate technological capacity? In the case of Argentina, the onset of relative peace influenced the pace of the weapons program. In the late 1970s, Buenos Aires was on the brink of war with Chile over the possession of islands off the southern coast of South America, but these two states concluded a peace treaty in 1984 that alleviated the perceived threats stemming from Santiago.[14] The 1982 Falklands/Malvinas War with Great Britain underscored Buenos Aires's conventional inferiority relative to Western powers.[15] This temporarily accelerated the nuclear program but Argentina soon realized that Britain did not constitute a substantial threat to its national security.[16] Buenos Aires did not experience a single militarized dispute with London after 1983 and bilateral diplomatic relations were restored in 1990.[17] A rivalry between Argentina and Brazil evolved over territorial disputes and water conflicts on the Rio de la Plata.[18] During the 1970s, Argentine general Alejandro Lanusse pursued a "diplomatic campaign" against Brazil that strained relations between the two regional powers in Latin America.[19] Yet bilateral relations with Brazil began to warm shortly after Argentina initiated its nuclear weapons program. In 1979 the two states resolved the dispute over water resources and Argentina began to view Brazil more favorably as the 1980s unfolded.[20] The evolution of Argentina's relationships with these three states minimized its political resolve to build nuclear weapons to such an extent that some have questioned whether Buenos Aires had any interest in nuclear weapons during the 1980s.[21] By most accounts, Argentina continued its weapons program until 1990 but proceeded without significant urgency due to the absence of protracted and sustained militarized conflict.[22]

The relative absence of militarized conflict also influenced Brazil's nuclear weapons program. If Brazil faced any security threats between 1978 and 1990 they most likely would have emanated from Argentina. As noted above, however, Argentine-Brazilian relations were far from adversarial during this period.[23] The lack of pressing strategic needs to nuclearize slowed the tempo of Brazil's program.[24] As Maximiano Fonseca, the Brazilian navy minister, indicated: "We don't need the bomb *now*, since there is no foreign enemy in sight."[25] Fonseca supported Brazil's efforts to build nuclear weapons but he did not believe it was necessary to acquire the bomb immediately due to the lack of an imminent security threat.[26] Brazil, like Argentina, ultimately abandoned its bomb program in 1990 largely because the absence of strategic incentives to proliferate led leaders to reevaluate the utility of nuclear weapons.[27]

We know from the statistical analysis that peaceful nuclear cooperation, on average, increases the likelihood of nuclear weapons acquisition. The Brazilian and Argentine cases underscore that there are exceptions to this general rule. Indeed, the technological capacity stemming from atomic aid does not guarantee that states will acquire the bomb, even if they have an interest in developing nuclear weapons. These cases also demonstrate that a reduction in the strategic incentives to proliferate arising from a relative absence of militarized conflict can reduce the probability that assistance contributes to weapons acquisition, an insight that is consistent with the logic of hypothesis 8.2.

Successfully Predicted Cases

Having discussed the outliers, I turn to the cases that are consistent with the prediction of my argument. I briefly analyze each case that appears in the right-hand side of table 8.1 and subsequently explore one case from the lower-right quadrant (Pakistan) and another from the upper-right quadrant (Libya) in greater detail to further tease out how and why nuclear cooperation influences the outcomes of nuclear weapons programs.

Those cases in the lower-right quadrant of table 8.1 successfully utilized peaceful nuclear cooperation to build nuclear weapons. India, for instance, used nuclear technology and materials obtained from Canada and the United States for peaceful purposes to conduct its 1974 nuclear test (see chapters 4 and 7).

The Israeli case is also illustrative. France provided military assistance to Israel during the late 1950s and early 1960s that probably enhanced its ability to assemble a nuclear weapon quicker than it would have been able to through solely indigenous means.[28] Yet Israel also benefited from peaceful nuclear aid. A research reactor exported by the United States in 1960 provided the Israelis with a legitimate front for its nuclear weapons programs.[29] Additionally—and more directly—heavy water supplied by the United States, the United Kingdom, and Norway helped enable Israel to acquire nuclear weapons.[30] Gary Milhollin highlights the importance of foreign-supplied heavy water for Israel's weapons when he notes that "the reactor at Dimona is Israel's only means of making plutonium, and plutonium is Israel's primary nuclear weapon material. When Dimona opened in 1963 . . . Israel was producing heavy water only in laboratory quantities. Therefore, it was physically impossible to start Dimona without U.S. or Norwegian heavy water."[31] The French military aid, therefore, was alone insufficient for Israeli to build nuclear weapons.

France benefited from peaceful nuclear assistance prior to acquiring nuclear weapons in 1960. The United States signed an NCA with France in November 1956 that led to cooperation in "power and research development" in addition to the transfer of 2,500kg of U-235 and significant quantities of heavy water.[32] Washington also signed NCAs with EURATOM, a European consortium that includes France, beginning in the late 1950s that authorized the development of large commercial reactors and the supply of enriched uranium for use in these facilities. Based on a 1955 NCA, Britain shared experiences with France on the construction and operation of graphite-moderated reactors—the type of facility ultimately used to produce plutonium for the first French bombs—and exported enriched uranium and other nuclear materials for use in the civil nuclear program.[33] Additionally, Canada provided France with know-how relevant to metallurgy and heavy water reactors during the late 1950s.[34] Although France indigenously produced reactors and a reprocessing plant at Marcoule, the materials and know-how transferred as a result of civilian nuclear assistance likely facilitated the progress of its military program.[35] This is especially true because France, unlike the United States, does not draw a red line between its civilian and military nuclear programs. According to Michael Pecqueur, a former chairman of the French Atomic Energy Commission (CEA), "France has two kinds of nuclear materials: those that are free for any use, and those that can be used only for peaceful purposes under international agreements. . . . France has the option to use any free materials for our military program."[36]

The North Korean and South African cases further underscore the significance of the technical base resulting from atomic assistance. The Soviet Union trained North Korean nuclear scientists beginning in the late 1950s and completed construction of a research reactor at Yongbyon in 1965. This technical aid provided a sufficient base of knowledge in nuclear matters to help the North Koreans build an "experimental nuclear installation" in the 1980s.[37] After constructing this facility and collecting technology and expertise under the auspices of a peaceful program, Pyongyang withdrew from the NPT in 2003, an event that I explore further in chapter 9. This made it difficult for international regulatory agencies such as the IAEA to monitor activities. Pyongyang subsequently used the experimental reactor to produce plutonium for a nuclear bomb that it tested in October 2006.[38] The South African case also illuminates the importance of indigenous know-how obtained as a result of nuclear cooperation. As I discussed in the introduction and in chapter 7, when Pretoria initiated a nuclear weapons program in 1974 it appointed J. Wynand de Villiers to head the project.[39] De Villiers was trained by the United States in the early 1960s and he later helped South Africa assemble a small nuclear arsenal.[40]

Peaceful nuclear cooperation played a more minor role in the Chinese case. China received military assistance from the Soviet Union in building the Lanzhou uranium enrichment facility during the late 1950s.[41] This aid proved to be critical in helping China build the bomb. The Soviets signed another agreement with China in the 1950s that had a civilian purpose and led to the training of Chinese nuclear engineers.[42] It is difficult to ascertain precisely what impact this had on Beijing's ability to produce the bomb based on the available evidence, although it likely contributed to China's ability to produce weapons-grade uranium.[43]

Those countries in the upper-right quadrant of table 8.1 are consistent with what my argument would predict because they signed relatively few NCAs and did not acquire nuclear weapons.[44] Iran remained highly motivated to build nuclear weapons between 1985 and 2000 yet it was unable to acquire the bomb during this period. The United States provided Iran with peaceful nuclear assistance between 1957 and 1979, as I discussed in chapter 4. This aid bolstered Iran's ability to build nuclear weapons but it was alone insufficient to quickly enable proliferation following the decision to pursue the bomb in 1985.[45] Iran's Supreme Leader, Ayatollah Ruhollah Khomeini, shunned the peaceful nuclear program when he came to power in 1979 because he viewed it as a symbol of Western influence. And when Tehran resumed its interest in the nuclear program in the mid-1980s, it had a very difficult time securing atomic assistance. As a result, Iran's scientific knowledge base was relatively small and it lacked civilian facilities that could be used directly to build nuclear weapons. This put the country in an unfavorable position compared to similarly developed states that benefited from greater levels of atomic assistance. Tehran planned to build nuclear weapons by producing weapons-grade uranium in covert military facilities and it began receiving military assistance from Pakistan in 1987 to support this effort.[46] Yet because it lacked an adequate knowledge base to draw on, Iran experienced substantial technical difficulties with the enrichment plants that delayed the progress of the weapons program. Iran did not master enrichment on a laboratory scale until 2006—nineteen years after it began receiving military assistance.[47] To put this in perspective, Brazil and India mastered this technology in ten years, both without receiving military assistance but benefiting from substantial peaceful nuclear cooperation.[48] If Iran remains politically motivated to build nuclear weapons, technical hurdles may not prevent it from doing so in the coming years. What this case shows, however, is that the absence of civilian nuclear assistance can considerably delay weapons acquisition, even if a country receives aid intended to promote proliferation.

As we saw in chapter 5, Brazil, France, and Italy aided Iraq's civilian nuclear program in the 1970s. These transfers enhanced Iraq's capacity to build nuclear weapons. The preponderance of the evidence suggests that Saddam Hussein planned to use the French-supplied Osiraq reactor to produce plutonium for bombs, even if the facility was not perfectly suited for this purpose.[49] Baghdad's plans were altered when Israeli destroyed the facility in a "bolt from the blue" raid on June 7, 1981.[50] Although there is a scholarly debate on the efficacy of the Osiraq strike, it at least temporarily foreclosed Iraq's ability to produce plutonium for weapons using a civilian facility.[51] Iraq signed only one additional NCA between 1981 and 2000; the lack of peaceful nuclear assistance during this period had a critical impact on its ability to produce nuclear weapons. Following the destruction of its principal reactor, Iraq planned to build nuclear weapons by producing HEU rather than attempting to rebuild a nuclear reactor. This created a practical problem because Iraq had not built up a sufficient technical base through its civilian nuclear activities to master sophisticated enrichment technologies. As one Iraqi nuclear scientist noted, "A centrifuge is like a delicate soufflé that will fall apart if anything is done incorrectly, and our chefs were woefully unprepared."[52] An International Atomic Energy Agency report issued in the aftermath of the 1990–91 Persian Gulf War reaffirmed this point, noting that a "lack of technical depth" imposed serious difficulties on the implementation of the enrichment program and it would have been "some years" before Iraq could have produced sufficient quantities of weapons-grade uranium.[53] Had Saddam Hussein refrained from invading Kuwait, Iraq might have eventually produced nuclear weapons but the lack of civilian nuclear assistance dramatically slowed the pace of the program after the shift to enrichment in 1981.

South Korea received a small amount of peaceful nuclear assistance compared to other proliferators, such as India and Pakistan. It did, however, accumulate a nontrivial amount of aid prior to terminating its weapons program in 1978. South Korea imported two research reactors, enriched uranium, and a small fuel fabrication plant from the United States.[54] Washington also provided training to South Korean scientists under the terms of a 1964 NCA.[55] Canada and the United States supplied nuclear power plants, but those did not come online until after the termination of the weapons program. South Korea hoped to import a plutonium reprocessing facility from France but this transaction was never consummated. Given developments in the civilian nuclear program, a U.S. intelligence estimate concluded in the mid-1970s that Seoul would not be able to produce a nuclear weapon prior to 1984.[56] The challenges associated with producing fissile material for bombs had an impact on the outcome of South Korea's weapons program.[57]

Yet it is important not to overstate the role of atomic assistance in this particular case. Seoul's technical base was sufficiently developed as a result of civilian aid that it could have indigenously built a covert reprocessing facility with sufficient time and resources.[58] Moreover, South Korea abandoned its weapons program primarily due to U.S. pressure.[59] Washington threatened to end its security relationship with Seoul if the weapons program continued and had plans to withdraw American ground troops and nuclear weapons from South Korea.[60] When President Jimmy Carter reaffirmed the U.S. defense commitment to South Korea in 1978, Seoul calculated that it had less utility for an independent nuclear deterrent and abandoned the bomb campaign. Although the outcome of South Korea's program is consistent with what my theory would predict given the amount of nuclear cooperation in the 1970s, a closer examination reveals that my argument only partially explains this case.

This historical analysis up to this point suggests that peaceful nuclear assistance facilitated nuclear weapons acquisition in several cases and the lack of atomic aid restrained proliferation in other instances. Below I analyze two cases in greater detail. First I analyze the Pakistani case to explore how peaceful assistance contributed to bomb acquisition and then I examine the Libyan nuclear program, which did not result in a single nuclear weapon despite persisting for more than thirty years.

PAKISTAN: SUCCESSFUL BOMB ACQUISITION

Pakistan's civil nuclear program began in the 1950s as a result of foreign assistance. In August 1955, the United States signed a nuclear cooperation agreement with Pakistan that led to the construction of a small research reactor at the Pakistan Institute of Nuclear Science and Technology (PINSTECH) and the supply of highly enriched uranium to fuel it. The PINSTECH reactor, which began operation in 1963, was used to train Pakistani technicians, produce isotopes, and conduct neutron physics experiments.[61] In the 1960s, Canada signed a nuclear cooperation agreement with Pakistan that enabled Ottawa to build the Karachi Nuclear Power Plant (KANUPP) and supply heavy water and uranium to fuel the reactor. This power reactor began operation in 1972. Canada also helped Pakistan develop a fuel fabrication facility at Chasma in the late 1970s.[62] Western European suppliers offered considerable amounts of atomic assistance to Pakistan as well. The United Kingdom, for example, provided hot cells capable of separating plutonium on a laboratory scale.[63] Similarly, Belgium and France assisted Pakistan in developing the "New Laboratories" at PINSTECH to reprocess spent nuclear fuel.[64] Brussels also provided Islamabad with a heavy water production facility that came online at Multan in 1980.[65] Paris agreed in 1976 to supply a large-scale reprocessing center at Chasma, but it suspended this deal in 1978, partially in response to U.S. pressure.[66]

In addition to transferring these materials and technology, many suppliers provided substantial know-how to Pakistan.[67] For instance, the United States trained promising young scientists from Pakistan at Argonne National

Laboratory between 1955 and 1961.[68] These scientists were trained in the design and construction of nuclear reactors, the handling of radioactive materials, chemistry and metallurgy, and other peaceful applications of atomic energy.[69] The United Kingdom, Belgium, and other countries in Western Europe provided similar training to Pakistani personnel.[70]

After Pakistan suffered a defeat at the hands of India in their 1971 war, it initiated a nuclear weapons program. Islamabad redoubled its efforts to get nuclear weapons after India tested a nuclear explosive device in May 1974.[71] When Prime Minister Zulfikar Ali Bhutto initiated the weapons program, he planned to develop reactors and reprocessing centers to produce plutonium for nuclear weapons. The prime minister tapped Munir Ahmad Khan, the chairman of the Pakistan Atomic Energy Commission (PAEC), to implement this plan. Khan was one of the Pakistanis trained by the United States more than a decade earlier.[72] Not only did Khan personally benefit from the training he received from the United States, as chairman of the PAEC he was able to share his expertise once he returned to Pakistan. Others who learned via peaceful nuclear cooperation were also able to share their experiences with Pakistani scientists. This accumulation of know-how enabled Pakistan to develop a technical base that was "equally adept" to India's scientific abilities in the early 1970s.[73] This indigenous knowledge base, in turn, increased the PAEC's confidence that it could deliver the bomb for Pakistan.[74]

Bhutto and Khan believed that Pakistan could use facilities obtained for peaceful purposes to develop nuclear weapons—just as India would do in 1974.[75] But ultimately, Islamabad chose a slightly different path. It focused instead on the uranium route to the bomb. The history of Pakistan's enrichment program is well known.[76] In September 1974, a young metallurgist named A. Q. Khan wrote a letter to Prime Minister Bhutto offering to help Pakistan build the bomb.[77] Khan had been working in the Netherlands for a subcontractor of the European enrichment consortium URENCO. Khan stole sensitive information dealing with centrifuge technology that can be used to enrich uranium while employed by URENCO. At the end of 1975, he suddenly left the Netherlands and returned to Pakistan with stolen blueprints for centrifuges and a Rolodex containing information on one hundred companies that supply enrichment technology.[78]

Pakistan used this information to purchase subcomponents from abroad and construct covert enrichment facilities dedicated to a bomb program.[79] As a result of Khan's shopping spree, Pakistan had virtually everything it needed to construct a centrifuge enrichment plant as early as 1979.[80] With this equipment in hand, Pakistan began to indigenously construct enrichment facilities at Sihala and Kahuta using stolen blueprints.[81] In the end, highly enriched uranium produced at these plants enabled Islamabad to produce at least one bomb by 1987 and conduct nuclear tests eleven years later.[82]

Pakistan was able to master sophisticated enrichment technology and produce HEU for bombs partially because of the peaceful nuclear assistance it received beginning in the mid-1950s. Atomic aid provided by the United States, Canada, and Western European countries established a technical knowledge base in Pakistan. Islamabad was able to draw on this know-how to construct and operate the enrichment centers at Sihala and Kahuta. For instance, as I noted above, Pakistani scientists received training in uranium metallurgy—the physical and chemical behavior of uranium and its alloys. Expertise in metallurgy is vital to enriching uranium using the gas centrifuge method. Without this know-how, Islamabad would not have known what to do with the technology and materials it procured from abroad. Pakistani scientists were well aware that their weapons program, including the Sihala and Kahuta plants, would not have advanced as it did in the absence of civil nuclear cooperation. Since it was a developing country, Pakistan could not have obtained the requisite expertise solely through indigenous means. Munir Khan illustrated this well:

> I have no place from which to draw talented scientists and engineers to work in our nuclear establishment. We don't have a training system for the kind of cadres we need. But, if we can get France or somebody else to come and create a broad nuclear infrastructure, and build these plants and these laboratories, I will train hundreds of my people in ways that otherwise they would never be able to be trained. And with that training, and with the blueprints and the other things that we'd get along the way, then we could set up separate plants that would not be under safeguards, that would not be built with direct foreign assistance, but I would not have the people who could do that. If I don't get the cooperation, I can't train the people to run a weapons program.[83]

Samar Mubarakmand, who headed the team of scientists that orchestrated Pakistan's 1998 nuclear tests, expressed similar sentiments. He argued that Pakistan was able to develop the bomb because it had a cadre of trained scientists and technicians. Samar suggested that any country can procure dual-use equipment relevant to a weapons program but states cannot go nuclear "unless there is a human resource available . . . which understands [nuclear-related] work to such an extent that it is able to develop and raise this program from zero to 100% all by itself."[84] He added that countries such as Libya were unable to develop the bomb because they lacked what Pakistan had: the requisite knowledge base.

LIBYA: PROGRAM INITIATION WITHOUT ACQUISITION

Libya is an interesting case because Tripoli was unable to produce the bomb despite incentives to proliferate that persisted for decades. In 1969 a military coup resulted in the overthrow of Libya's monarch, King Idris, and an eccentric revolutionary named Muammar Qaddafi rose to power. Shortly after seizing power Qaddafi initiated a nuclear weapons program. Tripoli initially planned to acquire the bomb by purchasing a complete warhead. In the 1970s, Qaddafi approached China, India, and possibly other countries in the hopes of buying nuclear explosives.[85] Libya quickly recognized that existing nuclear powers would not be willing to share such sensitive military technology and modified its proliferation strategy. Like many other countries, it came to understand that the best way to build nuclear weapons was to develop a nuclear infrastructure for peaceful purposes and subsequently draw on the relevant materials, technology, and know-how to augment the weapons program. However, Tripoli had difficulty securing atomic assistance under Qaddafi's tenure—even after ratifying the NPT in 1975. Libya received a modest amount of aid from India and Argentina but the Soviet Union was Libya's main nuclear supplier until the late 1990s. As I discussed in chapter 4, Tripoli and Moscow signed a nuclear cooperation agreement in 1975 that led to the transfer of a small reactor and the construction of an atomic research center at Tajoura.[86] Libya sought to expand its cooperation with Moscow by importing additional nuclear power plants. However, the Soviet Union was not enthusiastic about this idea and it substantially curtailed its peaceful nuclear assistance to Libya in the 1980s.[87]

Libya later turned to covert procurement activities. Between 1997 and 2003, Tripoli received military-related nuclear assistance from Pakistan as part of the expansive network that also contributed to the weapons programs of Iran and North Korea.[88] Unlike other beneficiaries of Pakistani nuclear assistance, Libya requested "turnkey" facilities, meaning it wanted to import an entire weapons infrastructure and draw very little on indigenous resources. In addition to centrifuges, Pakistan provided Tripoli with a proven bomb design and significant quantities of uranium hexafluoride.[89] Drawing on this assistance, Libya began covert construction of the al-Hashan uranium enrichment facility located just outside of Tripoli.[90] The military aid should have enabled Libya to produce the bomb.

Yet technical problems resulting from a lack of peaceful atomic assistance led to critical difficulties. The problem was that Tripoli lacked an indigenous knowledge base that it could draw on for military purposes. For example, a U.S. government report issued in 1984 noted that "weaknesses of its technical manpower base and lack of coherent planning and research programs" had substantially hampered Libya's nuclear weapons program.[91] Given these shortcomings, Libya placed some nuclear technology in storage for several years during the 1980s because it arrived without instructions for assembly and operation.[92] Things did not improve significantly during the 1990s and early 2000s. U.S. ambassador Donald Mahley, who frequently represented Washington in WMD-related negotiations, noted in 2004 that he interacted with Libyan nuclear scientists who were "knowledgeable, dedicated and in-novative," but he dealt with the same people repeatedly because Libya had "almost no bench."[93] Technical limitations were evident when it came to the operation of the al-Hashan enrichment facility. Between 2000 and 2002 Libya performed successful high-speed tests of the centrifuges at al-Hashan, but it was far from being able to enrich uranium for nuclear weapons at that time.[94] Tripoli also imported more advanced centrifuges beginning in 2000 but they were never in workable condition. Indeed, IAEA inspectors later discovered that many of these advanced centrifuge components were in unopened boxes.[95] The bottom line is that the lack of a civilian nuclear program upon which it could draw stymied Libya's weapons acquisition ef-forts—even after receiving military aid from Pakistan. This, more than any other single factor, explains why Tripoli did not produce the bomb during the period when its political incentives to nuclearize were the strongest.[96]

Technical limitations also contributed to the termination of Libya's weap-ons program in 2003. In chapter 7, I argued that leaders are sensitive to the costs of building nuclear weapons; a high price tag and the necessity of diverting resources from other national priorities can sometimes deter lead-ers from pursuing the bomb. Although I used this logic to explain weapons program initiation, it also partially sheds light on a nuclear reversal in the Libyan case. Qaddafi realized by 2003 that continuing the weapons pro-gram would impose economic costs without any guarantees that the bomb would be successfully produced, given that substantial technological hurdles remained even after more than three decades of work. According to an un-identified European diplomat who spoke with Libyan elites, Tripoli "had determined that it was too expensive to develop nuclear weapons."[97] Qad-dafi and Libyan premier Shukri Ghanem made similar arguments in public.[98] The Libyan leader began to realize that the long-standing nuclear weapons program was diverting too many resources away from other economic and military programs.[99] Other factors contributed to Libya's reversal but techni-cal limitations played a key role.[100]

Conclusion

The case studies presented above shed light on *how* peaceful nuclear as-sistance facilitates nuclear weapons acquisition. Atomic assistance matters in both direct and indirect ways. Countries can employ civilian nuclear technol-ogy and materials directly to build nuclear weapons. The Indian case is the best illustration of this but Israel also used heavy water provided for civilian use to aid in the production of plutonium for its nuclear weapons. It is more common for nuclear cooperation to facilitate bomb acquisition by bolstering an indigenous knowledge base that can be drawn on to construct covert military facilities. France, North Korea, Pakistan, and South Africa are among the proliferators that benefited from assistance in this manner. The Libyan case underscored that nuclear weapons programs can fail if countries lack the scientific and technical expertise that typically stems from peaceful nuclear cooperation. This finding has important implications for nonproliferation policies that I discuss further in chapter 9 and in the conclusion of the book.

Appendix 8.1: Data Analysis and Empirical Findings

Dataset, Variables, and Method

As noted above, I construct a standard time series–cross-sectional dataset that includes all states in the international system from 1945 to 2000. States exit the sample once they build the bomb and can only reenter if they give up their nuclear weapons, as South Africa did in the early 1990s. This is important because including states in the sample after they built nuclear weapons would bias the results in favor of my argument. There are 5,825 country-year observations in the dataset.

The dependent variable measures whether a state acquires nuclear weapons in a given year. Drawing on existing proliferation datasets, I code this variable 1 if a state produces a functional nuclear weapon in year t and 0 otherwise.[101] Proliferation is a rare event, occurring in around 0.14 percent of the observations in the dataset.

The explanatory variables used in the statistical analysis are similar to those employed in chapter 7, with a few exceptions (see appendix 7.1 for details). Including a variable measuring whether a state is an NPT member in statistical models of bomb acquisition is problematic because no states acquired nuclear weapons while part of the treaty during the time period analyzed in this chapter.[102] It is possible, however, to include a variable measuring the global strength of the NPT. This variable measures the percentage of countries in the international system that ratified the NPT in year t-1.[103] I also include a variable counting the number of years between 1945 and year t-1 along with three cubic splines to control for temporal dependence, as recommended by Nathanial Beck, Jonathan Katz, and Richard Tucker.[104] The independent variables are lagged one year behind the dependent variable to control for possible simultaneity bias.[105]

Given that the dependent variable is dichotomous and nuclear weapons acquisition is an uncommon occurrence, I use rare events logit to fit the statistical models. Recall that this estimator, which I also used in chapter 7, is appropriate when the dependent variable has thousands of times fewer ones than zeros.[106] The standard errors are clustered by country to account for heteroskedastic error variance.

Table 8.2 Determinants of nuclear weapons acquisition, 1945–2000

	(1) Nonsafety NCAs– baseline model	(2) Nonsafety NCAs– full model	(3) Nonsafety NCAs–limited sample
Nuclear cooperation agreements	0.033***	0.097**	0.058**
	(0.009)	(0.048)	(0.026)
Militarized interstate disputes		0.091	0.172
		(0.204)	(0.137)
Military assistance		2.901*	0.945*
		(1.742)	(0.530)
GDP per capita		−0.000	0.001
		(0.000)	(0.000)
GDP per capita squared		−0.000	−0.000
		(0.000)	(0.000)
Industrial capacity		1.742	3.333***
		(1.763)	(1.271)
Rivalry		1.517	1.823**
		(1.401)	(0.772)
Nuclear protection		−0.432	−0.536
		(1.018)	(0.656)
Polity		0.017	−0.039
		(0.092)	(0.038)
Democratization		−0.070	−0.017
		(0.097)	(0.036)
Openness		0.022	0.024*
		(0.022)	(0.014)
Liberalization		0.080***	−0.016
		(0.010)	(0.011)
Percent NPT		−0.097	−0.040***
		(0.067)	(0.013)
No proliferation years	0.069	−0.214	0.259
	(0.193)	(0.433)	(0.261)
Spline 1	0.001	0.001	−0.001
	(0.001)	(0.002)	(0.006)

(*Continued*)

Table 8.2—*cont.*

	(1)	(2)	(3)
	Nonsafety NCAs– baseline model	Nonsafety NCAs– full model	Nonsafety NCAs–limited sample
Spline 2	−0.001	−0.003	0.002
	(0.001)	(0.003)	(0.003)
Spline 3	0.000	0.003	−0.001
	(0.001)	(0.002)	(0.001)
Constant	−6.679***	−5.915	−10.189***
	(1.433)	(5.497)	(3.301)
Observations	7,308	5,834	399

Notes: Robust standard errors in parentheses; *** $p<0.01$, ** $p<0.05$, * $p<0.1$

Results

Table 8.2 presents the initial statistical results.[107] Model 1 is a baseline model that only includes the nonsafety NCA variable and controls for temporal dependence. Model 2 is a full model that adds all of the control variables. Model 3 is a censored model that limits the estimation sample to states that have an interest in nuclear weapons.[108] This explains why the number of observations in model 3 (399) is substantially less than the number of observations in model 2 (5,834).

Nonsafety NCAs are positive and statistically significant in all three models, indicating that countries receiving peaceful nuclear assistance are more likely to acquire the bomb compared to countries that do not receive aid. Peaceful nuclear assistance makes a difference when I examine all states (models 1–2) and only those that we might expect to develop the bomb at some point (model 3). As we saw in Figure 8.1, the relationship between this variable and nuclear weapons acquisition is also substantively significant. These findings support hypothesis 8.1.

Turning to the control variables, military assistance is positive and statistically significant in models 2 and 3. Not surprisingly, countries that receive aid explicitly intended to help them build nuclear weapons are more likely to proliferate than states that do not receive this type of assistance. Some of the other controls are significant in one—but not all—of the models presented in table 8.3. Liberalization is positive and statistically significant in model 2 but insignificant in model 3. Rivalry is statistically significant in model 3, providing some support for the argument that states in threatening security environments are more likely to build nuclear weapons. Notice, however, that this finding is sensitive to model specification and the MID variable, which is another proxy for threat environment, is statistically insignificant.

Table 8.3 Disaggregated peaceful nuclear assistance and the determinants of nuclear weapons acquisition, 1945–2000

	(4) Intangible NCAs	(5) Material NCAs	(6) Comprehensive research NCAs	(7) Comprehensive power NCAs	(8) Limited power NCAs	(9) Unlimited power NCAs
Nuclear cooperation agreements	0.186	0.234	−0.874	0.269***	0.300**	0.309***
	(0.121)	(0.209)	(0.745)	(0.076)	(0.139)	(0.118)
Militarized inter-state disputes	0.108	0.079	0.135	0.086	0.064	0.105
	(0.187)	(0.203)	(0.189)	(0.201)	(0.217)	(0.184)
Military assistance	2.657*	2.720	2.413	3.074*	3.138	2.868*
	(1.599)	(1.853)	(1.808)	(1.807)	(2.015)	(1.586)
GDP per capita	0.000	−0.000	0.000	−0.000	−0.000	−0.000
	(0.000)	(0.000)	(0.001)	(0.000)	(0.000)	(0.000)
GDP per capita squared	−0.000	0.000	−0.000	−0.000	0.000	0.000
	(0.000)	(0.000)	(0.000)	(0.000)	(0.000)	(0.000)
Industrial capacity	1.676	1.631	1.527	1.577	1.546	1.752
	(1.554)	(1.968)	(0.968)	(1.821)	(1.722)	(1.768)
Rivalry	1.457	1.294	1.128	1.598	1.612	1.425
	(1.549)	(1.337)	(1.383)	(1.358)	(1.463)	(1.465)
Nuclear protection	−0.497	−0.336	−0.393	−0.406	−0.305	−0.524
	(1.051)	(1.192)	(1.484)	(0.921)	(1.026)	(0.965)

(Continued)

Table 8.3—*cont.*

	(4) Intangible NCAs	(5) Material NCAs	(6) Comprehensive research NCAs	(7) Comprehensive power NCAs	(8) Limited power NCAs	(9) Unlimited power NCAs
Polity	0.005	0.034	0.043	0.024	0.028	0.046
	(0.093)	(0.105)	(0.147)	(0.090)	(0.088)	(0.101)
Democratization	−0.064	−0.064	−0.054	−0.076	−0.100	−0.072
	(0.100)	(0.101)	(0.121)	(0.096)	(0.119)	(0.087)
Openness	0.021	0.009	0.005	0.028	0.024	0.017
	(0.022)	(0.028)	(0.033)	(0.022)	(0.028)	(0.024)
Liberalization	0.087***	0.076***	0.092***	0.070***	0.036***	0.079***
	(0.011)	(0.014)	(0.030)	(0.010)	(0.013)	(0.012)
Percent NPT	−0.101	−0.088	−0.087	−0.090	−0.086	−0.093
	(0.067)	(0.069)	(0.065)	(0.065)	(0.078)	(0.059)
No proliferation years	−0.191	−0.229	−0.173	−0.234	−0.244	−0.204
	(0.419)	(0.423)	(0.370)	(0.416)	(0.418)	(0.412)
Spline 1	0.001	0.000	0.000	0.001	0.000	0.001
	(0.002)	(0.002)	(0.002)	(0.003)	(0.002)	(0.003)
Spline 2	−0.003	−0.002	−0.002	−0.003	−0.003	−0.003
	(0.003)	(0.003)	(0.003)	(0.003)	(0.003)	(0.003)
Spline 3	0.002	0.002	0.002	0.002	0.002	0.002
	(0.002)	(0.002)	(0.002)	(0.002)	(0.002)	(0.002)
Constant	−6.158	−4.998	−5.696	−5.754	−5.532	−5.702
	(4.786)	(5.748)	(3.582)	(5.707)	(5.478)	(5.513)
Observations	5,834	5,834	5,834	5,834	5,834	5,834

Notes: Robust standard errors in parentheses; *** $p<0.01$, ** $p<0.05$, * $p<0.1$

The coefficient on the variable measuring industrial capacity is positive and highly statistically significant in model 3, suggesting that technologically developed states are more likely to acquire nuclear weapons. Yet this finding is also not robust given that industrial capacity is insignificant in model 2.[109] Two other variables, openness and percent NPT, are statistically significant in model 3 but not model 2. GDP per capita, GDP per capita squared, nuclear protection (i.e., having an alliance with a nuclear weapons state), polity, and democratization do not achieve statistical significance in any of the models displayed here.[110]

ANALYSIS OF DISAGGREGATED NCAS

Table 8.3 presents the results of the analysis that disaggregates peaceful nuclear assistance. Each of the models estimated here uses a different independent variable to gauge whether the type of assistance matters when it comes to nuclear weapons acquisition. Mirroring procedures used to construct the aggregated variable, all of these measures count the total number of agreements a state signed between 1945 and year t-1. Aside from using alternate measures of nuclear cooperation, these models are identical to model 2 (see table 8.2). Model 4 measures nuclear assistance based on intangible NCAs. Model 5 uses nuclear material NCAs to code the key independent variable, while model 6 bases this measure on comprehensive research NCAs. Models 7–9 rely on comprehensive power NCAs, which are the most comprehensive type of agreement. Model 7 codes nuclear cooperation based on the total number of comprehensive power NCAs a state signed. Model 8 only codes limited power NCAs that explicitly prohibit transfers of enrichment facilities, reprocessing centers, and heavy water reactors. Model 9 codes unlimited comprehensive power NCAs that do not restrict transfers of these facilities.

Some interesting findings emerge from this analysis. In short, the type of assistance matters. Nuclear cooperation agreements are positive and statistically significant in models 7–9. This tells us that comprehensive power NCAs are closely connected with nuclear weapons acquisition, regardless of whether they prohibit transfers of "restricted" technology. Thus, the relationship between atomic assistance and bomb production not only exists but it is broader than the conventional wisdom suggests. Yet there is a limit to the proliferation potential of atomic assistance. Nuclear cooperation agreements are statistically insignificant in models 4–6, indicating that intangible NCAs, nuclear material NCAs, and comprehensive research NCAs are not correlated with nuclear weapons acquisition. Agreements that authorize (1) limited technical exchanges, (2) transfers of materials, or (3) assistance in developing a program for nuclear research do not appear to raise the likelihood of nuclear acquisition. The initial findings presented in table 8.3 appear to have been driven by comprehensive power NCAs.

Turning to the control variables, the findings are generally consistent but there is one noteworthy difference compared to the results from model

2. Military assistance is statistically significant and positive in three of the models presented in table 8.3 but insignificant in the other three. This suggests that the connection between proliferation aid and bomb acquisition is sensitive to model specification. This is not especially surprising given that this variable barely achieved conventional levels of statistical significance in model 2 (p = 0.096).

MILITARIZED CONFLICT, NUCLEAR ASSISTANCE, AND WEAPONS ACQUISITION

The results presented in table 8.4 allow me to assess whether militarized conflict conditions the connection between nuclear weapons acquisition and peaceful nuclear cooperation. What I expect to find is that this relationship is statistically significant even in the absence of militarized conflict but that it becomes substantively stronger as the number of militarized interstate disputes increase. To appropriately test this hypothesis, I need to include an interaction term between the NCA variable and militarized interstate disputes along with the two constituent parts.[111] For this analysis I use comprehensive power NCAs to operationalize civilian nuclear assistance because these treaties were closely associated with nuclear weapons acquisition in the preceding tests. Model 10 uses rare events logit to estimate the statistical relationships while model 11 uses probit.

As noted elsewhere in the book, it is often difficult to interpret interaction terms solely on the basis of the information presented in a standard regression table. The NCA variable is positive and statistically significant in models 10 and 11. All this tells us is that comprehensive power NCAs are statistically related to nuclear weapons acquisition when the number of militarized disputes equals zero. The interaction term is insignificant in both models but this does not necessarily imply that conflict does not condition the relationship between NCAs and bomb production. The findings previously illustrated in figure 8.2 were supportive of hypothesis 8.2; the marginal effect of assistance on weapons acquisition rose as the number of MIDs increased (and the 95 percent confidence interval did not include zero at any point).

I replicated model 10 using alternate measures of nuclear cooperation. I found some support for hypothesis 8.2 when using the aggregated NCA variable, but these findings were driven by the inclusion of comprehensive power NCAs. When I employed intangible NCAs and nuclear material NCAs to operationalize atomic assistance, neither the interaction term nor the constituent parts achieved conventional levels of statistical significance. There was little evidence that these agreements were correlated with nuclear weapons acquisition even at moderate and high levels of conflict. Comprehensive research NCAs also do not appear to raise the risk of proliferation in threatening security environments. In some tests, the interaction term between these treaties and militarized disputes was positive and statistically significant but this finding was sensitive to model specification.[112] When it comes to nuclear

Table 8.4 The conditional effect of peaceful nuclear assistance, 1945–2000

	(10) Comprehensive power NCAs– rare events logit	(11) Comprehensive power NCAs– probit
Nuclear cooperation agreements	0.236**	0.122***
	(0.105)	(0.040)
Militarized interstate disputes	0.066	0.029
	(0.214)	(0.064)
NCAs * MIDs	0.001	0.031
	(0.050)	(0.024)
Military assistance	2.952	2.128***
	(1.938)	(0.579)
GDP per capita	−0.000	0.000**
	(0.000)	(0.000)
GDP per capita squared	0.000	−0.000***
	(0.000)	(0.000)
Industrial capacity	1.584	1.567***
	(1.655)	(0.486)
Rivalry	1.531	0.653
	(1.433)	(0.445)
Nuclear protection	−0.472	−0.418
	(0.958)	(0.328)
Polity	0.026	0.026
	(0.089)	(0.026)
Democratization	−0.078	−0.049
	(0.100)	(0.032)
Openness	0.023	0.003
	(0.022)	(0.007)
Liberalization	0.075***	0.012**
	(0.014)	(0.006)
Percent NPT	−0.083	−0.057**
	(0.071)	(0.023)
No proliferation years	−0.219	−0.012
	(0.400)	(0.123)
Spline 1	0.000	0.001
	(0.003)	(0.001)
Spline 2	−0.002	−0.003**
	(0.003)	(0.001)
Spline 3	0.002	0.002**
	(0.003)	(0.001)
Constant	−5.515	−4.859***
	(5.284)	(1.528)
Observations	5,834	5,834

Notes: Robust standard errors in parentheses; *** p<0.01, ** p<0.05, * p<0.1

weapons acquisition, comprehensive power NCAs are the treaties that matter the most for both of the hypotheses evaluated in this chapter.

ALTERNATE DEPENDENT VARIABLE CODING

As discussed in appendix 7.1, there is scholarly disagreement regarding when certain countries pursued or acquired nuclear weapons. I code two alternate dependent variables to explore whether the findings are sensitive to my prior coding choices for the dependent variable.[113] The first alters the coding of North Korea. Pyongyang conducted its first nuclear test in 2006. I did not code it as acquiring weapons prior to 2000 in the analysis conducted above, although it is possible that North Korea acquired its first bomb in 1999.[114] I recode the dependent variable to account for this possibility. The second alternate measure uses the Jo and Gartzke proliferation data to code acquisition.[115] Using these alternate dependent variables, I replicate model 7, which measured nuclear cooperation using comprehensive power NCAs. Nuclear cooperation agreements remain positive and statistically significant (not reported), suggesting that my findings are not sensitive to changes in the coding of the dependent variable. In terms of the control variables, the findings are likewise similar to those reported in table 8.2.

Chapter 9

Have International Institutions Made the World Safer?

The international community, led by major powers such as the United States, has instituted policies to separate the peaceful and military uses of the atom. The 1968 nuclear Nonproliferation Treaty established a comprehensive system of safeguards to make it more difficult for countries to draw on peaceful nuclear assistance to build nuclear weapons; the Additional Protocol fortified the safeguards regime in the late 1990s. To what degree have these measures made a difference?

The nonproliferation regime could theoretically limit the proliferation potential of peaceful nuclear assistance by reducing uncertainty about states' intentions, and by raising the costs of violating a commitment not to build nuclear weapons. In practice, however, nuclear safeguards have had a relatively modest effect in reducing the dangers of atomic assistance for nuclear weapons proliferation. The relationship between nuclear cooperation and weapons pursuit remains statistically significant from 1945 to 2000. The substantive effect of nuclear assistance on weapons pursuit declines slightly during this period, but this is not necessarily because of the nonproliferation regime. Instead, we see this pattern because a smaller percentage of states in high-conflict environments benefited from assistance over time. As I argued in chapter 7, militarized conflict substantially magnifies the connection between peaceful cooperation and weapons program onset. Since fewer states receiving nuclear assistance have strategic incentives to develop nuclear weapons, we observe a probabilistic decline in the civil-military connection.

There is a limit to the constraining power of the NPT. Atomic assistance increases the likelihood of nuclear weapons pursuit regardless of a state's status in the nonproliferation regime. States that do not ratify the NPT are more likely to launch a military program following increases in atomic assistance and, surprisingly, so are states that commit to the treaty. Ratification reduces the likelihood that states will pursue nuclear weapons when they do not benefit from peaceful nuclear assistance, but this finding disappears as states accumulate aid. NPT members are no less likely to proliferate than nonmembers if they receive moderate levels of nuclear assistance. These findings make it hard to attribute much of a causal effect to the NPT.

Case studies of nuclear decision making in Syria and Japan confirm the insights from the statistical analysis, showing that the causal mechanisms of the safeguards argument do not operate as expected. The Syrian case demonstrates that countries can exploit the safeguards regime with relative ease. Japan is an "easy" case for the safeguards argument because we would expect to find evidence that the NPT played a critical role. The historical record reveals, however, that the treaty cannot account for Japanese nuclear restraint. There is no compelling evidence that Tokyo would have behaved any differently if the NPT was never created.

In terms of nuclear weapons acquisition, safeguards have not made it substantially more difficult for states with weapons programs to develop the bomb. Case studies of sixteen nuclear weapons programs show that proliferators repeatedly engaged in violations of their safeguards agreements, suggesting that they were not deterred by the prospect of detection. Moreover, when violations occurred, the International Atomic Energy Agency rarely detected them in a timely fashion. These trends continued even after the creation of the AP, suggesting that this protocol may not go far enough in augmenting the capabilities of the IAEA.

How Could Safeguards Limit the Perils of Peaceful Nuclear Cooperation?

International institutions that are designed to detect noncompliance with agreements can promote cooperation by increasing transparency and reducing uncertainty.[1] The NPT regime is one example of such a "monitoring arrangement."[2] To verify that states do not use civilian nuclear assistance for military purposes, the IAEA institutes a system of material control and accountancy, surveillance, and verification known as safeguards.[3] The purpose of IAEA safeguards is to *detect* the diversion of materials from the civilian nuclear fuel cycle and *provide warning* to the international community in the event that such an event occurs. To detect diversions, the IAEA examines records provided by individual countries and conducts inspections of facilities to verify the declared inventories of nuclear materials.[4]

The IAEA's inspections regime could "work" in two respects. First, the enhanced likelihood of detection could deter states from drawing on civilian nuclear programs for military purposes. The assumption here is that verified cheating results in the imposition of costs that states generally want to avoid. Second, safeguards could provide countries with peace of mind that their rivals will not use peaceful aid to build nuclear weapons. If countries believe their cooperation will not be exploited, they are more likely to refrain from developing the bomb.[5]

IAEA safeguards represent the bedrock of the nonproliferation regime. Without a mechanism for identifying cheaters, the NPT would lack an element that is widely believed to be important in promoting cooperation and compliance.[6] Moreover, other nonproliferation institutions depend on the viability of nuclear safeguards. The Nuclear Suppliers Group, for instance, encourages states to require safeguards when transferring nuclear technology, materials, or know-how.[7] We already saw in chapter 3 that NSG members are no more likely to behave "responsibly" than nonmembers and that patterns of nuclear cooperation were similar before and after the creation of the group. Nevertheless, it is plausible that the NSG has weakened the connection between nuclear assistance and proliferation—but only if safeguards effectively detect violations and deter states from drawing on civilian nuclear activities for military purposes. In other words, the efficacy of the NSG in mitigating the perils of peaceful nuclear cooperation is tied to the utility of nuclear safeguards. If safeguards have not made a difference, we can reasonably conclude that the NSG had little effect on the connection between the peaceful and military uses of the atom.

Safeguards were first applied to the transfer of uranium fuel from Canada to Japan in 1959 but the IAEA did not institute a robust safeguards regime until 1965.[8] The NPT fortified the safeguards regime. As noted elsewhere in the book, the treaty requires nonnuclear weapons states to forgo the development of nuclear weapons and accept IAEA safeguards to verify this commitment. NPT-based safeguards are "full scope" in nature, meaning that they apply to all of the declared facilities possessed by nonnuclear weapons states.[9] If NNWS do not conclude full-scope safeguards agreements with the IAEA, they are prohibited from receiving peaceful nuclear assistance. This requirement represents the first linkage of atomic assistance and safeguards under international law. Yet as we saw in chapter 3, this bargain goes unrealized; NPT members are no more likely, on average, to receive atomic aid than non-NPT members.

Later efforts sought to further weaken the relationship between nuclear aid and the spread of nuclear weapons. In response to Iraq's nuclear activities, which are discussed further below, IAEA director general Hans Blix noted in 1991 that his agency's safeguards system needed "more teeth."[10] The IAEA initiated a program in 1993 known as "93+2" to identify ways to enhance the agency's safeguards and propose a more comprehensive system to the Board of Governors by 1995. Program 93+2 led to the establishment of the IAEA Additional Protocol in 1997. The AP provides the agency with greater authority to verify compliance with the NPT. Perhaps most important, it allows the IAEA to inspect both declared and undeclared facilities on short notice and collect environmental samples at sites when there are reasons to believe that suspicious activities have taken place. This means that the IAEA can visit any installation that could be involved in a secret nuclear weapons program—even if a country does not declare the facility to the agency. In the words of one expert, this authority "transform[s] IAEA inspectors from accountants to detectives."[11] The AP also requires states to provide more information to the IAEA regarding their activities.[12] Note, however, that in order for the requirements discussed above to become legally binding, states must conclude AP agreements with the IAEA on a bilateral basis.[13] This is a critical point that I revisit below and in the conclusion.

Observable Implications

Has the strengthening of the safeguards regime weakened the relationship between atomic assistance and nuclear proliferation over time? How would we know whether it had or had not? There are observable implications of the safeguards argument for nuclear weapons program onset and bomb acquisition.

When it comes to nuclear weapons program initiation, the marginal effect of peaceful nuclear assistance should fluctuate over time, consistent with the illustration in figure 9.1. There should be a steep increase in the likelihood of assistance contributing to weapons pursuit from the early 1950s until the mid-1960s.[14] During this period, many states began to accrue civilian nuclear assistance and there were few checks on their ability to use this aid for military purposes. We should observe a decline in the relationship between atomic assistance and weapons program onset beginning in 1965 when the IAEA instituted a robust safeguards regime. This relationship should decline even more sharply after the NPT entered into force in 1970. Throughout the 1970s and 1980s we should witness a further decline in the civil-military connection as more states commit to the treaty. A final sharp decrease in the marginal effect of nuclear cooperation should occur after the institution of the AP in 1997.

If the observed relationship between nuclear assistance and weapons program initiation mirrors the pattern discussed above, we would also expect to find evidence from historical cases of the NPT and AP deterring countries from diverting nuclear materials or building covert facilities due to the enhanced likelihood of detection. We should also see evidence that these institutions provided states with peace of mind that their rivals would not proliferate.

International treaties are not legally binding unless states ratify them, and the NPT is no exception.[15] States that remain outside the NPT are generally not legally bound to foreswear nuclear weapons or accept comprehensive

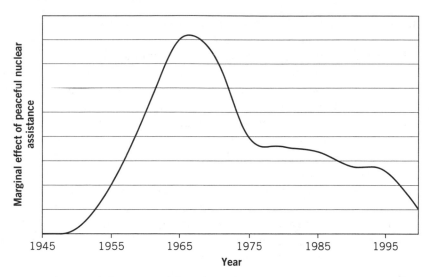

Figure 9.1 Theoretical illustration of the safeguards argument

IAEA safeguards.[16] A second observable implication of the safeguards argument, therefore, is that NPT membership conditions the relationship between nuclear cooperation and weapons program onset. In particular, peaceful nuclear cooperation should raise the likelihood that non-NPT members begin weapons programs but there should be no relationship between assistance and proliferation for members of the treaty. On average, states that commit to the NPT should be dissuaded from pursuing the bomb—even if they receive atomic assistance—because the treaty's safeguards regime raises the risk that they will be caught if they draw on civilian nuclear programs for military purposes. Testing this second implication is important because many states remained outside the NPT for years after the treaty was established.[17] Algeria and Turkey, for instance, waited twenty-seven and twelve years, respectively, to ratify the NPT. The time trends discussed above might not influence these states to the same extent that they would affect countries that committed to the treaty when it opened for signatures in 1968.

The safeguards argument has two observable implications for nuclear weapons acquisition. First, states pursuing nuclear weapons should be deterred from drawing on civilian nuclear activities to augment their military programs because safeguards raise the risk of detection. In other words, we should not observe frequent violations of nuclear safeguards agreements. Second, when violations occur, the IAEA should detect them in a timely fashion and provide advance warning to the international community, permitting individual states to take action as appropriate. As noted above, the timely detection of violations is critical to the overall effectiveness of the NPT monitoring regime. It would be hard to credit the NPT for stymieing weapons acquisition efforts in cases where proliferators are able to frequently break the rules without getting caught.

The argument outlined above represents a rational theory of safeguards effectiveness that is rooted in the neoliberal tradition. Before proceeding to an analysis of the historical record, it is important to mention that constructivists offer two additional theories of NPT effectiveness. One suggests that the NPT stigmatized proliferation and made nuclear weapons less attractive for "responsible" states that seek to play by the rules.[18] The other contends that the nonproliferation regime helped establish a "taboo" against the use of nuclear weapons, minimizing the strategic benefits that accompany bomb possession.[19] Although the focus of this chapter is on a neoliberal theory of safeguards, both of the constructivist arguments suggest that the pattern between atomic assistance and nuclear weapons program initiation should mirror the relationship depicted in figure 9.1; they imply that proliferation was rampant prior to the creation of the NPT and the regime reversed this trend. If the historical record reveals a different pattern, this would cast doubt on the neoliberal theory as well as the two constructivist arguments. Additionally, I explore the empirical validity of the constructivist mechanisms when examining the case of Japan.

The following sections test the pro-regime argument against the historical record. I begin by analyzing the role of safeguards in limiting nuclear weapons program initiation. Subsequently I explore whether safeguards made it more difficult for states to successfully develop the bomb.

Safeguards, Weapons Program Initiation, and Peaceful Nuclear Cooperation

Have nuclear safeguards weakened the relationship between peaceful nuclear assistance and nuclear weapons program initiation? I employ quantitative and qualitative methods to answer this question. As I discussed throughout the book, scholars face challenges when using large-N statistical analysis to model rare events such as nuclear proliferation. Yet, with appropriate caution, quantitative analysis can be useful for determining whether NPT-backed safeguards made a difference while controlling for other factors that are thought to influence proliferation. I analyze two cases of nuclear restraint to determine whether the logic driving the safeguards argument operates correctly in particular cases. The criteria for case selection are discussed further below.

Statistical Analysis

I begin by analyzing whether the relationship between nuclear cooperation and weapons program initiation changes over time. For the purposes of this analysis, I focus on the connection between comprehensive power NCAs and program onset. Recall that these agreements were closely associated with weapons program initiation in the analysis conducted in chapter 7.

Figure 9.2 plots the marginal effect of comprehensive power NCAs on weapons program initiation as the number of years since 1945 increases.[20] The first thing to observe is that the marginal effect is statistically significant across the full range of the time variable. The 95 percent confidence interval (denoted by the dotted lines) does not include zero at any point in figure 9.2. Thus, improvements to the nonproliferation regime have not eliminated the relationship between atomic assistance and nuclear weapons program onset. In terms of substantive significance, the marginal effect declines slightly from 1945 to 2000. It is not clear that we can credit the nonproliferation regime for this trend, however. The marginal effect displayed in figure 9.2 does not mirror the theoretical illustration of the safeguards argument (see figure 9.1). The effect begins to weaken well before there is a major improvement to the regime. This is in stark contrast to the prediction that the marginal effect should increase in the early stages of the nuclear age when there were few institutional checks on civilian nuclear activities. The effect of nuclear assistance on bomb programs dropped by 66 percent between 1945 and 1960—five years before the IAEA developed a system of safeguards for nuclear power reactors. For comparative purposes, consider that the marginal effect dropped by 55 percent between 1970 and 1985, a period where two

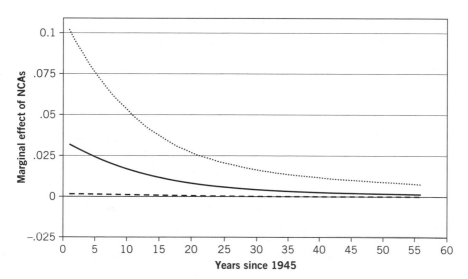

Figure 9.2 Marginal effect of nuclear cooperation agreements on nuclear weapons program onset as the number of years since 1945 increases

Note: Dotted lines represent 95 percent confidence interval. Calculations are based on model 2 (see appendix 9.1).

of the most significant improvements in the nonproliferation regime were made. Thus, the decline in the strength of the relationship between nuclear assistance and proliferation between 1945 and 1960 is greater than the decline in this connection in the fifteen years following the entry into force of the NPT. This is problematic for the safeguards argument, which predicted a steep decline in the marginal effect beginning in 1970.

My argument that peaceful nuclear assistance is especially likely to lead to weapons program initiation if countries experience militarized conflict (see chapter 7) helps account for the slight decline in the marginal effect of nuclear cooperation. Figure 9.3 illustrates why this is the case. The figure plots the number of nonnuclear weapons states receiving peaceful nuclear assistance over time along with the percentage of countries benefiting from aid that experienced militarized conflict.

Notice that the number of nonnuclear weapons states receiving modest peaceful nuclear assistance has increased over time.[21] Yet the percentage of nonnuclear weapons states receiving modest nuclear assistance in high conflict environments declined over time.[22] Roughly one-quarter of all states receiving nuclear assistance faced above-average levels of conflict in the 1960s. By the 1990s, this figure dropped to around 8 percent; only five states that were not already pursuing nuclear weapons in this decade—Japan, Greece, Thailand, Turkey, and Serbia—faced a significant number of militarized disputes with their adversaries after accumulating peaceful nuclear assistance. Given that my proliferation argument is probabilistic, a handful of cases that satisfy these two key conditions by no means guarantee that weapons program initiation will occur. The likelihood that we will observe at least

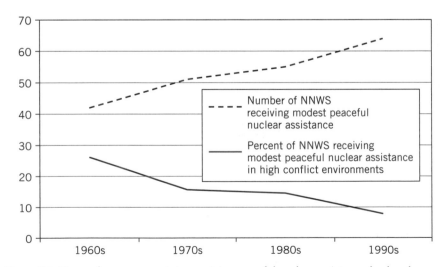

Figure 9.3 Nonnuclear weapons states receiving peaceful nuclear assistance, by decade

one new nuclear weapons program rises, however, as the percentage of states receiving nuclear assistance in high conflict environments increases.

I turn now to a second testable implication of the safeguards argument. Does NPT membership condition the relationship between peaceful nuclear cooperation and nuclear weapons program initiation? The statistical findings reveal that, on average, it does not. Comprehensive power NCAs are statistically related to weapons program initiation for both NPT and non-NPT members. Figure 9.4 illustrates the substantive effect of nuclear cooperation on the likelihood of pursuing the bomb for these two groups of countries. For non-NPT states, increases in nuclear assistance (measured by comprehensive power NCAs) raises the probability of initiating a weapons program by 99 percent, all else being equal. Atomic aid has a larger substantive impact on states that ratify the NPT; these states are 150 percent more likely to pursue nuclear weapons with increases in assistance. These findings should be interpreted cautiously since only a few countries have initiated a weapons program while part of the treaty.[23] Nevertheless, the evidence presented in figure 9.4 contradicts the safeguards argument.

Another way to assess the impact of the NPT is to examine whether nuclear cooperation conditions the relationship between treaty membership and beginning a weapons program. NPT membership was negatively correlated with pursuing the bomb in some of the tests conducted in chapter 7 (recall that the variable was insignificant in other tests). There are two plausible explanations for the negative and significant correlation.[24] The first is that the NPT "screens" participants and merely reinforces existing

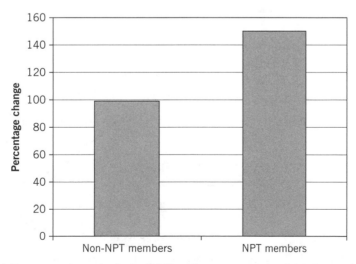

Figure 9.4 Percentage change in the probability of weapons program initiation resulting from increases in peaceful nuclear assistance, NPT members vs. non-NPT members
  Note: Calculations are based on model 4 (see appendix 9.1)

nonproliferation commitments. In other words, states ratify the treaty when they expect to comply. The second possibility is that the treaty "constrains" states, causing those that would otherwise have decided to begin a weapons program not to do so.[25] The safeguards argument would predict that the NPT reduces the likelihood of weapons pursuit for all states—even if they accrue large amounts of nuclear assistance. If this is not the case, it might be more appropriate to think of the NPT as a screen rather than a constraint.

Figure 9.5 illustrates the findings of an empirical analysis designed to address this issue. The figure plots the marginal effect of NPT membership on nuclear weapons program initiation as the number of comprehensive power NCAs increase.[26] In the absence of nuclear cooperation, the NPT has a strong negative effect on this stage of proliferation. Importantly, the statistical significance of the NPT washes away when the number of comprehensive power NCAs rises above fourteen for a country.[27] These results tell us that the NPT does not have the same effect on all states. Countries that do not receive nuclear assistance are highly unlikely to pursue nuclear weapons when they enter the treaty. But things change once they accrue atomic aid. Peaceful nuclear cooperation creates conditions in the recipient country that increase the likelihood of nuclear weapons pursuit, as I argued in chapter 7. These conditions appear to be powerful enough to swamp the effect of the NPT on proliferation decisions. To be clear, this does not mean that the NPT has not affected the way that states think about nuclear weapons programs.

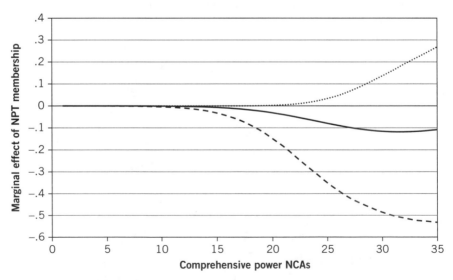

Figure 9.5 Marginal effect of NPT membership on nuclear weapons program initiation as comprehensive power NCAs increase

Note: Dotted lines represent 95 percent confidence interval. Calculations are based on model 4 (see appendix 9.1)

But on average, states that accrue nuclear aid are no less likely to begin a weapons program if they enter the NPT. This casts further doubt on the argument that safeguards have eliminated (or substantially reduced) the perils of peaceful nuclear cooperation for nuclear weapons program onset.

Case Studies

This section explores the causal logic of this argument by examining the decision-making context in the cases of Syria and Japan. The Syrian case is meant to illustrate the weaknesses of IAEA safeguards and underscore why the regime has not weakened the connection between peaceful nuclear cooperation and nuclear weapons programs to a greater extent. My objective in analyzing the Japanese case, on the other hand, is to determine whether there is support for the causal logic of the pro-NPT argument in an "easy" context. From a methodological standpoint, it is difficult to show that the nonproliferation regime has *caused* nuclear restraint. As Richard Betts points out, to show that safeguards and the NPT have prevented proliferation scholars must show that a country "would have sought nuclear weapons . . . but refrained from doing so, or was stopped, because of [the] treaty."[28] If such a country exists, Japan is a likely candidate.[29] Both neoliberals and constructivists routinely cite the case of Japan as evidence that the NPT has made a difference.[30]

SYRIA: THE INADEQUACY OF SAFEGUARDS

Syria ratified the NPT in 1969 but it did not conclude a comprehensive safeguards agreement with the IAEA until 1992; Damascus has yet to accept the AP as of December 2011. With one Chinese-supplied research reactor and relatively few trained scientists and technicians, its civilian nuclear development today is roughly on par with Vietnam's but substantially less than South Korea's in the 1970s or Pakistan's in the 1980s. Syria has attempted to secure additional atomic aid over the last few decades but many suppliers have not been enthusiastic about assisting Damascus in developing a civil nuclear program.[31]

Some analysts suspected that Syria would build nuclear weapons in response to Israel's acquisition of the bomb in the late 1960s. However, based on the dataset I employed in chapter 7, Syria did not initiate a nuclear weapons program from 1945 to 2000.[32] There is some debate regarding whether Syria pursued the bomb in more recent years. Certainly many of Syria's recent nuclear activities are consistent with the behavior of a proliferator. But it is unclear whether Syria's post-2000 behavior meets the threshold for nuclear weapons pursuit that I adopted in chapter 7; I was unable to locate definitive evidence in the available public record that officials at the cabinet level or above made a political decision to pursue the bomb.[33] This comports with recent proliferation research and declassified assessments made by the U.S. government since the 1990s. For instance, during an intelligence briefing in April 2008, an official indicated that the United States only had "low confidence" that nuclear material produced in the Syrian reactor was intended for a weapons program.[34]

Either way, there is strong evidence that NPT-based safeguards cannot account for Syria's nuclear decision making. On at least two occasions since 1992, Damascus violated its comprehensive safeguards agreement with the IAEA. Syria secretly built the al-Kibar reactor with North Korean assistance, in violation of its commitment under the NPT to notify the IAEA of nuclear facilities under construction within its borders.[35] The IAEA was unaware that the reactor existed until it was bombed by Israel in September 2007, before construction was complete. The al-Kibar raid reveals that Israel had little confidence that the current safeguards regime would deter Syrian nuclearization. Other countries likely shared this view, as evidenced by the silent international response to the strike. Unlike in 1981, when Israel bombed a safeguarded Iraqi reactor, few leaders openly condemned the 2007 military raid and the UN Security Council did not take action against the attacking state.[36] Recognizing that the strike represented a lack of confidence in safeguards, the IAEA reacted negatively to Israel's use of military force. Director General Mohamed ElBaradei said after the attack: "It is deeply regrettable that information concerning this installation was not provided to the Agency in a timely manner and that force was resorted to unilaterally before the Agency was given an opportunity to establish the facts, in accordance with its responsibilities under the NPT and Syria's Safeguards Agreement."[37]

Following the al-Kibar raid, the IAEA immediately demanded to inspect the site. Efforts were complicated, however, by Syria's obstinacy. To begin, Syria bulldozed what was left of the facility immediately after the strike and refused to acknowledge that Israel had destroyed a nuclear reactor. Damascus claimed that al-Kibar was a military installation that had no nuclear-related purpose, implying that the IAEA did not have the authority to visit the site. President Bashar al-Assad proclaimed that evidence suggesting that a nuclear facility had been destroyed was "fabricated 100 percent."[38] Despite this initial resistance, Syria eventually invited a team of IAEA inspectors to visit the country in June 2008. Environmental samples taken at al-Kibar revealed uranium particles that had been altered as a result of chemical processing, suggesting the presence of a nuclear facility.[39]

Irregularities were also detected at the Chinese-supplied research reactor. In particular, samples at this facility revealed the presence of uranium particles that had been modified by human activities; Damascus' prior declarations to the IAEA did not suggest that this type of uranium existed at the site.[40] These revelations suggest that Syria may have used some of the natural uranium intended for the al-Kibar reactor to conduct reprocessing experiments at the research facility, in violation of the country's safeguards agreement.[41] This is especially telling given that Syria's research reactor is under IAEA safeguards. The reporting discrepancies indicate that Damascus believed that the agency would not detect the altered uranium or that it would not enforce the safeguards agreement in the event that the material was uncovered.

After evidence of these suspicious activities emerged, Syria stonewalled the IAEA. The agency has requested visits to three additional sites that might be related to al-Kibar and has repeatedly called on Syria to provide documentation on the past use of the North Korean-supplied facility. It has also requested information related to Syria's procurement activities.[42] On all of these fronts Syria has been uncooperative. As stated in an August 2009 report issued by the IAEA Board of Governors, the agency's ability to verify compliance with the NPT "is severely impeded because Syria has not provided sufficient access to information, locations, equipment, or materials."[43] Not surprisingly, Damascus has continued to deny any wrongdoing and to assert that international condemnation of its actions is unfounded. As Suleiman Haddad, chairman of the Syrian parliament's foreign affairs committee, stated: "America has tried and tried to put pressure on Syria. . . . We will not respond to this pressure. We are in full cooperation with the IAEA to prove to the world that Syria has nothing to hide."[44]

I do not assume that Syria's violation of its safeguards agreement indicates beyond a reasonable doubt that it had an active nuclear weapons program. But these events do show that IAEA safeguards did not deter Syria from misreporting activities that took place at its research reactor or from building a covert nuclear facility. Damascus's uncooperative behavior toward the IAEA following the Israeli strike further underscores the perception that safeguards were not a real constraint in the eyes of Syrian decision makers. The IAEA depends on cooperation from the state it is investigating and lacks the power to enforce the NPT in the absence of such support.[45] In the end, enforcement responsibilities typically fall to individual countries such as the United States, Russia, or Israel. Since Damascus had already experienced a military raid of its principal nuclear facility, there was little the IAEA—or any state—could do to coerce Syrian cooperation.

It is telling that the events discussed above occurred after the IAEA revamped its safeguards regime with the AP in 1997. The problem from the standpoint of nonproliferation is that Syria has refused to sign this protocol. Consequently, the IAEA does not have the authority to inspect Damascus's undeclared facilities.[46] Recall that the AP allows IAEA inspectors to behave like detectives, piecing together information on a state's nuclear program from a variety of sources. In theory this increases the likelihood that covert nuclear facilities will be detected. The detection of these facilities often depends, however, on member countries sharing information with the IAEA that is obtained by intelligence agencies. States may understandably be reluctant to share sensitive intelligence with an international organization because doing so could compromise their sources and methods. The United States, for instance, shared information with the IAEA on the al-Kibar site—but it did so more than six months after the Israeli strike. In the aftermath of the attack, the United States and Israel reportedly "went to great lengths to prevent others from finding out where the site was," presumably because they wanted to protect their intelligence sources and methods.[47] The lack of information sharing stymied the IAEA's efforts to obtain a complete picture of Syria's nuclear activities.[48] This example underscores that the agency might not have detected the al-Kibar site even if Syria had ratified the AP, although it would likely have increased the probability of this outcome.[49] Even if we assume that the AP increased the likelihood of detection, the protocol would only make a difference if this reduced Syria's willingness to build a covert facility. Given Syria's recent defiance of the IAEA, it is by no means obvious that the possibility of getting caught would have deterred Syria from violating the NPT. Damascus would have feared the prospect of a military strike from Israel in the event that suspicious activities were exposed. But this ultimately happened after Israeli and American intelligence discovered al-Kibar, regardless of Syria's status outside the AP.

If safeguards and the NPT had little effect on Syria's nuclear decision making, what explains nuclear restraint prior to 2000? One factor that played a critical role was Syria's possession of chemical weapons. Nuclear weapons and chemical weapons are not equivalent; clearly the former are far more destructive then the latter. But in the 1970s Damascus began to view chemical weapons as a substitute for the bomb, especially in light of the technical limitations associated with producing nuclear weapons. President Hafez al-Assad thought that chemical weapons could reduce Israel's willingness to initiate militarized disputes by neutralizing its superior military capabilities.[50] In other words, chemical weapons could provide benefits that often come from possessing the bomb. With this in mind, Syria began manufacturing chemical weapons in the late 1970s.[51] There is some evidence that Syria's possession of chemical weapons produced the intended strategic payoffs. Former Israeli prime minister Yitzhak Rabin noted in 1994, for instance, that a strategic stalemate had emerged between Israel and Syria and that this was partially attributable to the latter state's chemical weapons capability.[52] Syrian president Bashar al-Assad, the son of the former leader, reaffirmed this when he stated in 2003 that "Syria is entitled to defend itself by acquiring a chemical and biological deterrent . . . it is not difficult to get most of these weapons anywhere in the world and they can be obtained any time."[53] Implicit in the younger Assad's statement is the notion that chemical weapons possession can lead to some of the same benefits as nuclear weapons for substantially lower costs.

JAPAN: AN NPT POSTER CHILD?

A secret Japanese study conducted in 1967 concluded that "Japan could produce the material necessary to make an atomic bomb by extracting plutonium from its civilian nuclear power plants."[54] The argument I advanced in
chapter 7 suggests that Japan should have drawn on its civil nuclear program
to augment its military capabilities after the Chinese nuclear test in 1964 or
the North Korean nuclear crisis in the 1990s. Although Tokyo has debated
the nuclear option at various points since the late 1960s, it never made a political decision to build nuclear weapons, making this case an outlier for my
theory of weapons program initiation.[55] This case is also an outlier for other
proliferation arguments, which is one reason why it has attracted considerable scholarly attention.[56]

Japan signed the NPT in 1970 and ratified it in 1976; it also became one
of the first states to ratify the AP in 1999. Given these commitments, Tokyo
submits to demanding international inspections of its nuclear facilities that
reportedly account for nearly one-third of the IAEA's budget.[57] Some have
argued that nuclear safeguards and the NPT more generally can explain why
atomic assistance did not trigger a proliferation decision. Yet, as I will show
below, the causal mechanisms driving the safeguards argument do not operate correctly in the Japanese case.

The first mechanism outlined above suggests that the IAEA safeguards
regime deterred Tokyo from drawing on its expansive civilian nuclear infrastructure for military purposes because there were perceived costs associated with cheating. Japanese policymakers occasionally make statements
that seem to support this assertion.[58] A former senior Japanese official indicated in a July 2003 interview, for example, that Japan has not pursued
nuclear weapons because "we would have to break our agreements with the
NPT [and] the IAEA" and this would cause us to "lose all credibility in the
international community."[59] Scholars likewise suggest that the NPT played
a decisive role in the case of Japan because of the IAEA's monitoring regime. John Endicott exemplified this well when he stated: "There is no other
peaceful nuclear power program so thoroughly examined by the IAEA as
Japan. . . . I do not know how fissile materials in Japan would be channeled
into weapons program without the world knowing."[60]

The evidence cited above, however, does not necessarily support the causal
logic of this mechanism. Drawing too heavily on pro-NPT statements made
by Japanese officials is also problematic because it is politically prudent to
praise the treaty in public, even if officials do not believe that it constrains
their behavior. Indeed, known violators of the NPT make statements nearly
identical to the ones cited above. For instance, Ali Reza Moaiyeri, a senior
Iranian diplomat, highlighted the importance of the NPT in an April 2008
speech and went on to state that his country would not build nuclear weapons because "Iran's support for the NPT is unequivocal and we continue to

honor our obligation."[61] Since officials from *all* states routinely praise the treaty, it is questionable that such statements offer definitive support for pro-regime arguments.

Moreover, praise for the NPT is not universal among Japanese elites. When Japan signed the NPT in 1970 (Tokyo ratified the treaty in 1976), it explicitly acknowledged its right to withdraw in the event that "the supreme interests of the nation" necessitated the pursuit of nuclear weapons.[62] More recent statements from policymakers underscore that the proliferation barriers associated with IAEA safeguards are not the principal reason for Japanese nuclear restraint. In 2002, for example, Ichiro Ozawa, an influential opposition leader, noted that Japan has "plenty of plutonium in our nuclear power plants, so it's possible for us to produce 3,000 to 4,000 nuclear warheads," implying that safeguards would not necessarily deter nuclearization.[63] This is consistent with U.S. assessments reached after Japan ratified the NPT. For example, a declassified National Security Council (NSC) document on the potential proliferation risks of Japan's civilian nuclear program stated, "Even technically perfect safeguards cannot fully meet the problems posed by reprocessing."[64] This evidence does not prove that the NPT has not mattered in the Japanese case, of course. It does, however, suggest that scholars should be cautious when interpreting pro-NPT rhetoric.

There are three additional reasons why this first mechanism did not play a decisive causal role in the Japanese case. First, Japan's key decision to exercise nuclear restraint preceded the entry into force of the NPT. Prime Minister Eisaku Sato announced Japan's "Three Non-Nuclear Principles" in December 1967—nearly a decade before Japan ratified the NPT. These principles, which hold that Japan will not manufacture, possess, or deploy nuclear weapons on its soil, remain the foundation of the country's nuclear policy today. In early 1968, Prime Minister Sato commissioned a study to "explore the costs and benefits of Japan's nuclearization in a comprehensive way."[65] This study concluded that it would not be prudent for Japan to pursue nuclear weapons and solidified Tokyo's decision to remain nonnuclear. Given the timing of these events, it would be difficult to conclude that things would have played out any differently if the NPT had never been created. As Etel Solingen notes, "The decision to remain non-nuclear was prior to, not a consequence of, the decision to ratify the NPT."[66]

Second, other factors are more salient than the NPT in explaining why Japan has not initiated a weapons program. Most important, the U.S.-Japan Peace Treaty, which was concluded in September 1951, provides Tokyo with nuclear protection. When China's 1964 nuclear test and the escalation of the Vietnam War fueled a domestic debate on the benefits of an independent nuclear deterrent, Prime Minister Sato repeatedly asserted that the U.S. nuclear umbrella provided sufficient defense of Japan.[67] The 1968 classified study similarly concluded that Japanese nuclear weapons were unnecessary to defend against Chinese aggression in light of Tokyo's alliance with Washington.[68] Moreover, Japan calculated that nuclear weapons would make the country vulnerable to attack because its high population density and relatively small geographic size undermined the logic of mutually assured destruction (MAD).[69] The pursuit of the bomb would also strain Tokyo's relationship with Washington and perhaps fuel an arms race in East Asia. As a classified 1995 study conducted by the Japanese Defense Agency indicated, "[if Japan builds nuclear weapons,] the reliability of the U.S. nuclear umbrella would be undermined and Japan would be viewed as distrustful of its military alliance with the United States; [and] neighbors would fear that Japan was taking a more independent defense policy stance."[70] Given these dangers, the report concluded that Japan could best maximize its security by continuing to rely on the U.S. nuclear deterrent.[71]

Domestic constraints have also contributed to Japan's nuclear policies, although to a lesser degree. The Japanese constitution places limits on the country's development of "offensive" weapons, and some have argued that this helps explain Tokyo's nuclear restraint.[72] However, Japanese officials have maintained that possessing a low-yield nuclear weapon would not violate the constitution if such a device was required for self-defense and if it did not pose an offensive threat to other states.[73] Perhaps a more significant domestic constraint is the lack of public support for a weapons program, which emerges partially because Japan is the only country to have atomic bombs used against it.[74] Domestic support for proliferation has increased slightly in recent years but the overwhelming majority of the population continues to oppose nuclear weapons. In the immediate aftermath of North Korea's 2006 nuclear test, one poll found that roughly 18 percent of the respondents believed that Japan should rethink its commitment to nonproliferation.[75]

Third, it is by no means obvious that NPT-based safeguards would have deterred Japan from initiating a weapons program in the event that strategic and domestic conditions had been more conducive to the development of a nuclear arsenal.[76] Tokyo has emphasized that international and domestic constraints do not prevent proliferation, even if they raise the political and economic costs of nuclearization. Indeed, recent statements from policymakers underscore that the IAEA safeguards or the NPT more generally would not necessarily impede a proliferation decision if the public became more supportive of an independent nuclear deterrent and/or if the security environment worsened. Chief Cabinet Secretary Fukuda Yasuo stated, for instance, that "under Japanese law, there is no reason to prevent Japan from arming itself with nuclear weapons. . . . If public opinion agrees with nuclear armament, the denuclearization principle can be revised."[77] Even Takemi Keizo, a legislator who opposes nuclear proliferation, argued in 2004 that "Japan's possession of nuclear weapons would be in sight" in the event that the country's alliance with the United States deteriorated.[78] These statements do not imply that Japan will build nuclear weapons any time soon; most analysts and policymakers conclude that nuclearization would not be a prudent option for Tokyo under existing conditions.[79] But if these conditions change, it is unclear whether the safeguards regime would stand in the way of Japanese proliferation. Rather than violate its NPT commitments, Japan would likely withdraw from the treaty if it decided to pursue the bomb, as the country's 1970 policy statement implies.

The second causal mechanism emphasized above implies that the safeguards regime provides countries with assurances that others would not draw on civil nuclear programs to pursue nuclear weapons. There is scant evidence to support this in the Japanese case. When Tokyo debated the merits of the NPT in the early 1970s, some policymakers believed that the treaty could effectively identify and punish cheaters—but this perception quickly eroded.[80] Safeguards did not deter Japan's principal rivals from developing nuclear weapons. China and the Soviet Union acquired the bomb prior to the creation of the NPT and they are entitled to possess weapons under the terms of the treaty.[81] South Korea pursued nuclear weapons in the 1970s, even after it committed to the NPT. Perhaps most important, North Korea drew on peaceful nuclear cooperation to acquire the bomb despite being a member of the NPT until 2003, as I will discuss later in this chapter. Pyongyang's activities deeply damaged Japan's confidence in the IAEA's safeguards regime and the NPT more generally. After North Korea withdrew from the treaty, Japanese officials pointed to "the necessity of enhancing the IAEA's capability of safeguards activities," implying that they believed that the current regime was inadequate.[82] This belief was only magnified following Pyongyang's nuclear tests in 2006 and 2009.

In light of this evidence, it would be difficult to argue that safeguards alleviated Japan's security concerns with respect to its greatest adversaries. Even scholars who argue that the NPT made a difference for Japan acknowledge that its "faith in the NPT is not because policymakers believe that it will definitely keep other states from pursuing a nuclear option."[83]

Japan wants to believe in the NPT but it does not appear to have much confidence in the critical grand bargains of the treaty. It was disappointed with the lack of sustained international pressure against India and Pakistan following their 1998 nuclear tests. Iran's development of secret uranium enrichment facilities also alarmed policymakers in Tokyo. These events reinforced a fear that states could defy norms embodied in the NPT with few consequences.[84] U.S. actions have also contributed to this perception, although Japanese officials are sometimes reluctant to criticize Washington publicly due to the important alliance the two countries share. Particularly damaging is the view that the United States and the other four NPT nuclear weapons states have not done enough to meet their obligations under Article

VI to make "good faith efforts" towards the eventual elimination of nuclear weapons. The first point that Japan emphasized during its opening statement at the 2010 NPT Review Conference is that "the nuclear-weapon states [must] reaffirm an unequivocal undertaking to accomplish the total elimination of their nuclear arsenals."[85] If this does not happen the legitimacy of the treaty may be further undermined, which could cause Japan to reconsider its nuclear policy. As Joseph Cirincione argues, "Japan will seriously consider its nuclear options if it comes to believe that the United States and other nuclear-weapon states no longer have any intention of pursuing [its obligations] under Article VI of the NPT."[86]

The discussion up to this point has focused on the causal mechanisms rooted in neoliberalism. Recall that constructivists offer another reason for Japanese nuclear restraint. The nonproliferation regime could have weakened the perils of nuclear cooperation in this case by stigmatizing the pursuit or use of nuclear weapons. Based on the available evidence, however, this cannot account for Japanese nuclear decision-making. This argument suffers from many of the same problems as the neoliberal theory of NPT effectiveness.

Perhaps most damaging is the fact that Japan solidified its nonnuclear policy before signing the NPT, at a time when nuclear weapons were widely believed to enhance national prestige.[87] A related problem is that the constructivist argument implies that norms altered the way that Japan viewed nuclear weapons over time. Maria Rublee argues, for example, that Japan was "persuaded . . . that nuclear weapons are morally wrong and thus can never be considered as a legitimate political or military tool."[88] Yet Tokyo's rationale for not pursuing nuclear weapons remained remarkably consistent over time. Japan secretly considered pursuing nuclear weapons in the first time in the 1960s, before the NPT entered into force and prior to the stigmatization of nuclear weapons internationally. Tokyo again debated the merits of an independent nuclear deterrent in the mid-1990s, following the North Korean nuclear crisis. At this point, the "taboo" against nuclear weapons pursuit and use were well developed and the NPT had been in existence for twenty-five years. But the same factors drove Japan's nuclear policy in both the 1960s and the 1990s. The conclusion reached by the Japanese government in each instance was that it should not pursue nuclear weapons "so long as the credibility of the U.S. nuclear umbrella was maintained."[89] Since the principal justification for nuclear restraint remained constant even as the taboo emerged, norms stemming from the NPT could not have been a driving force behind Japan's decision making. Moreover, there is not compelling evidence that Japan's nuclear policy would be different if the NPT was never created, as I argued above. At best, the taboo mattered at the margins in a limited respect.

Safeguards, Weapons Acquisition, and
Peaceful Nuclear Cooperation

Has the safeguards regime made it more difficult for states pursuing nuclear weapons to successfully build the bomb? Given the small number of states that have pursued nuclear weapons, it is feasible to briefly examine each case and evaluate whether safeguards "worked." We already saw in chapter 8 that atomic assistance facilitated proliferation in many of the nuclear weapons programs that were active between 1945 and 2000. My aim here is to revisit these cases to assess whether safeguards constrained nuclear weapons programs. Specifically, I look for evidence that safeguards deterred countries from diverting nuclear materials from a peaceful program to a weapons program or from building covert nuclear facilities. I also evaluate whether the IAEA detected violations in a timely fashion if and when they occurred.

For the purposes of this analysis it is conceptually useful to categorize proliferators in three groups: (1) states that built nuclear weapons before the NPT entered into force; (2) states that successfully built nuclear weapons after the entry into force of the NPT; and (3) states that failed to produce nuclear weapons after the emergence of the NPT-based safeguards regime.

Case Studies

PRE-NPT NUCLEAR POWERS
Five countries—the United States, the Soviet Union, the United Kingdom, France, and China—built nuclear weapons before the emergence of the NPT. We obviously would not expect the safeguards regime to have influenced these cases. When China conducted its first nuclear test in 1964, for instance, the IAEA had not even established a facility-specific safeguards regime that covered power reactors. Moreover, as noted above, these five countries are entitled to possess nuclear weapons under the terms of the NPT and are subject to different obligations than nonnuclear weapons states.

POST-NPT NUCLEAR POWERS
Other states benefited from peaceful nuclear cooperation and managed to build the bomb after policies were put in place to separate the peaceful and military uses of the atom.[90] These cases are more troubling from the standpoint of nonproliferation because it was theoretically possible for the safeguards regime to have limited weapons acquisition. There are five countries that fall into this category: India, Israel, Pakistan, North Korea, and South Africa.[91]

In four of these cases, we would not expect that safeguards played a major role because the respective states were not part of the NPT and did not accept full-scope safeguards. Some of these non-NPT members did not even accept facility-specific safeguards on technology that ultimately aided a weapons program. Safeguards did not yet exist when Canada supplied the CIRUS reactor India, which New Delhi ultimately used to produce plutonium for its 1974 nuclear test (see chapters 4 and 7). The principal Israeli reactor at Dimona remains unsafeguarded and Britain and other suppliers exported critical nuclear materials without requiring safeguards.[92] South Africa's covert uranium enrichment program, which ultimately helped produce a small nuclear arsenal, was unsafeguarded until 1991, although Pretoria did accept safeguards on certain research and power reactors while it had an active bomb program. Facility-specific safeguards were more relevant in the Pakistani case. Islamabad accepted safeguards on the nuclear reactor it imported from Canada in the 1970s but these safeguards had little deterrent effect. Although Munir Khan initially hesitated to violate Pakistan's safeguards agreement, President Bhutto was prepared to divert plutonium from the civil reactor for use in nuclear weapons (see chapter 8).[93]

North Korea was part of the NPT while it had an active nuclear program but safeguards played little role in curtailing its weapon-related activities. Pyongyang's NPT commitment was disingenuous from the beginning; when North Korea entered the treaty in 1985 it was already pursuing nuclear weapons and had no intention of terminating the bomb program.[94] After dragging its feet, Pyongyang finally concluded a treaty-mandated safeguards agreement with the IAEA in January 1992.[95] It soon became clear that North Korea had an illicit reprocessing program that it had not declared to the IAEA. When the IAEA called for special inspections in February 1993 to investigate Pyongyang's reprocessing activities, North Korea threatened to withdraw from the NPT. The United States persuaded North Korea to remain in the NPT but it continued to defy the IAEA. In March 1994, Pyongyang refused to allow an inspection team to visit a plutonium reprocessing plant at Yongbyon and the IAEA Board of Governors passed a resolution demanding that Pyongyang comply with its safeguards agreement.[96] Three months later, Hans Blix, the IAEA's director general, frustratingly stated that North Korea's uncooperative actions had "seriously eroded" the agency's ability to determine whether North Korea remained in compliance with the NPT.[97]

On October 21, 1994, the United States reached a deal with North Korea known as the Agreed Framework that temporarily resolved the nuclear crisis. Pyongyang pledged to eliminate all of its nuclear facilities, including the reprocessing plant at Yongbyon, and accepted IAEA special inspections to verify the program's dismantlement. For its part, Washington provided limited security assurances to North Korea and agreed to transfer light water reactors for power production.[98]

Shortly after the conclusion of the Agreed Framework, Pyongyang initiated a covert uranium enrichment program. By mid-2002 North Korea had procured the materials for an enrichment facility, although this does not necessarily imply that it had a significant enrichment capacity at that time.[99] Plans to construct this site were not disclosed to the IAEA, but U.S. intelligence uncovered the facility and confronted the North Koreans about their enrichment efforts in October 2002. When the IAEA asked Pyongyang to explain its uranium enrichment program, North Korea expelled all IAEA inspectors from the country and restarted the plutonium production program. On January 11, 2003, North Korea officially withdrew from the NPT. At the time of its withdrawal from the treaty North Korea had likely produced a small nuclear arsenal; Pyongyang left little doubt about its capabilities when it conducted nuclear weapons tests in 2006 and 2009 using plutonium produced in the Yongbyon reactor, which was indigenously constructed drawing on training and other civil nuclear assistance from the Soviet Union (see chapter 8). North Korea's actions made it clear that it was not deterred by the prospect of its weapons program being revealed by the IAEA. Moreover, U.S. intelligence—not IAEA inspections—provided the most useful information about the country's noncompliance with its NPT obligations. An IAEA report issued in August 2003 summarized the limitations of safeguards in the North Korean case: "Since 1993, the Agency has been unable to fully implement the comprehensive safeguards agreement with [North Korea]. . . . The Agency has never been allowed . . . to verify the correctness and completeness of the DPRK's initial declaration of nuclear material subject to safeguards under that agreement."[100]

Not only were safeguards not helpful in constraining Pyongyang's nuclear weapons program, the nonproliferation regime unintentionally facilitated North Korea's ability to produce nuclear weapons. As Andrew Semmel, a U.S. representative to the NPT 2005 review conference, noted, North Korea used

> "apparent compliance with the Treaty to present a peaceful public image. That image and the claimed 'right' to a peaceful nuclear program were used to mask access to foreign help in building fissile material production facilities that could support a nuclear weapons capability. . . . [North Korea] was able to use its status as an NPT member in 'good standing' to divert attention from its real motivations and to facilitate foreign nuclear assistance."[101]

POST-NPT FAILED WEAPONS PROGRAMS

We might be more likely to find evidence in favor of the safeguards argument when analyzing post-NPT cases that ended in failure. There are five such cases: Argentina, Brazil, Iraq, Libya, and South Korea. One other state— Iran—still has active weapons ambitions but has yet to build the bomb.

Argentina and Brazil were not part of the NPT when they pursued nuclear weapons. Since they were not legally bound by the treaty's requirements, the NPT could not have influenced either state's behavior. As non-NPT members, both countries were eligible for facility-specific safeguards agreements but there is little evidence that this sheds light on their weapons program outcomes. Argentina's key nuclear facility, the Pilcaniyeu uranium enrichment plant, was not subject to facility-specific safeguards while Buenos Aires pursued nuclear weapons.[102] The flip side of the argument that safeguards deter states from drawing on civil facilities for military purposes is that countries pursuing nuclear weapons are more likely to use facilities to build nuclear weapons if they are unsafeguarded. That this did not occur is a bit of a puzzle for the safeguards argument. Brazil similarly possessed unsafeguarded nuclear facilities that were not used to build nuclear weapons.[103] Moreover, both Argentina and Brazil terminated their bomb programs primarily because the strategic incentives to proliferate subsided and domestic factors were less conducive to weaponization.

The other countries in this category pursued nuclear weapons while they were part of the NPT. Iran ratified the treaty in 1970 and reached a safeguards agreement with the IAEA in 1974. During the 1980s, Iran began pursuing nuclear weapons. Iranian elites did not believe that the NPT-backed safeguards regime could stymie its weapons program. Former president Akbar Hashemi Rafsanjani made his feelings about the treaty clear when he stated that "international laws are only scraps of paper."[104] These sentiments resulted in repeated violations of Iran's safeguards agreement, which were rarely discovered by the IAEA in a timely fashion. Tehran began a gas centrifuge uranium enrichment program in 1985 and a laser enrichment program in 1991 but it did not declare either program to the IAEA.[105] Iran invited the IAEA to inspect six nuclear facilities in the 1990s but Tehran's cooperation with the agency was halfhearted. The inspections were "inconclusive" and did not turn up evidence of either illicit enrichment program in part because of "Iran's refusal to allow full access" and permit environmental sampling.[106] U.S. intelligence revealed a secret enrichment facility at Natanz and a heavy water production facility at Arak in 2002. The IAEA later began an investigation and concluded in a June 2003 report that "Iran has failed to meets its obligations under its Safeguards Agreement with respect to the reporting of nuclear material, the subsequent processing and use of that material and the declaration of facilities where the material was stored and processed."[107]

In the immediate aftermath of these revelations Iran cooperated with the IAEA in a limited fashion to avoid possible political and economic sanctions. Iran pledged to suspend all uranium enrichment activities in February 2003 and it signed the AP ten months later. However, Iran later reversed the enrichment suspension and refused to ratify the AP. The IAEA Board of Governors voted to refer Iran to the U.N. Security Council in November 2005 because of Tehran's repeated safeguards violations. A February 2006 IAEA report aptly summarized the IAEA's sentiments: "It is regrettable, and a matter of concern, that . . . uncertainties related to the scope and nature of Iran's nuclear programme have not been clarified after three years of intensive Agency verification"[108] The report went on to state that without Iran's active cooperation, "the Agency's ability to reconstruct the history of Iran's past programme and to verify the correctness and completeness of the statements made by Iran . . . will be limited, and questions about the past and current direction of Iran's nuclear programme will continue to be raised."[109] The UN Security Council passed four resolutions between 2006 and 2011 sanctioning Iran for its nuclear weapon-related activities.

In September 2009 it was publicly revealed that Iran had a second covert uranium enrichment facility known as the Fordow Fuel Enrichment Plant (FFEP), located outside the city of Qom. Although this site had been under construction since 2002 it had not been declared to the IAEA. U.S. intelligence detected the construction of the FFEP in 2006 and Iran acknowledged the presence of the facility only after it became clear that Western governments could confirm that it existed.[110] These actions further underscore that Tehran was not deterred by the prospect of getting caught as a result of Agency safeguards. As of the end of 2011, Iran had yet to build nuclear weapons but there is little evidence that safeguards contributed to this result.

Iraq no longer has an active nuclear weapons program but the NPT safeguards regime played very little role in this outcome.[111] As I discussed in chapter 8, Iraq procured a small reactor from France known as Osiraq but the facility was destroyed by Israel in a surprise preventive strike in 1981. Baghdad's plan to use an IAEA-inspected facility to produce plutonium for nuclear weapons demonstrates that it did not view safeguards as a meaningful constraint.[112] A former Iraqi official indicated this when he noted that Baghdad had a "hidden agenda" with respect to the Osiraq facility and planned to "deceive and manipulate the [IAEA]."[113] In destroying the reactor, Israel also made it clear that it did not believe that safeguards were adequate to limit Iraq's nuclear weapons ambitions. Indeed, the strike represented a vote of no confidence in the safeguards regime. As Sigvard Edlund, the director general of the IAEA, stated in the aftermath of the raid, "The Israeli attack on Iraq's nuclear research center was also an attack on the Agency's safeguards."[114]

Iraqi elites did not believe that its safeguards obligations would affect its post-1981 plans to build covert uranium enrichment facilities. Saddam Hussein asked Ja-afar Dhiya Ja'afar, his senior nuclear advisor, "If we stay in the NPT, will it in any way hinder the clandestine nuclear program?" Ja'afar response was a vehement "no."[115] Most of the undeclared facilities that Iraq built throughout the 1980s were unknown to the IAEA and the

United States. Alarmingly, Iraq even operated an undeclared installation at the Tuwaitha Nuclear Research Center that was located within yards of an IAEA-monitored research reactor.[116] After invading Kuwait in August 1990, Saddam initiated a "crash program" to divert safeguarded highly enriched uranium for use in nuclear weapons.[117] Again, this demonstrates that Iraq was not deterred from violating its safeguards agreement.

In January 1991, the United States, backed by a broad international coalition, began a bombing campaign designed to expel Iraq from Kuwait. Following the U.S. coalition's quick and decisive victory, the U.N. Security Council passed Resolution 687, which created a special commission known as UNSCOM to verify the destruction of Saddam's chemical, biological, and nuclear weapons programs. UNSCOM destroyed many weapon-related facilities and made it more difficult for Iraq to reconstitute its nuclear weapons program.[118] At the same time, Saddam Hussein's intentionally created ambiguities about the weapons program in order to preserve his domestic political position.[119] Although the Iraqi case seems to suggest that inspections coupled with economic sanctions can be effective, it is important to remember that what happened in Iraq in the 1990s was a far cry from the normal NPT-based safeguards regime. It took an interstate war and an unprecedented postwar inspections regime to curtail the Iraqi nuclear weapons program.

Iraq's efforts exposed critical weaknesses in the NPT safeguards regime and led to the creation of the AP in the late 1990s. UNSCOM made it clear to the world that Iraq's nuclear program was far more sprawling than anyone—including the United States and the IAEA—realized prior to the war. As the Gulf War Air Power Survey alarmingly stated following the war, "we now know that the Iraqis' program to amass enough enriched uranium to begin producing atomic bombs was more extensive, more redundant, further along, and considerably less vulnerable to air attack than was realized at the outset of Desert Storm."[120] The NPT-backed safeguards regime was insufficient to expose covert weapon-related facilities in this case.

On the surface, we might expect to find the most support for the safeguards argument in the Libyan case. One seemingly obvious explanation for Libya's inability to produce nuclear weapons despite a thirty-three-year weapons program is that the nonproliferation regime frustrated Tripoli's efforts to successfully build the bomb. Yet we already saw in chapter 8 that Tripoli failed to produce the bomb primarily because it lacked an indigenous knowledge base, which typically is created as a result of civilian nuclear cooperation. There is evidence that IAEA safeguards cannot account for Libya's inability to produce nuclear weapons. For more than twenty years Tripoli engaged in violations of its safeguards agreement. In the early 1980s, Libya initiated research and development on uranium enrichment at the Tajoura nuclear complex and gained experience in the design and operation of centrifuge equipment. Between 1984 and 1990, it also separated small amounts of plutonium and conducted laboratory-scale uranium conversion experiments at Tajoura. None of these activities were reported to the IAEA, in violation of Libya's safeguards agreement.[121] These events underscore that the prospect of getting caught did not deter Tripoli from engaging in illicit activities at a safeguarded facility. Later, Libya began constructing a covert nuclear facility at al-Hashan without reporting its activities. All told, twelve sites were involved in Tripoli's undeclared nuclear activities.[122] The existence of so many undeclared installations demonstrates that Tripoli believed it could conceal facilities from the IAEA and/or that it did not fear that the agency would punish it if an installation was detected. Perhaps Qaddafi believed it could keep the al-Hashan site secret because he had not yet ratified an AP agreement, meaning that the IAEA did not have the authority to inspect undeclared facilities.

Despite the fact that many of the transgressions discussed above occurred at a safeguarded facility, they were never detected by the IAEA. What we know about these violations comes from Libya's own revelations after Qaddafi terminated the weapons program in December 2003. At that point, Tripoli declared its previous activities and the IAEA opened an investigation to explore the veracity and completeness of these disclosures. A report issued in February 2004 summarized the IAEA's key findings: "Over an extended period of time, Libya was in breach of its obligation to comply with the provisions of the Safeguards Agreement. . . . [These violations] are matters of the utmost concern."[123] This conclusion underscores that IAEA safeguards are not foolproof.[124] If a country is determined to build nuclear weapons, the Libyan case suggests that it is very difficult to ensure the peaceful uses of nuclear technology and materials.[125]

South Korea pursued nuclear weapons between 1970 and 1978. Like Iran, Iraq, and Libya, South Korea was part of the NPT while it had an active weapons program. Seoul signed the treaty in 1968 and ratified it seven years later.[126] South Korea's pursuit of nuclear weapons violated its principal NPT obligation, but the country did not commit wholesale violations of its safeguards agreement in the 1970s. When Seoul ratified the AP in 2004 it revealed that undeclared chemical enrichment experiments had taken place between 1979 and 1981 and laser enrichment experiments occurred in 2000. Further, South Korea indicated that it experimented with plutonium separation in the early 1980s.[127] These revelations were alarming but all of the activities took place after Seoul terminated the weapons program in 1978.

The South Korean violations highlight one element of my proliferation argument, even though this case is less damaging for the IAEA safeguards regime than some of the others that I previously discussed. The case reaffirms the notion that the accumulation of peaceful nuclear assistance can empower scientists and members of atomic energy commissions who might conduct weapon-related experiments and put pressure on political leaders to nuclearize. As one cogent analysis of South Korea's nuclear activities noted, "Once gained, it is hard to lose technical capacity or the taste for acquiring scientific knowledge, and technologies with weapons applications were now part of the institutional memory of the South Korean nuclear establishment."[128]

Why did South Korea refrain from drawing on peaceful nuclear assistance to augment its weapons acquisition efforts prior to 1979? One plausible explanation is that IAEA safeguards deterred Seoul from mimicking the behavior of countries such as India. Close examination of South Korean decision making, however, reveals that safeguards cannot fully account for nuclear restraint in this case. Political elites made it clear that South Korea could build nuclear weapons "whenever we want to," and that the safeguards agreement would not stand in the way of these efforts.[129] And as I underscored in chapter 8, Seoul refrained from severe safeguards violations and ultimately gave up the program altogether primarily because of U.S. pressure.[130]

The evidence presented in this chapter reveals that nuclear safeguards and the NPT more generally have had a fairly limited effect in reducing the perils of atomic assistance for nuclear weapons proliferation. Why has the regime not made more of a difference? What implications does this have for the looming renaissance in nuclear power? I address these important questions in the conclusion of the book.

Appendix 9.1: Data Analysis and Empirical Findings

The statistical analysis employed in this chapter examined whether nuclear safeguards reduced the connection between atomic assistance and nuclear weapons program initiation over time. The analysis of nuclear weapons acquisition relied solely on qualitative analysis.

Dataset, Variables, and Method

The dataset used for the analysis in this chapter is similar to the one used in chapter 7 (see appendix 7.1 for details). The dependent variable measures whether a state made a political decision to begin a nuclear weapons program based on the Singh and Way dataset.[131] As noted previously, I measure nuclear cooperation as the total number of comprehensive power NCAs a state has signed as of year $t - 1$. Note, however, that I occasionally code this variable differently to assess the sensitivity of the findings. I include all of the independent variables from table 7.3.

To test the two implications of the safeguards argument, it is necessary to code three additional variables. The first measures the difference between year t and 1945. The second is an interaction term between this time variable and comprehensive power NCAs. This variable, along with the two constituent parts, allows me to assess whether the relationship between atomic assistance and weapons program onset has changed over time.[132] The third new variable is an interaction term between NPT membership and power agreements, which is included to test whether atomic aid affects treaty members and nonmembers differently. Mirroring procedures from chapter 7, I use probit and rare events logit to estimate the statistical models.

Results

Table 9.1 displays the results of the statistical analysis designed to test whether the effect of atomic assistance on weapons programs declines over time. These findings were previously illustrated in figure 9.2.[133] Models 1–3 include the interaction term between time and comprehensive power NCAs, along with all of the relevant independent variables. Model 1 is estimated using rare events logit, while model 2 is estimated using probit. Model 3 uses an aggregated dependent variable that includes all nonsafety NCAs a state has signed between 1945 and year $t - 1$.[134]

Nuclear cooperation agreements are statistically related to nuclear weapons program initiation in all three models and the sign is in the positive direction. This indicates that when the year count is set to zero, NCAs increase the likelihood of nuclear weapons pursuit. On the basis of the information presented in table 9.1, we do not know whether this relationship is statistically significant after 1945.[135] We can turn to the information presented in figure 9.2 to obtain a more complete interpretation of the interaction term; the figure allows us to assess the statistical significance of the NCA variable across the entire range of the time variable.[136] The marginal effect of comprehensive power NCAs was statistically significant from across the entire range of the time variable (i.e., from 1945 to 2000), as we saw above. These findings are inconsistent with the implication of the safeguards argument, as I previously discussed.[137]

In terms of the control variables, the findings are substantively similar to those presented in chapter 7 (see especially table 7.3). The principal difference is that the two GDP variables behave as expected in models 1–3; these variables were often insignificant in chapter 7. It is also worth noting that the NPT variable is statistically significant in models 2–3 but insignificant in model 1. These findings are consistent with the conclusion that NPT membership reduces the likelihood of proliferation in some tests but that this result is sensitive to model specification.

Table 9.2 displays the findings with the interaction term between NPT status and NCAs. These results allow me to evaluate whether NPT membership conditions the relationship between atomic aid and weapons pursuit

Table 9.1 Determinants of nuclear weapons program initiation, with time interactions

	(1)	(2)	(3)
	Comprehensive power NCAs–rare events logit	Comprehensive power NCAs–probit	Nonsafety NCAs
Nuclear cooperation agreements	0.561**	0.245**	0.069*
	(0.254)	(0.111)	(0.040)
Years since 1945	−0.081	−0.029	−0.004
	(0.074)	(0.024)	(0.016)
NCAs * years since 1945	−0.001	−0.001	−0.001
	(0.005)	(0.002)	(0.001)
Military assistance	1.982	0.649	0.263
	(1.213)	(0.557)	(0.579)
Militarized interstate disputes	0.280***	0.167***	0.165***
	(0.090)	(0.035)	(0.034)
GDP per capita	0.001**	0.000***	0.000***
	(0.000)	(0.000)	(0.000)
GDP per capita squared	−0.000***	−0.000***	−0.000***
	(0.000)	(0.000)	(0.000)
Industrial capacity	0.812	0.214	0.311
	(1.014)	(0.318)	(0.271)
Rivalry	2.589**	0.982**	0.747**
	(1.302)	(0.399)	(0.292)
Nuclear allies	−0.294	0.010	−0.007
	(0.830)	(0.282)	(0.311)
Polity	0.074	0.026	0.019
	(0.054)	(0.020)	(0.018)
Democratization	−0.089	−0.031	−0.022
	(0.074)	(0.025)	(0.021)
Openness	0.008	0.005	−0.000
	(0.020)	(0.004)	(0.005)
Liberalization	−0.008	−0.003	−0.004
	(0.019)	(0.005)	(0.006)
NPT	−1.282	−0.766**	−0.588*
	(0.860)	(0.298)	(0.305)
No program years	0.011	0.017	0.010
	(0.304)	(0.105)	(0.097)

(*Continued*)

Table 9.1—*cont.*

	(1) Comprehensive power NCAs–rare events logit	(2) Comprehensive power NCAs–probit	(3) Nonsafety NCAs
Spline 1	0.001	0.000	0.000
	(0.003)	(0.001)	(0.001)
Spline 2	−0.002	−0.001	−0.001
	(0.003)	(0.001)	(0.001)
Spline 3	0.001	0.001	0.001
	(0.001)	(0.000)	(0.000)
Constant	−8.088***	−4.414***	−4.321***
	(2.107)	(0.784)	(0.741)
Observations	5519	5519	5519

Notes: Robust standard errors in parentheses; *** p<0.01, ** p<0.05, * p<0.1

Table 9.2 Determinants of nuclear weapons program initiation, with NPT interactions

	(4) Comprehensive research NCAs	(5) Non-safety NCAs
Nuclear cooperation agreements	0.309**	0.081**
	(0.142)	(0.037)
NPT	−1.527*	−1.355
	(0.856)	(0.955)
NCAs * NPT	0.086	0.058**
	(0.068)	(0.027)
Military assistance	1.083	0.868
	(1.237)	(1.270)
Militarized interstate disputes	0.343***	0.308***
	(0.096)	(0.087)
GDP per capita	0.001	0.001
	(0.000)	(0.000)
GDP per capita squared	−0.000***	−0.000**
	(0.000)	(0.000)
Industrial capacity	0.568	0.825
	(0.884)	(0.810)

Table 9.2—*cont.*

	(4) Comprehensive research NCAs	(5) Non-safety NCAs
Rivalry	2.418*	2.041*
	(1.361)	(1.113)
Nuclear allies	−0.222	−0.284
	(0.831)	(0.925)
Polity	0.078	0.063
	(0.055)	(0.053)
Democratization	−0.079	−0.063
	(0.064)	(0.056)
Openness	0.002	−0.003
	(0.017)	(0.016)
Liberalization	−0.003	−0.007
	(0.016)	(0.017)
No program years	−0.004	−0.006
	(0.315)	(0.270)
Spline 1	0.000	0.001
	(0.002)	(0.002)
Spline 2	−0.001	−0.001
	(0.002)	(0.002)
Spline 3	0.001	0.001
	(0.001)	(0.001)
Constant	−9.050***	−8.312***
	(2.239)	(1.938)
Observations	5519	5519

Notes: Robust standard errors in parentheses; *** $p<0.01$, ** $p<0.05$, * $p<0.1$

and vice versa. Both models displayed in the table are estimated using rare events logit. Model 4 measures nuclear cooperation based solely on comprehensive power NCAs while model 5 uses the aggregated NCA variable that includes all nonsafety agreements.

Some interesting findings emerge from this analysis. The NCA variable is positive and statistically significant in both models, indicating that atomic assistance is positively correlated with weapons program onset when a state is not part of the NPT. The coefficient on the variable measuring whether a state has ratified the NPT is negative and statistically significant in both

models. Thus, when states have not signed a single NCA, the NPT reduces
the likelihood of nuclear weapons pursuit. The interaction term is positive
and statistically significant in model 5 but insignificant in model 4. The information presented in figures 9.4 and 9.5 allowed us to obtain a more complete understanding of these findings. We saw in figure 9.4 that NPT states
are more likely to initiate weapons programs following increases in nuclear
cooperation, compared to states that remain outside the treaty. I calculated
the marginal effect of comprehensive power NCAs on weapons pursuit as the
value of the NPT variable increases from 0 (nonmembership) to 1 (membership), along with the 95 percent confidence interval. The confidence interval
did not include zero at either data point, indicating that nuclear assistance is
correlated with proliferation among both NPT members and nonmembers of
the treaty. Figure 9.5 illustrated how this effect changed as nuclear assistance
increases.[138] Recall that the statistical significance of the NPT washes away
after the number of comprehensive power NCAs rises. In other words, as
states accumulate nuclear assistance over time, NPT membership no longer
predicts whether or not it will pursue nuclear weapons. The findings with
respect to the control variables are substantively similar to the results presented in table 7.3.

I noted above that we should interpret these results cautiously because a
small number of countries pursued nuclear weapons while part of the NPT.
Iran and Iraq both accumulated foreign assistance and made political decisions to pursue the bomb after they had entered the treaty. It is reasonable to
wonder whether one of these cases is driving the findings presented in table
9.2. I replicated model 4 with Iraq excluded from the estimation sample
and the findings (not reported) were substantively similar. The results were
likewise similar when I removed Iran from the dataset.[139]

Conclusion

What Peaceful Nuclear Assistance Teaches Us about International Relations

This book has broadly addressed the use of economic statecraft to achieve foreign policy objectives and the ways in which attempts to influence the behavior of other states can have unintended consequences for international security. It has analyzed three specific questions relating to civilian nuclear cooperation: Why do nuclear suppliers provide peaceful nuclear assistance to other countries? Does peaceful nuclear assistance raise the likelihood of nuclear weapons proliferation? And, have international institutions influenced the nuclear marketplace and mitigated the potential perils of atomic assistance?

I argued that politico-strategic factors drive the nuclear marketplace. Countries provide atomic assistance to enhance their political influence by strengthening recipient countries and improving their bilateral relationships with those states. In particular, suppliers use aid to reinforce their allies and alliances, to forge partnerships with enemies of enemies, and to strengthen existing democracies (if the supplier is also a democracy). Suppliers also barter nuclear technology for oil when they are worried about their energy security. Statistical tests and qualitative historical analysis lent support to this argument.

The export practices of nuclear suppliers inadvertently contribute to the spread of nuclear weapons. I argued that states generally seek atomic aid with benign intentions but circumstances change over time as civilian nuclear programs advance. Civilian assistance leads to technological breakthroughs that reduce the expected costs of a weapons program and empower members of bureaucracies, who sometimes lobby elites to develop the bomb. Atomic aid is especially likely to contribute to a proliferation decision when states experience militarized conflict with other countries. In the end, states are able to draw on dual-use technology and know-how to augment their nuclear weapons programs. A multimethod assessment of this argument revealed that countries that receive assistance in developing a civilian nuclear program are more likely than states that do not receive assistance (or receive relatively less aid) to initiate nuclear weapons programs and acquire the bomb.

International institutions have mattered less in shaping the connection between the peaceful and military uses of the atom than many people believe. The Nuclear Suppliers Group has played a fairly limited role in shaping states' export practices. And while the nuclear Nonproliferation Treaty and the International Atomic Energy Agency have achieved some successes, they have done less to curb proliferation than is commonly assumed.

The arguments and evidence made in the book challenge conventional thinking about peaceful nuclear assistance and nuclear weapons proliferation. More generally, this book has implications for central issues in international politics relating to economic statecraft, international institutions, the diffusion of military power, and the nexus between energy and security.

Economic Statecraft and Foreign Policy

Scholars have maintained for hundreds of years that international commercial activities affect war and peace. Plutarch wrote in AD 100, for example, that trade brings "cooperation and friendship" while the termination of economic relations causes life to be "savage and destitute."[1] Yet relatively few books have systematically linked economics and security, in part because the former is sometimes dismissed as "low politics." This book underscores the value of linking economic statecraft and international security. Indeed, it shows that peaceful nuclear assistance—which is sometimes studied within political economy—is an important foreign policy tool that governments employ to manage their relationships with friends and foes.

The instrument of economic statecraft studied in this book is unique because of the linkages between the peaceful and military uses of the atom. Many people think that countries provide civilian nuclear assistance to promote nonproliferation objectives. Simply put, countries that abide by nonproliferation rules are thought to be rewarded with help in developing a peaceful nuclear program. States that break the rules, on the other hand, are denied this benefit. International reaction to the recent U.S.-India nuclear deal (see chapter 4) illustrates that this conception of nuclear cooperation is firmly ingrained in the minds of academics and practitioners. Promoting nonproliferation is not a major reason for providing peaceful nuclear assistance, however. Nuclear suppliers have repeatedly demonstrated that they are willing to export technology to states with poor nonproliferation records if doing so allows them to achieve political or strategic objectives such as strengthening their alliances, countering the influence of threatening states, or enhancing their energy security. This behavior undermines one of the critical grand bargains of the NPT.

Suppliers are not oblivious to the rules of the nuclear marketplace or the proliferation risks associated with their export practices. They are apt to downplay these things, however, because the perceived benefits of atomic assistance trump the possible dangers for nonproliferation. Governments might also be risk acceptant when it comes to civilian nuclear assistance partially because of their time horizons—that is, how far they look into the future when making decisions. They may be willing to engage in actions that seem reckless from a nonproliferation standpoint because they can reap politico-strategic benefits in the short term. Indeed, aid can help achieve goals of this nature fairly quickly. The relationship between proliferation and peaceful nuclear assistance, in contrast, is more like a marathon than a sprint. The accidental contribution to the spread of nuclear weapons, if it occurs, is more likely to materialize in the distant future perhaps well after the leader who agreed to provide assistance is out of office. Consider Canada's nuclear cooperation with India, which I discussed in chapter 4. Canada transferred a reactor to India in the 1950s and reaped political benefits shortly after consummating the deal. Then, in 1974, India used this civilian assistance to test a nuclear explosive device. Although aid ended up being devastating for nonproliferation, Canada (and the international community more generally) did not experience strategic costs until nearly two decades after it first occurred.

Similar to a poker table, the nuclear marketplace is characterized by calculated gambles. Sometimes these gambles pay off. Yet, on average, things do not work out well for nuclear exporters.[2] The Faustian bargain that governments accept when they enter the nuclear marketplace adds to our understanding of risk taking in international politics. In the end, peaceful nuclear assistance is a classic example of accepting long-term losses in pursuit of short-term gains.[3]

Nuclear Weapons Proliferation and the Diffusion of Military Power

This book, like some before it, analyzes the effects of economic statecraft. Many scholars judge whether tools "work" based on their intended outcomes. Economic sanctions that were intended to end South Africa's apartheid policies, for example, might be considered successful if they curtailed racial discrimination in that country. On the contrary, I do not evaluate the intended effects of peaceful nuclear assistance. Instead, this book explores the unintended effects of economic statecraft. It shows that while nuclear suppliers may reap politico-strategic benefits from providing peaceful nuclear assistance they also inadvertently contribute to the spread of nuclear weapons.

This insight adds to our understanding of nuclear proliferation dynamics. Peaceful nuclear assistance usually does not lead to proliferation immediately but, over time, aid creates conditions in the recipient country that raise the risk of weapons pursuit and acquisition. This happens even though virtually all countries capable of transferring nuclear technology are committed to nonproliferation and many find the spread of nuclear weapons the most pressing threat to their national security. In this respect, the way that nuclear weapons spread is tragically ironic.

The connection between atomic assistance and nuclear proliferation is surprisingly broad. To the extent that policymakers worry about nuclear cooperation as a proliferation risk, they focus on uranium enrichment centers and plutonium reprocessing facilities. NSG guidelines, for instance, recommend restricting exports of these plants because they can be used directly to produce fissile material for weapons. This book's findings suggest that this narrow conception of the linkage between atomic assistance and proliferation is misguided. Indeed, NCAs that explicitly prohibit transfers of enrichment centers, reprocessing facilities, and heavy water reactors are about as proliferation prone as similar agreements that do not prohibit this type of cooperation.

There is, however, a limit to the proliferation potential of peaceful nuclear cooperation. The international community should be most worried about the treaties that are designed to help a state operate or expand a nuclear energy program. Assistance geared toward nuclear research is less worrisome, but it is not entirely innocuous. As I noted above, atomic aid often contributes to proliferation by developing an indigenous knowledge base that enables states to build weapon-relevant facilities. Historically, comprehensive power agreements—the broadest type of NCA—have been the most efficient in fostering scientific competence in nuclear engineering and related fields. But more limited agreements could be effective in promoting nuclear knowledge in the future. Moreover, research reactors could be a "gateway technology" in that having a nuclear research program could cause states to seek more sophisticated technology and know-how relevant to electricity production. Indeed, many of the states currently operating nuclear power plants imported research reactors before receiving more advanced technology from foreign suppliers.

Most research downplays the supply side of nuclear proliferation or ignores it altogether. Scholars should take greater stock of the supply side. By concentrating on foreign assistance and technological capacity, we can obtain a more complete understanding of how and why nuclear weapons spread. Atomic aid can contribute to proliferation in both obvious and nuanced ways. Yet the supply side is just part of the proliferation equation. This book shows that the connection between atomic assistance and proliferation is especially strong if external security threats arise or persist. It is fruitful, therefore, to integrate the supply side and the demand side of nuclear proliferation.[4]

Scholars and policymakers who focus on the supply side emphasize trade activities intended to foster proliferation, like those associated with the Pakistani-based A. Q. Khan network.[5] These may raise cause for concern but a heavy focus on this type of assistance is problematic. It leaves policymakers with the impression that a few rogue countries such as Pakistan and North Korea are the only suppliers contributing to the spread of nuclear weapons. This is, unfortunately, an incomplete perception. Nuclear transactions intended exclusively for peaceful purposes also raise the risk of proliferation. These activities are conducted under the auspices of the NPT by suppliers that are generally thought to play by the rules, such as the United States, Canada, and Germany. Peaceful atomic assistance is substantially more common than illicit military aid. As noted throughout the book, countries signed nearly fifteen hundred NCAs from 1945 to 2000, compared to eight cases where suppliers transferred technology or materials explicitly for military purposes. Furthermore, military assistance is correlated with nuclear weapons acquisition but not weapons program onset, whereas peaceful assistance is associated with both of these outcomes.

To be clear, this does not mean that policymakers should ignore illicit proliferation assistance. The Khan network had consequences for international security and a similar network could emerge in the future. But states should not obsess about military assistance while neglecting the dangers of peaceful nuclear cooperation.

Beyond nuclear proliferation, this book adds to our understanding of the diffusion of military power in international politics.[6] Scholars and policymakers have long recognized that commercial technologies can be used to augment military capabilities (and vice versa).[7] Indeed, defense planners in the United States and elsewhere often seek to integrate civilian and military industrial bases in order to reduce the costs of developing major weapons systems.[8] Yet civilian technologies can provide the basis for military power even when states do not intend to intentionally exploit the dual use dilemma. This book serves as a reminder that military technologies may spread unintentionally as a result of the international spread of civilian technology and knowledge.

Energy and Security

The intersection between energy and security received considerable attention in the aftermath of the energy crisis of the 1970s. In the ensuing decades, however, scholars in security studies devoted much less attention to the causes and consequences of states' energy policies. Research on this subject continued, but this work occurred in other fields such as economics and in other subfields in political science. This book has shown that we can better understand issues of critical importance for international security by examining peaceful nuclear assistance. In doing so, it provides one example of how exploring the nexus between of energy and security in international relations can be fruitful. But this book just scratches the surface when it comes to this linkage; future research on the connections between energy and international security would be most welcome.

International Institutions

The book speaks to important debates in international relations regarding international institutions. An early debate focused on whether institutions matter. Realist scholars asserted that institutions are epiphenomenal, meaning they have little independent effect on state behavior.[9] Institutions, in other words, cannot prevent states from behaving as "power maximizers." Alternatively, proponents of liberal institutionalism argued that institutions can change state behavior because they alter incentives for states to cheat, lower transaction costs, and provide focal points for cooperation.[10] This makes cooperation possible in circumstances where the anarchic international system otherwise would not allow it.

Scholars later added nuance to this debate by elucidating the conditions under which institutions influence state behavior. One perspective, known as the "management approach," contends that states generally comply with their international commitments but noncompliance may emerge due to limited state capacity.[11] Oran Young exemplified this argument when he argued that "the effectiveness of international institutions varies directly with the capacity of the governments of members to implement their provisions."[12] Others see compliance as an enforcement problem. States make decisions regarding compliance based on a rational calculus and choose to cheat when the benefits of noncompliance exceed the costs associated with detection. Therefore, compliance hinges on the ability of institutions to identifying cheaters through monitoring and the possibility of coercive sanctions.[13] Institutions that are designed poorly and are incapable of identifying or punishing cheaters will likely not affect the way that countries behave.[14] On the other hand, institutions that are highly effective when it comes to monitoring and enforcement can heavily influence outcomes in international politics.[15] A third perspective focuses on domestic politics rather than the design of international institutions.[16] States may comply with their commitments because victims of noncompliance provide enhanced monitoring capacity by sharing information about treaty violations, and because domestic constituencies raise the costs of cheating. The implication of this approach is that institutions can make a difference even when they are "weak" and lack the power to levy incentives and sanctions.[17] A final camp posits that commitment and compliance are inexorably linked.[18] Treaties merely "screen" participants—they do not "constrain" them. We observe compliance with treaty commitments because states generally commit when they expect to comply. Like the standard realist argument, this perspective questions whether institutions have an independent causal effect on state behavior. But unlike realists, these scholars emphasize selection effects rather than power.

This book weighs in on these debates by examining the efficacy of the nuclear nonproliferation regime. This regime—particularly its anchor, the NPT—has received substantial praise over the last four decades. Some experts have hailed the NPT as the "most successful treaty ever devised."[19] It earned this title in large part because the number of states that possess nuclear weapons today is much fewer than the number predicted by many observers prior to the creation of the NPT. In the early 1960s, President Kennedy famously warned that fifteen or twenty nations would have nuclear weapons by 1970. Since Kennedy's warning only six additional countries have crossed the nuclear weapons threshold—China, Israel, India, South Africa, Pakistan, and North Korea. The NPT has also been considered a success due to the number of states that have ratified the treaty; today only India, Israel, Pakistan, and North Korea remain outside the regime.

The findings presented here challenge this conventional view of the non-proliferation regime. I examined four elements of the regime: (1) the energy assistance provision of the NPT (Article IV); (2) the NSG's guidelines; (3) the no acquisition provision of the NPT (Article II); and (4) the safeguards requirements embodied in the NPT (Article III) and implemented by the IAEA. In these four respects, the nonproliferation regime has failed or been largely ineffective. This does not mean that the regime has failed in all respects, of course. I did not examine every article of the NPT because some of them fell outside the scope of this book. For instance, Article VI of the treaty requires the nuclear weapons states to "pursue negotiations in good faith on effective measures relating to cessation of the nuclear arms race at an early date and to nuclear disarmament." It is possible that this disarmament provision influenced state behavior given that a number of arms control agreements have been signed and the overall number of nuclear weapons in the world has been reduced considerably. This book also did not examine other nonproliferation institutions such as regional Nuclear Weapon Free Zone (NWFZ) treaties, the Proliferation Security Initiative (PSI), or UN Security Council Resolution 1540. Future research should examine these institutions so that we can obtain a more complete picture of commitment and compliance in the nonproliferation regime. It would also be useful to further scrutinize whether the NPT constrains or screens. One way to accomplish this is to systematically examine why states ratify the NPT and analyze whether the treaty remains correlated with nuclear weapons pursuit once we account for the conditions that motivated countries to commit in the first place.[20] If the findings presented in this book are any indication, we would expect the statistical significance of the NPT variable to wash away once we have accounted for this selection effect.

Why has the nonproliferation regime exerted such a limited effect on state behavior? The ineffectiveness of the regime does not necessarily stem from an inability of institutions to prevent states from behaving as "power maximizers," as a standard realist explanation might imply. It is more likely that we observe this outcome because the IAEA, NPT, and NSG are poorly designed with respect to monitoring and enforcement.

Two critical design-related weaknesses in the safeguards regime undermine its effectiveness. First, the IAEA—unlike some effective institutions— is not set up to levy harsh penalties against cheaters. The agency's mandate does not include enforcement; if the agency finds evidence that nuclear materials have been diverted it can report the violation to the agency's Board of Governors. The Board, in turn, has the authority to report noncompliance to the UN Security Council and call for the return of IAEA-provided materials and equipment.[21] The Security Council considers how to deal with noncompliant states on an ad hoc basis because there are no established norms related to enforcement.[22] The fifteen-member body can pass resolutions sanctioning states that violate safeguards agreements. Yet these resolutions are only effective when they have broad support from the international community, and this rarely happens. The United States, for instance, has had a difficult time convincing others—especially Russia and China—to support robust sanctions against Iran despite the evidence that emerged after 2003 suggesting that Tehran had violated its safeguards agreement. Compounding matters even further, some have argued that safeguards agreements are not always enforced evenhandedly at the Security Council.[23] Ultimately, the burden is on individual countries to enforce NPT safeguards agreements. The range of unilateral enforcement options includes the imposition of economic sanctions, the withdrawal of existing foreign aid, the termination of defense commitments, or the use of military force against nuclear installations. These policies can be effective but they are often applied inconsistently with varying degrees of political commitment.[24]

The second design-related weakness of the safeguards regime is that states must voluntarily submit to IAEA safeguards.[25] The agency has little authority to monitor the activities of states that remain outside the nuclear nonproliferation regime. This is less problematic today than in earlier decades because almost all states have committed to the NPT. Nevertheless, non-NPT members, including Israel, Pakistan, India, and North Korea, are not currently subject to full-scope safeguards.[26] The Additional Protocol substantially fortifies the safeguards regime but only enhances the IAEA's monitoring capabilities when states ratify the protocol. This is potentially problematic because many countries thinking about developing nuclear energy programs have yet to ratify the AP. Accordingly, it is unclear whether this arrangement will be sufficient to reverse historical trends.

A related problem is that Article X of the NPT gives states the authority to withdraw from the treaty if extreme circumstances arise as long as they provide ninety days' notice. This "escape clause" was included in the treaty to increase states' willingness to make an NPT commitment in the first place. The downside is that a state could join the NPT, legally accumulate peaceful nuclear assistance, and then withdraw from the treaty, as North Korea did in 2003. This is problematic because states can extract the benefits of NPT membership and later use atomic aid to augment a weapons program in the absence of IAEA safeguards. Today, leaders around the world fear that Iran might exercise its Article X right and withdraw from the NPT once it produces adequate material in its civilian nuclear program to build nuclear weapons, leaving the IAEA incapable of detecting a diversion. There are, of course, possible costs associated with withdrawal from the NPT. The point here is that the IAEA is institutionally incapable of monitoring a state's activities when it invokes the exit option. Moreover, it cannot prevent states from withdrawing from the NPT or impose penalties on those that do so, although the Board of Governors could refer the case to the UN Security Council.

These weaknesses represent a lethal combination. If states enter the NPT, they can conceivably commit violations without detection and penalty. Countries can also remain outside of the NPT/AP or withdraw from the treaty simply by providing three months' notice, depriving the international community of what little monitoring and enforcement capabilities it has. Both of these paths create problems that undermine the effectiveness of the NPT. Interestingly, the former path was taken most frequently by nondemocratic states such as Iran, Iraq, Libya, and Syria (see chapters 8 and 9). Democratic states that pursued nuclear weapons from 1970 to 2000 (e.g., India and Israel) pursued nuclear weapons outside of the NPT. This may be because democracies have greater respect for the rule of law, their domestic institutions are more transparent, or violating international commitments imposes greater audience costs.[27] Regardless, the institutional weaknesses of the nonproliferation regime created problems that applied to both democracies and authoritarian regimes, although perhaps to varying extents. Future research should devote greater attention to the issue, focusing on the factors that motivate states to violate the NPT and the conditions under which countries might withdraw from the treaty in the future.

The NSG's institutional weaknesses go a long way toward explaining why the cartel did not have a greater effect on the behavior of suppliers in the nuclear marketplace. As noted throughout the book, the NSG is an informal association of nuclear suppliers.[28] Its guidelines are advisory, meaning they are not legally binding, and the group does not have any authority with respect to verification or enforcement. The requirements of NSG membership are also fairly lax. Initially, the group did not even require full-scope safeguards as a precondition of atomic assistance; it only called on exporters to "obtain formal government assurances from recipients stating that items . . . will not be diverted to unsafeguarded nuclear fuel cycle or explosive activities."[29] In 1992 the organization revised its guidelines to discourage nuclear cooperation in the absence of full-scope safeguards, but even then such transfers were allowed in "exceptional cases."[30] These institutional characteristics make it relatively easy for states to change or ignore the NSG's guidelines, as the controversy surrounding the U.S.-India nuclear deal illustrates. It is impossible to know whether the NSG would have had a bigger effect on nuclear suppliers if it had been designed differently. But the group was not set up in a way that maximized this likelihood.

It is important to point out that the NSG probably would not have mitigated the perils of peaceful nuclear cooperation even if it had changed the way that exporters conducted their business. Requiring safeguards as a precondition of peaceful nuclear cooperation will only minimize proliferation risks if safeguards weaken the connection between atomic assistance and the spread of nuclear weapons. Yet safeguards had little independent effect on either nuclear weapons program onset or bomb acquisition. Ultimately, the shortcomings of the IAEA weakened the effectiveness of the NSG.

A simple implication emerges from the preceding discussion. Institutions that are weak with respect to monitoring and enforcement are likely to be ineffective in shaping state behavior. Soft law might matter in the areas of human rights and the environment, but we should not count on this same outcome when it comes to international security. Not only might ineffectively designed institutions fail to promote compliance, they might actually exacerbate the problems they were designed to address. Indeed, this book shows that institutions can have unintended consequences. A driving force behind the U.S. Atoms for Peace program and the creation of the IAEA was the idea that peaceful nuclear cooperation would reduce the risk of nuclear weapons proliferation. This belief turned out to be wrong. Encouraging the diffusion of nuclear technology and know-how may have, paradoxically, undermined one of the IAEA's core missions. Just as Albert Wohlstetter and his colleagues suggested in 1979, policymakers would do well to remember Florence Nightingale's 150-year-old advice to designers of hospitals: "The very first requirement in a hospital [is that] it should do the sick no harm."[31]

What if Eisenhower never delivered his Atoms for Peace address at the UN? What would the world look like from a nonproliferation standpoint if suppliers decided to embargo nuclear exports? It is impossible to know for certain, but this book provides some clues about the likely answer. Some proliferation would still have occurred since nuclear cooperation is not the only thing that enables nuclear weapons to spread. Yet many states that initiated a weapons program would have struggled to produce the bomb.[32] The United States and the Soviet Union demonstrated that it is possible to acquire the bomb in the absence of peaceful nuclear cooperation. Doing so usually requires high levels of development and a robust political commitment. Libya, for instance, pursued the bomb for more than thirty years and did not produce even one operational weapon, mostly because it only marginally benefited from atomic assistance. With sufficient political will, the most determined proliferators may have found a way to acquire the bomb—but it would have taken much longer without peaceful nuclear aid.

Policy Implications

Countries are exploring the development or expansion of nuclear energy for the first time in decades, a movement that some have labeled the "nuclear renaissance."[33] There is only one state in Africa with an operational nuclear power plant (South Africa), but eleven others are seeking to harness the peaceful uses of the atom for energy production. Bolivia, Chile, El Salvador, Peru, and Venezuela are among the nuclear energy aspirants in Latin America. Several European states, including Belarus, Croatia, Greece, and Poland, are considering the development of a civilian nuclear program. East Asian countries such as Indonesia, Malaysia, Myanmar, and Vietnam are likewise exploring this possibility. In the Middle East, just about every state has expressed interest in nuclear power. As I noted in the introduction of the book, it is unclear what will come of this renaissance in the aftermath of the March 2011 Japanese nuclear accident at Fukushima. Some countries will probably be deterred from building nuclear power plants as a result of the accident but others are likely to move forward. The problems discussed in this book are therefore unlikely to go away as a result of Fukushima.

States seeking to develop civil nuclear programs today universally maintain that they desire nuclear energy for civilian purposes. For example, Saudi Arabia's foreign minister, Prince Saud al-Faisal, asserted that his country's aim "is to obtain the technology for peaceful purposes, no more."[34] At this point, there may be little reason for the international community to doubt these claims. This book shows that, on average, nuclear programs begin with peaceful intentions and fears of nuclear hedging are overblown. The problem is that things change over time. Once a country accumulates foreign nuclear assistance, technological advances and the empowerment of nuclear bureaucracies can raise the risk of nuclear weapons program initiation. This is especially true if a country experiences international crises or militarized conflict with one of its neighbors.

If the status quo persists, the findings of this book suggest that the nuclear energy renaissance will probably lead to the further spread of nuclear weapons. It is impossible to predict with any certainty how many new nuclear powers will emerge as a result of the renaissance. But the evidence compiled in the book provides insights into the answer. Peaceful nuclear cooperation contributed to nuclear proliferation in the past, but it did not create a world in which dozens of countries possessed the bomb. States such as South Africa may not have initiated nuclear weapons programs had they not benefited from peaceful nuclear assistance beginning in the 1950s. Others, such as Pakistan, probably would have pursued nuclear weapons even in the absence of civilian nuclear assistance but aid allowed them to produce the bomb sooner than would otherwise have been possible. Yet the worst-case predictions made by President Kennedy and others did not materialize even as civilian nuclear assistance continued throughout the nuclear age.

So, the nuclear renaissance may produce a handful of additional states with nuclear weapons over the next few decades but it is unlikely to result in wholesale nuclear proliferation. It is instructive to consider the prospect of future nuclear proliferation in the Middle East. The region is highly unlikely to produce ten nuclear weapons states if the United States and other suppliers move forward with plans to provide civilian nuclear assistance. If history is any indication, however, it would not be surprising if large-scale atomic aid in the Middle East contributed to the onset of two or three new nuclear weapons programs. This is especially the case given that many of these countries face external threats and could experience wars or other types of militarized conflict in the relatively near future. This book has shown that the combination of peaceful nuclear assistance and security threats can be a recipe for nuclear proliferation.

What should governments do as more countries express interest in nuclear energy? Should they alter their export practices? If so, in what ways? Answers to these questions are far from simple, making it difficult to offer unequivocal recommendations.

How policymakers should respond depends, in part, on the degree to which they are willing to tolerate the further spread of nuclear weapons. John Mueller and others have suggested that the threat posed by nuclear proliferation is overblown and that nuclear weapons have mattered very little in international politics.[35] If a government official shares this perspective, he or she may conclude that urgent action is not needed to defend against the perils of peaceful nuclear cooperation. On the other hand, it would not be unreasonable to argue that the introduction of nuclear weapons in just one additional country would have adverse effects for national and international security. Policymakers who share this view may think it wise to reconsider some of their current policies on nonproliferation and peaceful nuclear cooperation, even though proliferation has proceeded at a relatively slow rate since the 1950s.

Given a choice between living in a world with nine nuclear weapons states or a hypothetical world with twelve nuclear powers, it is hard to imagine that U.S. policymakers would prefer the latter. The critical question is: How much time, energy, and resources, officials are willing to devote to curbing the spread of nuclear weapons?

My own view is that it would be unwise to allow the status quo to persist. There is a limit to what governments can and should do to alleviate the dangers of peaceful nuclear assistance, but there are some modest actions they could take to address the problem. Action is particularly desirable because the failures of the NPT identified in this book are eroding faith in the nonproliferation regime. Perceived noncompliance with Article IV is a source of dismay for the nonnuclear weapons states. At the 2010 NPT Review Conference, for example, Marty Natalegawa, the Indonesian minister of foreign affairs who spoke on behalf of the Nonaligned Movement said: "Article IV is explicit . . . and the NAM States Parties do not see any room for reinterpretation or setting of conditions for the peaceful uses of nuclear energy. The undue restrictions currently being applied to many developing countries Parties to the NPT is regrettable, and should be removed."[36] Venezuelan president Hugo Chavez echoed these sentiments when he recently stated, "It cannot be that the countries that have developed nuclear energy prohibit those of the third world from developing it."[37] Even U.S. allies, such as Turkey, are losing confidence in the NPT because suppliers have failed to aid the civilian nuclear programs of treaty members.[38] There is likewise a growing fear that the provisions of the treaty dealing with the nonacquisition of nuclear weapons (Article II) and safeguards (Article III) are untenable. It is easy to understand why this is the case given the findings presented in this book.

Peaceful nuclear cooperation aided proliferation in many historical cases by establishing an indigenous technological base that states could draw on to build covert facilities dedicated to weapons programs. The AP should help counter this strategy by enhancing the IAEA's ability to inspect undeclared facilities. But the protocol does not go far enough, as the previous discussion makes clear.

There are six measures that could further alleviate the problems identified in this book.[39] Most of these recommendations have been proposed by other scholars and policymakers. They are worth revisiting here because they flow from the book's central conclusions. These recommendations are strong enough to make a difference but modest enough to avoid some of the risks of strengthening international institutions.[40] In the end, they could reduce—although probably not eliminate—the perils of atomic assistance.

Preserve the Article IV Bargain

Peaceful nuclear assistance is an effective means of statecraft that suppliers can employ to strengthen their political and strategic interests. They should not necessarily cease using atomic aid in this manner. Yet they should avoid choosing grand strategy over nonproliferation when the two conflict. This practice is exemplified by the U.S.-India deal, but this book shows that it was alarmingly common throughout the nuclear age. Not only can this strategy lead to proliferation, it also undermines one of the central bargains of the NPT.

Limit Indigenous Knowledge

Nuclear cooperation inevitably produces trained scientists and technicians in the recipient country. Yet suppliers worried about proliferation might limit the extent to which an indigenous knowledge base blossoms as a result of atomic assistance. One way to accomplish this is to encourage countries to purchase "turnkey" facilities that are built exclusively with foreign assistance. From a nonproliferation standpoint, this is preferable to recipient states building reactors and other fuel cycle facilities indigenously using foreign supplied designs, technical support, and training. States that pursued the latter option (e.g., India) ended up with a more adept scientific community than those that primarily relied on the latter path (e.g., Libya). By promoting nuclear cooperation on a turnkey basis, suppliers might be able to frustrate the progress of a nuclear weapons program in the event that the recipient later makes a political decision to proliferate.[41]

Link the AP with Peaceful Nuclear Cooperation

The IAEA does not have the enhanced authority it needs if states refuse to ratify AP agreements with the agency. One way to entice countries to commit to the AP is by making ratification a precondition for the supply of nuclear technology, materials, and know-how. In other words, states would only receive peaceful nuclear assistance if they accepted enhanced IAEA inspections; those that refused to do so would be denied atomic aid. This might be effective in encouraging ratification of the AP because most states require foreign assistance to reap the benefits of a civilian nuclear program. Moreover, this policy would keep technology away from states that pose the greatest proliferation risks.

The challenge associated with this recommendation is to ensure that nuclear suppliers abide by it. A shared interest in limiting the spread of nuclear weapons would likely be insufficient to produce widespread compliance among all exporting countries. Indeed, history suggests that informal guidelines are not effective in changing the behavior of nuclear suppliers. Nor are commitments enshrined in treaties if they are vaguely worded and loosely enforced. If the international community were to adopt this recommendation, it should consider enacting measures to monitor the behavior of suppliers and levy penalties against those that provide nuclear assistance to a state that has not ratified the AP. Otherwise, suppliers might continue to supply nuclear technology to states with weak nonproliferation credentials if they are able to obtain strategic or political benefits.

Bolster Penalties for Safeguards Violations

One reason that states were not deterred from violating their safeguards agreements with the IAEA is that penalties are lax. This book showed that many proliferators believed that they could draw on peaceful nuclear activities—either directly or indirectly—and escape with little more than a slap on the wrist (assuming the IAEA even detected the violation). As U.S. Secretary of State Hillary Clinton said during her opening remarks at the 2010 NPT Review Conference, "Potential violators must know that they will pay a high price if they break the rules, and that is certainly not the case today. The international community's record of enforcing compliance in recent years is unacceptable."[42] Penalties for violations must be enhanced. Countries could, for instance, immediately suspend nuclear cooperation if the IAEA discovers a breach and the agency could similarly end all technical assistance to the violating state. They could also levy harsh economic sanctions against countries that flout their commitments. As the contemporary case of Iran illustrates, gaining multilateral support for such a response can be difficult. But finding ways to impose meaningful penalties is of utmost importance for the viability of the nonproliferation regime.

The perception that the rules are applied in a discriminatory fashion has undermined enforcement measures in the past. Therefore, the penalties outlined above must be applied evenhandedly and it would be best if they were automatically instituted following a violation. If a U.S. adversary (e.g., Venezuela) and a U.S. ally (e.g., Japan) commit the same violation, they should face the same penalty. This could quell fears that the rules can be ignored at the discretion of powerful countries. It would also eliminate uncertainty surrounding the expected costs of drawing on a civilian nuclear program for military purposes. If the penalties for violations are universally known in advance, it is more likely that the prospect of detection would deter transgressions (assuming the penalties are costly).

Make NPT Exit More Difficult

The "exit option" discussed earlier in this chapter is a well-known problem with the design of the NPT. The international community should not necessarily eliminate the escape clause from the treaty, as many states view it as a crucial protection of their sovereignty. Countries should, however, devise policies to weaken the dangers associated with Article X of the treaty by raising the costs of exit. Two policies are especially worth considering in this regard. First, states could require that those withdrawing from the treaty must return all technology and materials it received as a result of civilian nuclear cooperation.[43] Suppliers should be legally entitled to verify that reactors and other facilities they exported have been dismantled and rendered useless by conducting on-site inspections. Some bilateral NCAs signed in recent years include this type of provision. For example, the NCA signed by the United States and the United Arab Emirates (UAE) in December 2009 states that Washington can recover any "material, equipment or components . . . and any special fissionable material produced through their use" if the UAE abrogates its commitments.[44] This type of requirement is not typically included in NCAs, but it should become the universal standard and NSG guidelines should be revised accordingly. Although this policy could raise the costs of exit from the NPT, it would not solve issues dealing with nuclear know-how. Once scientists are trained in nuclear engineering and related fields, the knowledge they obtain cannot easily be taken away.

Second, withdrawals should automatically trigger a response from the UN Security Council. The expectation is that the UN would take swift and harsh action with broad multilateral support. Minor punishments and/or verbal reprimands would be insufficient to raise the costs of exit. This policy has growing international support, particularly from the IAEA and the United States. IAEA director-general Mohamed ElBaradei said in a 2004 speech, "I still believe that we need to have some response mechanism by the Security Council in case of withdrawal."[45] More recently, Secretary of State Clinton made the same point with greater enthusiasm: "We should . . . find ways to dissuade states from utilizing the treaty's withdrawal provision to avoid accountability. . . . we cannot stand by when a state committing treaty violations says it will pull out of the NPT in an attempt to escape penalties and even pursue nuclear weapons."[46] One way to prevent this, she went on to say, is "toughening enforcement against proliferation violations at the UN Security Council."[47]

Enhance the IAEA's Budget

The IAEA has failed to fully achieve its mandate, in part, because it is underfunded. ElBaradei recently hinted that this is the case when he stated that the agency's budget "does not by any stretch of the imagination meet our basic, essential requirements" and "our ability to carry out our essential functions is being chipped away."[48] This is troubling since the rise in demand for nuclear energy will increase the IAEA's requirements for safeguards and inspections. Countries must ensure that the agency has adequate resources to fulfill its mission.[49] Throwing money at a problem is rarely sufficient but in this case it should be part of the solution. The total budget for the IAEA should be at least doubled over the next ten years.[50]

Notes

INTRODUCTION

1. It is unknown exactly when de Villiers traveled to Argonne National Laboratory, but it was sometime between March 14, 1955, and September 1, 1961.

2. I use the words "civil" and "civilian" interchangeably throughout the book.

3. I explore this decision further in chapter 7.

4. Author's interview with Walter Kato, Cambridge, Mass., November 20, 2008.

5. Quoted in Frank Barnaby, *How to Build a Nuclear Bomb: And Other Weapons of Mass Destruction* (New York: Nation Books, 2004), 68.

6. In this book I use the terms "civilian nuclear cooperation," "peaceful nuclear assistance," and "atomic assistance" interchangeably.

7. Dwight D. Eisenhower, "Address by Mr. Dwight D. Eisenhower, President of the United States of America, to the 470th Plenary Meeting of the United Nations General Assembly," December 8, 1953.

8. Quoted in Bryan Walsh, "The Green Politics behind Nuclear Power," *Time*, February 16, 2010.

9. Steven Miller and Scott Sagan, "Nuclear Power without Nuclear Proliferation?" *Daedalus* 138, no. 4 (Fall 2009): 10.

10. Ibid.

11. See Bryan Early, "Strategies for Acquiring Foreign Nuclear Assistance in the Middle East: Lessons from the United Arab Emirates," *Nonproliferation Review* 17, no. 2 (2010): 259–80.

12. Matthew Fuhrmann, "Splitting Atoms: Why Do Countries Build Nuclear Power Plants?" *International Interactions* 38, no. 1 (2012). See also Matthew Fuhrmann, "Nuclear Inertia: How Do Nuclear Accidents Affect Nuclear Power Plant Construction? I Built a Giant Database to Find Out," *Slate*, April 26, 2011.

13. Jay Solomon, "Turmoil, Disasters Cloud Atomic Energy Pacts," *Wall Street Journal*, May 4, 2011.

14. See, for example, Kim Tae-gyu, "KEPCO Puts Safety First in Building UAE Nuclear Plant," *Korea Times*, March 30, 2011.

15. See, for example, Chaim Braun and Christopher Chyba, "Proliferation Rings: New Challenges to the Nuclear Nonproliferation Regime," *International Security* 29, no. 2 (Fall 2004): 5–49; Alexander Montgomery, "Ringing in Proliferation: How to Dismantle an Atomic Bomb Network," *International Security* 30, no. 2 (Fall 2005): 153–87; Gordon Corera, *Shopping for Bombs: Nuclear Proliferation, Global Insecurity, and the Rise and Fall of the A. Q. Khan Network* (Oxford: Oxford University Press, 2006); Matthew Kroenig, *Exporting the Bomb: Technology Transfer and the Spread of Nuclear Weapons* (Ithaca: Cornell University Press, 2010); and David Albright, *Peddling Peril: How the Secret Nuclear Trade Arms America's Enemies* (New York: Free Press, 2010).

16. Michael Mastanduno, *Economic Containment: CoCom and the Politics of East-West Trade* (Ithaca: Cornell University Press, 1992). See also, Lisa Martin, *Coercive Cooperation: Explaining Multilateral Economic Sanctions* (Princeton: Princeton University Press, 1992).

17. See, for example, Michael Wesley, "The Strategic Effects of Preferential Trade Agreements," *Australian Journal of International Affairs* 62, no. 2 (2008): 214–28.

18. David Baldwin, *Economic Statecraft* (Princeton: Princeton University Press, 1985), 4.

19. See, for example, Joanne Gowa, *Allies, Adversaries, and International Trade* (Princeton: Princeton University Press, 1994); Edward Mansfield, *Power, Trade and War* (Princeton: Princeton University Press, 1994); and Stephen Brooks, *Producing Security: Multinational Corporations, Globalization, and the Changing Calculus of Conflict* (Princeton: Princeton University Press, 2005).

20. See especially Jonathan Kirshner, *Currency Coercion: The Political Economy of International Monetary Power* (Princeton: Princeton University Press, 1995); Daniel Drezner, *The Sanctions Paradox: Economic Statecraft and International Relations* (Cambridge: Cambridge University Press, 1999); Lars Skalnes, *Politics, Markets, and Grand Strategy: Foreign Economic Policies as Strategic Instruments* (Ann Arbor: University of Michigan Press, 2000); Glenn Palmer and T. Clifton Morgan, *A Theory of Foreign Policy* (Princeton: Princeton University Press, 2006); and William Long, *Economic Incentives and Bilateral Cooperation* (Ann Arbor: University of Michigan Press, 1996).

21. See chapter 4 for further details.

22. Quoted in Allison Macfarlane, "Where, How, and Why Will Nuclear Happen? Nuclear 'Renaissance' Discourses from Buyers and Suppliers," unpublished manuscript, George Mason University, Fairfax.

23. On the value of foreign aid more generally, see Hans Morgenthau, "A Political Theory of Foreign Aid," *American Political Science Review* 56, no. 2 (June 1962): 301–9. See also Stephen Walt, *The Origins of Alliances* (Ithaca: Cornell University Press, 1987), 41.

24. See, for example, John Mueller and Karl Mueller, "Sanctions of Mass Destruction," *Foreign Affairs* 78, no. 3 (May–June 1999): 43–53.

25. See, for example, Scott Sagan, *The Limits of Safety: Organizations, Accidents and Nuclear Weapons* (Princeton: Princeton University Press, 1993); Scott Sagan and Kenneth Waltz, *The Spread of Nuclear Weapons: A Debate Renewed* (New York: W. W. Norton, 2002); and John Mueller, *Atomic Obsession: Nuclear Alarmism from Hiroshima to Al-Qaeda* (Oxford: Oxford University Press, 2009).

26. The full text of the speech is available on the White House website: http://www.whitehouse.gov/the_press_office/Remarks-By-President-Barack-Obama-In-Prague-As-Delivered/.

27. Albert Wohlstetter, Thomas Brown, Gregory Jones, David McGarvey, Henry Rowen, Vince Taylor, and Roberta Wohlstetter, *Swords from Plowshares: The Military Potential of Civilian Nuclear Energy* (Chicago: University of Chicago Press, 1979).

28. Other noteworthy work includes Gloria Duffy, "Soviet Nuclear Exports," *International Security* 3, no. 1 (1978): 83–111; Robert Boardman and James Keeley, eds., *Nuclear Exports and World Politics* (New York: St. Martins, 1983); William Potter, ed., *International Nuclear Trade and Nonproliferation: The Challenge of the Emerging Suppliers* (Lexington,

Mass.: Lexington Books, 1990); and Duane Bratt, *The Politics of CANDU Exports* (Toronto: University of Toronto Press, 2006).

29. See, for example, Scott Sagan, "Why Do States Build Nuclear Weapons? Three Models in Search of a Bomb," *International Security* 21, no. 3 (Winter 1996–97), 54–86; T. V. Paul, *Power versus Prudence: Why Nations Forgo Nuclear Weapons* (Montreal: McGill University Press, 2000); Jacques Hymans, *The Psychology of Nuclear Proliferation: Identity, Emotions, and Foreign Policy* (Cambridge: Cambridge University Press, 2006); and Etel Solingen, *Nuclear Logics: Contrasting Paths in East Asia and the Middle East* (Princeton: Princeton University Press, 2007).

30. Other supply side research includes Stephen Meyer, *The Dynamics of Nuclear Proliferation* (Chicago: University of Chicago Press, 1984); Matthew Fuhrmann, "Spreading Temptation: Proliferation and Peaceful Nuclear Cooperation Agreements," *International Security* 34, no. 1 (2009): 78–41; Matthew Fuhrmann, "Taking a Walk on the Supply Side: The Determinants of Civilian Nuclear Cooperation," *Journal of Conflict Resolution* 53, no. 2 (2009): 181–208; and Kroenig, *Exporting the Bomb*.

31. This does not imply that other research totally ignores peaceful nuclear assistance. The point is that this type of aid is rarely the principal focus of other work.

32. I discuss these cases further in chapter 1.

33. See, for example, Virginia Page Fortna, *Peace Time: Cease-Fire Agreements and the Durability of Peace* (Princeton: Princeton University Press, 2004); Xinyuan Dai, *International Institutions and National Policies* (Cambridge: Cambridge University Press, 2007); and Beth Simmons, *Mobilizing for Human Rights: International Law in Domestic Politics* (Cambridge: Cambridge University Press, 2009).

34. For recent examples of systematic research on the efficacy of the nonproliferation regime, see Hymans, *Psychology of Nuclear Proliferation*; Solingen, *Nuclear Logics*; Karthika Sasikumar and Christopher Way, "Paper Tiger or Barrier to Proliferation? What Accessions Reveal about NPT Effectiveness," unpublished manuscript, San Jose State University; and Daniel Verdier, "Multilateralism, Bilateralism, and Exclusion in the Nuclear Proliferation Regime," *International Organization* 62, no. 3: 439–76.

35. Quoted in Tom Gjelten, "Military Officers Tie Energy to National Security," *National Public Radio*, May 18, 2009.

36. Recent attempts to link energy and security include Michael Klare, *Blood and Oil: The Dangers and Consequences of America's Growing Dependence on Imported Petroleum* (New York: Metropolitan, 2004); Adam Stulberg, *Well Oiled Diplomacy: Strategic Manipulation and Russia's Energy Statecraft in Eurasia* (Albany: State University of New York Press, 2007); and Jeff Colgan, "Oil and Revolutionary Governments: Fuel for International Conflict," *International Organization* 64, no. 4 (2010): 661–94.

37. See, for example, Leonard Weiss, "Atoms for Peace," *Bulletin of the Atomic Scientists* 59, no. 6 (2003): 34–44.

38. Some scholars have referred to this dilemma as the "nonproliferator's quandary." William Lowrance, "Nuclear Futures for Sale: To Brazil from West Germany, 1975," *International Security* 1, no. 2 (1976): 147–66.

39. U.S. Congress, Office of Technology Assessment, *Nuclear Safeguards and the International Atomic Energy Agency*, OTA-ISS-615 (Washington, D.C.: U.S. Government Printing Office, June 1995), 27.

40. The other grand bargain is that the nuclear powers will make "good faith efforts" to work towards eventual nuclear disarmament.

41. Note, however, that Germany and Japan had nuclear weapons programs during World War II.

42. See chapter 8 for further details.

43. For an argument about how international commerce more generally serves as an instrument of alliance management, see Skalnes, *Politics, Markets, and Grand Strategy*.

44. Soviet nuclear cooperation with Libya in the 1980s and some of the other cases discussed in the book illustrate this point.

45. Initial findings reveal a positive correlation between nuclear cooperation and pursuing nuclear weapons. Later analysis shows, however, that this result emerges because some states in the dataset have little interest in developing a peaceful nuclear program.

CHAPTER 1

1. For instance, some have alleged that 100kg of highly enriched uranium was stolen from a facility in the United States in the 1960s and eventually transported to Israel. See Seymour Hersh, *The Samson Option: Israel's Nuclear Arsenal and American Foreign Policy* (New York: Random House, 1991).

2. Charles Perrow, *Normal Accidents* (Princeton: Princeton University Press, 1999).

3. See, for example, Samuel Walker, *Three Mile Island: A Nuclear Crisis in Historical Perspective* (Berkeley: University of California Press, 2004).

4. "Pakistan Nuclear Plant to Shut for Inspection," *Reuters*, October 1, 1993.

5. "China's Nuclear Cooperation Agreements," *Nuclear Threat Initiative*. Available at http://www.nti.org/db/China/nca.htm.

6. "Israel, U.S. to Share Nuclear Safety Research," *Reuters*, April 14, 2008.

7. See James Keeley, "A List of Bilateral Civilian Nuclear Cooperation Agreements." University of Calgary, 2003.

8. Ibid.

9. See United States, Office of Technology Assessment, *Technologies Underlying Weapons of Mass Destruction* (Washington, D.C.: U.S. Government Printing Office, 1993).

10. Gary Gardner, *Nuclear Nonproliferation: A Primer* (London: Lynne Rienner, 1994), 16.

11. Toni Johnson, "Global Uranium Supply and Demand," (Washington, D.C.: Council on Foreign Relations, 2007). Available at http://www.cfr.org/publication/14705/global_ura nium_supply_and_demand.html.

12. David Albright, "Civil Inventories of Highly Enriched Uranium." (Washington, D.C.: Institute for Science and International Security, 2004). Available at http://www.isis-online. org/global_stocks/old/civil_inventories_heu.html#table4.

13. Sam Roe, "U.S. Cold War Gift: Iran Nuclear Pact," *Chicago Tribune*, August 24, 2006.

14. Nuclear Threat Initiative, "Why Highly Enriched Uranium Is a Threat," Available at http://www.nti.org/db/heu/index.html#fn4; Ann MacLachlan, "Operators of Small Reactors to Meet to Discuss Conversion to LEU Fuel," *Nuclear Fuel*, April 25, 2005, 5; Judith Miller, "U.S. Is Holding Up Peking Atom Talks," *New York Times*, September 19, 1982; Michael Brenner, "People's Republic of China," in William Potter, ed., *International Nuclear Trade and Nonproliferation: The Challenge of the Emerging Suppliers* (Lexington, Mass.: Lexington Books, 1990), 253.

15. David Albright and Kimberly Kramer, "Stockpiles Still Growing," *Bulletin of the Atomic Scientists* 60, no. 6 (2004): 14–16.

16. David Albright, Frans Berkhout, and William Walker, *World Inventory of Plutonium and Highly Enriched Uranium, 1992* (Oxford: Oxford University Press, 1993).

17. Plutonium is one of two materials (the other being weapons-grade HEU) that can be used for the production of nuclear weapons.

18. United States, Energy Information Administration, "New Commercial Reactor Designs," November 2006. Available at http://www.eia.doe.gov/cneaf/nuclear/page/analysis/nu cenviss2.html.

19. See Gary Milhollin, "Heavy Water Cheaters," *Foreign Policy*, no. 69 (1987–88): 100–119.

20. Gardner, *Nuclear Nonproliferation*, 26.

21. M. D. Zentner, G. L. Coles, and R. J. Talbert, *Nuclear Proliferation Technology Trends Analysis* (Richland, Wash.: United States Department of Energy, Pacific Northwest National Laboratory, 2005), 60.

22. World Nuclear Association, "Research Reactors," May 2007. Available at http://www. world-nuclear.org/info/inf61.html.

23. *Nuclear Power Reactors in the World* (Vienna: International Atomic Energy Agency, 2007).

24. For a list of these components, see World Nuclear Association, "Nuclear Power Reactors," June 2008. Available at http://www.world-nuclear.org/info/inf32.html.

25. The two most common enrichment methods are the gaseous diffusion and gas centrifuge methods. Joseph Cirincione, Jon Wolfsthal, and Miriam Rajkumar, *Deadly Arsenals: Nuclear, Biological, and Chemical Threats*. 2nd ed. (Washington: Carnegie Endowment for International Peace, 2005); and Zentner, Coles, and Talbert, *Nuclear Proliferation Technology Trends Analysis.*

26. Gordon Corera, *Shopping for Bombs: Nuclear Proliferation, Global Insecurity, and the Rise and Fall of the A. Q. Khan Network* (Oxford: Oxford University Press, 2006), 23, 62.

27. David M. Spooner, Assistant Secretary of Commerce for Import Administration, Testimony before the Senate Committee on Energy and Natural Resources, March 5, 2008. Available at http://www.ita.doc.gov/press/speeches/spooner_030508.pdf.

28. Matthew Kroenig, *Exporting the Bomb: Technology Transfer and the Spread of Nuclear Weapons* (Ithaca: Cornell University Press, 2010).

29. Ibid, 10–11.

30. On the France-Israel case, see Avner Cohen, *Israel and the Bomb* (New York: Columbia University Press, 1998). On the Soviet Union–China case, see John Lewis and Litae Xue, *China Builds the Bomb* (Stanford: Stanford University Press, 1988).

31. David Albright and Corey Hinderstein, "Algeria: Big Deal in the Desert?" *Bulletin of the Atomic Scientists* 57, no. 3 (2001): 45–52.

32. Corera, *Shopping for Bombs*; David Albright and Corey Hinderstein, "Unraveling the A. Q. Khan and Future Proliferation Networks," *Washington Quarterly* 28, no.2 (2005): 111–28.

33. "Background Briefing with Senior U.S. Officials on Syria's Covert Nuclear Reactor and North Korea's Involvement," April 24, 2008. Washington: Office of the Director of National Intelligence.

34. Mark Mazzetti and David E. Sanger, "Israeli Airstrike Reignites Debate on Syrian Nuclear Ambitions," *New York Times,* September 23, 2007.

35. The United States also provided covert assistance to Pakistan beginning in 2001 to improve the safety and security of Islamabad's nuclear arsenal. See David Sanger and William Broad, "U.S. Secretly Aids Pakistan in Guarding Nuclear Arms," *New York Times*, November 18, 2007.

36. Margaret Gowing, *Independence and Deterrence: Britain and Atomic Energy, 1945–1952* (London: Macmillan, 1974); and John Simpson, *The Independent Nuclear State: The United States, Britain, and the Military Atom*, 2nd ed. (London: Macmillan, 1986).

37. Richard Ullman, "The Covert French Connection," *Foreign Policy*, no. 75 (1989): 3–33.

38. See Keeley, "Bilateral Civilian Nuclear Cooperation Agreements."

39. Matthew Kroenig in Christoph Bluth, Matthew Kroenig, Rensselaer Lee, William Sailor, and Matthew Fuhrmann, "Correspondence: Civilian Nuclear Cooperation and the Proliferation of Nuclear Weapons," *International Security* 35, no. 1 (2010): 189.

40. There were roughly fifteen hundred agreements signed from 1945 to 2000. So, even if one could identify one hundred treaties that were cancelled, this would represent less than 7 percent of all agreements signed during this period. For further details, see Matthew Fuhrmann in Bluth et al., "Correspondence," 195–96.

41. Kroenig in Bluth et al., "Correspondence," 189–90.

42. Kroenig, *Exporting the Bomb*, 162.

43. Jamaica received little more than a research reactor from the United States while India received research reactors, power plants, nuclear materials, extensive training, and fuel cycle assistance from a host of different suppliers.

44. A small number of NCAs authorize cooperation in more than one of the following areas: safety, intangibles, and nuclear materials. I adopt the following coding rules to classify such agreements. I classify agreements that authorize both safety and intangibles as intangible NCAs. I code agreements that authorize the transfer of nuclear materials and call for cooperation in nuclear safety as nuclear materials NCAs. Likewise, I classify agreements that deal with both materials and intangibles as nuclear materials NCAs. Finally, a small number of deals authorize the exchange of materials or equipment as part of a technical exchange program. I also code these deals as intangible NCAs since the "recipient" may not maintain the materials and/or equipment following the conclusion of the technical exchanges. Note that I employ these coding rules in fewer than thirty cases because most of these NCAs deal exclusively with one area of assistance.

45. Virginia Page Fortna, *Peace Time: Cease-Fire Agreements and the Durability of Peace* (Princeton: Princeton University Press, 2004).

46. Keeley, "Bilateral Civilian Nuclear Cooperation Agreements." See also James Keeley, "Coding Treaties: Examples from Nuclear Cooperation," *International Studies Quarterly*, 29, no. 1 (1985): 103–8.

47. Another appendix that lists the sources I consulted to code the variables included in this dataset is not included in the book due to space constraints. But it is available from the author upon request.

48. If one were to adopt a loose definition of peaceful nuclear assistance there may be upwards of two thousand NCAs. See Keeley, "Bilateral Civilian Nuclear Cooperation Agreements."

49. This number does not count cases where the same two countries signed more than one NCA in a given year.

50. This figure codes NCAs based on whether they entitle a state to receive nuclear assistance. It does not represent NCAs where a state commits to supply—but not receive—atomic aid.

51. I divide states into one of these regions based on the standard Correlates of War codings.

52. Keeley, "Bilateral Civilian Nuclear Cooperation Agreements."

53. Keeley excludes the following: (1) agreements that are explicitly defense-related; (2) financial agreements; (3) agreements dealing solely with agricultural or industrial agreements not related to nuclear power; (4) agreements dealing with the leasing of nuclear material; (5) liability agreements; and (6) multilateral agreements.

54. I exclude agreements for this reason only if an independent source casts doubt on whether an NCA was signed. The absence of information on a particular agreement in primary and secondary sources is not sufficient grounds for exclusion.

55. The list of excluded NCAs is available from the author on request.

56. NCAs that involve EURATOM or the Belgo-Luxembourg Economic Union are classified as bilateral agreements between each state party to these arrangements and the other party to the NCA. From 1952 to 1972, EURATOM members included Belgium, France, West Germany, Italy, Luxembourg, and the Netherlands. Denmark, Ireland, and the United Kingdom joined in 1973. Greece, Portugal, and Spain joined in 1986. Austria, Finland, and Sweden joined in 1995. As the name implies, the Belgo-Luxembourg Economic Union includes Belgium and Luxembourg.

57. For more on this and related challenges associated with coding NCAs, see Keeley, "Coding Treaties."

58. Potter, *International Nuclear Trade and Nonproliferation*.

59. Dong-Joon Jo and Erik Gartzke, "Determinants of Nuclear Weapons Proliferation," *Journal of Conflict Resolution* 51, no. 1 (2007): 1–28.

CHAPTER 2

1. See, for example, T. V. Paul, "Chinese-Pakistani Nuclear/Missile Ties and the Balance of Power," *Nonproliferation Review* 10, no. 2 (2003): 21–29; and Matthew Kroenig, *Exporting the Bomb: Technology Transfer and the Spread of Nuclear Weapons* (Ithaca: Cornell University Press, 2010).

2. Recall that by definition peaceful assistance is meant strictly for nonmilitary purposes.

3. For a theoretical discussion on foreign policy signaling see James Fearon, "Signaling Foreign Policy Interests: Tying Hands Versus Sinking Costs," *Journal of Conflict Resolution* 4, no. 1 (1997): 68–90.

4. The Correlates of War's Composite Index of National Capability (CINC) is based in part on a state's energy consumption. See J. David Singer, Stuart Bremer, and John Stuckey, "Capability Distribution, Uncertainty, and Major Power War, 1820–1965," in *Peace, War, and Numbers*, ed. Bruce Russett (Beverly Hills, Calif.: Sage, 1972), 19–48.

5. Mohamed ElBaradei, "Nuclear Power and the Global Challenges of Energy Security," World Nuclear Association Annual Symposium, London, September 6, 2007. Available at http://www.iaea.org/NewsCenter/Statements/2007/ebsp2007n012.html.

6. Steve Herman, "India's Dilemma: How to Continue Economic Growth While Reducing Carbon Emissions," *Voice of America,* November 22, 2007. Available at http://www.voanews.com/english/archive/2007-11/2007-11-26-voa18.cfm.

7. Joanne Gowa makes a similar argument regarding international trade and military power. See Joanne Gowa, *Allies, Adversaries and International Trade* (Princeton: Princeton University Press, 1994).

8. Quoted in Daniel Poneman, *Nuclear Power in the Developing World* (London: George Allen and Unwin, 1982), 85.

9. Hans Morgenthau, "A Political Theory of Foreign Aid," *American Political Science Review* 56, no. 2 (June 1962): 301–9. See also Stephen Walt, *The Origins of Alliances* (Ithaca: Cornell University Press, 1987), 41.

10. Walt, *Origins of Alliances,* p. 43.

11. Nuclear power also reduces fossil fuel emissions, making it an environmentally friendly energy source that could provide a partial solution to the problem of global warming. See, for example, William C. Sailor, David Bodansky, Chaim Braun, Steve Fetter, Bob van der Zwaan, "Nuclear Power: A Nuclear Solution to Climate Change?" *Science* 288, no. 5469 (2000): 1177–78.

12. Daniel Poneman, "Nuclear Policies in Developing Countries," *International Affairs* 57, no. 4 (1981): 568–84.

13. Quoted in Duane Bratt, *The Politics of CANDU Exports* (Toronto: University of Toronto Press, 2006), 32–33.

14. Larry Rohter and Juan Forero, "Venezuela's Leader Covets a Nuclear Energy Program," *New York Times*, November 27, 2005.

15. Boris Johnson, "We Need Nuclear Power and a New Generation of Boffins," *Telegraph* [London], March 2, 2006.

16. Quoted in Daniel Yergin, "Ensuring Energy Security," *Foreign Affairs* 85, no. 2 (2006): 69.

17. Erica Strecker Downs, *China's Quest for Energy Security* (Santa Monica, Calif.: RAND, 2000).

18. Quoted in Robert Lieber, "Energy, Economics and Security in Alliance Perspective," *International Security* 4, no. 4 (1980): 142.

19. United Arab Emirates, "Policy of the United Arab Emirates on the Evaluation and Potential Development of Peaceful Nuclear Energy," April 2008.

20. Anya Loukianova, "Belarus Takes a Second Look at Nuclear Energy," *Bulletin of the Atomic Scientists* (online edition), July 9, 2008. Available at http://www.thebulletin.org/web-edition/features/belarus-takes-a-second-look-nuclear-energy.

21. Matthew Fuhrmann, "Splitting Atoms: Why Do Countries Build Nuclear Power Plants?" *International Interactions* 38, no. 1 (2012).

22. See, for example, Ariel Levite, "Never Say Never Again: Nuclear Reversal Revisited," *International Security* 27, no. 3 (Winter 2002–03): 59–88.

23. Dwight D. Eisenhower, "Address by Mr. Dwight D. Eisenhower, President of the United States of America, to the 470th Plenary Meeting of the United Nations General Assembly," December 8, 1953. Available at http://www.un.org/depts/dhl/dag/docs/apv470e.pdf.

24. "Visit of Prime Minister Shastri of India, Briefing Book." *Library and Archives of Canada (LAC)* RG 25 Vol. 3494 (File 18-1-H-IND-1965/1).

25. On the potential consequences of nuclear proliferation, see Scott Sagan and Kenneth Waltz, *The Spread of Nuclear Weapons: A Debate Renewed* (New York: W. W. Norton, 2002).

26. Countries may calculate that one transfer is unlikely to singlehandedly carry a state over the nuclear tipping point. Nuclear weapons production is complex and it usually requires assistance from several foreign suppliers in more than one aspect of the fuel cycle.

27. See Eugene Wittkopf, *Western Bilateral Aid Allocations: A Comparative Study of Recipient State Attributes and Aid Received* (Beverly Hills, Calif.: Sage, 1972); and James Meernik, Eric Krueger, and Steven Poe, "Testing Models of U.S. Foreign Policy: Foreign Aid during and after the Cold War," *Journal of Politics* 60, no. 1 (1998): 63–85.

28. Walt, *Origins of Alliances.*

29. Randall Schweller, "Bandwagoning for Profit: Bringing the Revisionist State Back In," *International Security* 19, no. 1 (1994): 72–107.

30. Christopher Gelpi, "Alliances as Instruments of Intra-Allied Control', in *Imperfect Unions: Security Institutions over Time and Space*, ed. Helga Haftendorn, Robert O. Keohane, and Celeste A. Wallander (Oxford: Oxford University Press, 1999): 107–39; and John Vasquez and Colin Elman, eds., *Realism and the Balancing of Power: A New Debate* (Upper Saddle River, N.J.: Prentice-Hall, 2003).

31. Christopher Sprecher and Volker Krause, "Alliances, Armed Conflict, and Cooperation: Theoretical Approaches and Empirical Evidence," *Journal of Peace Research* 43, no. 4 (2006): 363–69.

32. John Miglietta, *American Alliance Politics in the Middle East, 1945–1992* (Lanham, Md.: Lexington Books, 2002).

33. Lars Skalnes, *Politics, Markets, and Grand Strategy: Foreign Economic Policies as Strategic Instruments* (Ann Arbor: University of Michigan Press, 2000), 16–17.

34. Alan Sabrosky, "Interstate Alliances: Their Reliability and the Expansion of War," in *The Correlates of War II: Testing Some Realpolitik Models*, ed. J. David Singer (New York: Free Press, 1980), 161–98.

35. Brett Ashley Leeds, Andrew Long, and Sarah McLaughlin Mitchell, "Reevaluating Alliance Reliability: Specific Threats, Specific Promises," *Journal of Conflict Resolution* 44, no. 5 (October 2000): 686–99.

36. United States National Security Council, "U.S. Policy Toward Japan," NSC 6008, May 20, 1960. Accessed from the *Digital National Security Archive*. Available at http://nsarchive.chadwyck.com/marketing/index.jsp.

37. See, for example, Richard Lester, "U.S.-Japanese Nuclear Relations: Structural Change and Political Strain," *Asian Survey* 22, no. 5 (May 1982): 417–33.

38. Skalnes, *Politics, Markets, and Grand Strategy.*

39. Gary Clyde Hufbauer, Jeffrey Schott, and Kimberly Elliott, *Economic Sanctions Reconsidered* (Washington, D.C.: Institute for International Economics, 1990).

40. See Michael Mastanduno, *Economic Containment: COCOM and the Politics of East-West Trade* (Ithaca: Cornell University Press, 1992).

41. See, for example, John Mearsheimer, "The False Promise of International Institutions," *International Security* 19, no. 1 (1994–95): 5–49.

42. Walt, *Origins of Alliances*.

43. See Robert Pape, "Soft Balancing against the United States," *International Security* 30, no. 1 (2005): 7–45; and T. V. Paul, James Wirtz, and Michael Fortmann, eds., *Balance of Power: Theory and Practice in the 21st Century* (Stanford: Stanford University Press, 2004).

44. Paul, "Chinese-Pakistani Nuclear/Missile Ties," 21–22.

45. Nuclear assistance to enemies of enemies may have another added benefit for suppliers. Since nuclear technology is dual use in nature, the threatening state may worry that nuclear trade for peaceful purposes could enhance the recipient state's ability to build nuclear weapons. By definition, this is never the explicit intent of civilian nuclear cooperation, but third parties may not be aware of this. Consequently, peaceful atomic aid under such circumstances can divert the threatening state's attention toward the recipient state's nuclear energy program and away from other power-maximizing objectives.

46. See William Thompson, "Identifying Rivals and Rivalries in World Politics," *International Studies Quarterly* 45, no. 4 (2001): 557–86.

47. Russian nuclear cooperation with Iran is another case aimed at countering American influence. See Pape, "Soft Balancing against the United States," p. 10.

48. Quoted in Rory Carroll and Luke Harding, "Russia to Build Nuclear Reactor for Chavez," *Guardian*, November 19, 2008.

49. John Holdren, "Nuclear Power and Nuclear Weapons: The Connection is Dangerous," *Bulletin of the Atomic Scientists* 39, no. 1 (January 1983): 42.

50. See, for example, Michael Doyle, "Kant, Liberal Legacies, and Foreign Affairs," *Philosophy and Public Affairs* 12, no. 3 (1983): 205–35; Bruce Russett and John Oneal, *Triangulating Peace* (New York: W. W. Norton, 2001); and Bruce Bueno de Mesquita, Alastair Smith, Randolph Siverson, and James Morrow, *The Logic of Political Survival* (Cambridge: MIT Press, 2003).

51. Brett Ashley Leeds and David Davis, "Beneath the Surface: Regime Type and International Interaction, 1953–78," *Journal of Peace Research* 36, no. 1 (1999): 5–21.

52. G. John Ikenberry and Anne-Marie Slaughter, *Forging a World of Liberty Under Law: U.S. National Security in the 21st Century* (Princeton: Woodrow Wilson School of Public and International Affairs, September 2006), 8.

53. George W. Bush, *The National Security Strategy of the United States of America* (Washington, D.C.: White House, March 2006), 7.

54. Martha Finnemore and Kathryn Sikkink, "International Norm Dynamics and Political Change," *International Organization* 52, no. 4 (1998): 891.

55. Tanisha Fazal, *State Death: The Politics and Geography of Conquest, Occupation, and Annexation* (Princeton: Princeton University Press, 2007).

56. See, for example, Finnemore and Sikkink, "International Norm Dynamics and Political Change"; Thomas Risse, Stephen Ropp, and Kathryn Sikkink, *The Power of Human Rights: International Norms and Domestic Change* (New York: Cambridge University Press, 1999).

57. See, for example, Beth Simmons, "International Law and State Behavior: Commitment and Compliance in International Monetary Affairs," *American Political Science Review* 94, no. 4 (2000): 819–35.

58. See, for example, Richard Price, *The Chemical Weapons Taboo* (Ithaca: Cornell University Press, 1997).

59. William Lowrance, "Nuclear Futures for Sale: To Brazil from West Germany, 1975," *International Security* 1, no. 2 (Autumn 1976): 147–66.

60. Etel Solingen, "Brazil: Technology, Countertrade, and Nuclear Exports," in *International Nuclear Trade and Nonproliferation*, ed. William Potter (Lexington, Mass.: Lexington Books, 1990), 111–52.

61. Sara Tanis, and Bennett Ramberg, "Argentina," in *International Nuclear Trade and Nonproliferation*, ed. Potter, 95–110.

62. See, for example, Steven Erlanger, "Russia Says Sale of Atom Reactors to Iran Is Still On," *New York Times*, April 4, 1995.

63. Duane Bratt, *Politics of CANDU Exports*, 20; Sam Roe, "U.S. Cold War Gift: Iran Nuclear Pact," *Chicago Tribune*, August 24, 2006.

64. Norman Aitken, "The Effect of EEC and EFTA on European Trade: A Temporal Cross-Section Analysis," *American Economic Review* 63, no. 5 (December 1973): 881–92; and James Anderson, "A Theoretical Foundation for the Gravity Equation," *American Economic Review* 69, no. 1 (1979): 106–15.

65. Anderson, "A Theoretical Foundation for the Gravity Equation."

66. Alfred Marshall, *Principles of Economics* (London: Macmillan and Co., 1890). See also Kar-iyn Wong, "External Economies of Scale and International Trade: Further Analysis," Seattle: University of Washington, 2000. Available at http://faculty.washington.edu/karyiu/papers/ext-ext.pdf.

67. Paul Krugman, "Scale Economies, Product Differentiation, and the Pattern of Trade," *American Economic Review* 70 (1980): 950–59.

68. Paul Krugman and Robin Wells, *Macroeconomics* (New York: Worth Publishers, 2005), 413.

69. Ibid.

70. Bratt, *Politics of CANDU Exports*, 21–22.

71. Claus Hofhansel, *Commercial Competition and National Security: Comparing U.S. And German Export Control Policies* (Westport, Conn.: Praeger, 1996); and Erwin Hackel, "The Domestic and International Context of West Germany's Nuclear Energy Policy," in *Nuclear Policy in Europe: France, Germany, and the International Debate*, ed. Erwin Hackel, Karl Kaiser, and Pierre Lellouche (Bonn: Europa Union Verlag, 1980), 108–9.

72. One could argue that economies of scale affect peaceful nuclear assistance in a different manner. Increased domestic demand generates economies of scale that in turn enable countries to be competitive in the international nuclear marketplace. I thank an anonymous reviewer for this insight. This is not the argument that is typically made in the extant literature on nuclear cooperation, however.

73. Evan Lieberman, "Nested Analysis as a Mixed-Method Strategy for Comparative Research," *American Political Science Review* 99, no. 3 (2005): 435–52.

74. Will H. Moore, "Synthesis V. Purity and Large-N Studies: How Might We Assess the Gap between Promise and Performance," *Human Rights and Human Welfare* 6 (2006): 89–97.

75. I thank Steven Miller for providing this example.

76. Andrew Bennett, and Bear Braumoeller, "Where the Model Frequently Meets the Road: Combining Statistic Formal and Case Study Methods," unpublished manuscript, Georgetown University, Washington; Alexander George and Andrew Bennett, *Case Studies and Theory Development in the Social Sciences* (Cambridge: MIT Press, 2005); and James Mahoney and Gary Goertz, "A Tale of Two Cultures: Contrasting Quantitative and Qualitative Research," *Political Analysis* 14, no. 3 (2006): 227–49.

CHAPTER 3

1. "Directed" means that each pair of states is included twice in a given year.

2. If a dyad signs more than one NCA in the same year, only the first agreement is included.

3. Note that when I test the democracy argument I only include democratic suppliers in the dataset.

4. The total number of observations in the dataset is smaller for the analysis presented in table 3.5 because only democratic suppliers are included in the sample.

5. I begin by calculating the predicted probability of assistance in each directed dyad year when a dummy explanatory variable is set to 0 and all other factors are held constant at their mean. Next, I calculate the probability of nuclear aid when the same variable takes on a value of 1, holding all other factors constant. To produce the percentage change, I subtract one from the ratio of the latter probability to the former and multiply this number by 100. I calculate the substantive effects of continuous variables by increasing their values from the sample mean to one standard deviation above the mean. The figures reported in figure 3.1 are based on models 3 and 4.

6. This calculation is based on model 4.

7. This could support an alternative argument having to do with economics of scale. Suppliers with larger domestic programs per capita might be in a better position to be competitive in the international market because of the reduced costs resulting from increased production.

8. Each dependent variable is coded 1 if two states signed a given treaty type in a particular year and 0 otherwise.

9. In other words, conflict often predicts failure perfectly. This creates a problem known as "separation" and causes the variable to be dropped from the statistical model.

10. As I discussed in chapter 1, there were eight cases of proliferation assistance from 1945 to 2000.

11. I included the same independent variables discussed above and coded a dependent variable 1 if countries engaged in military assistance in year t and 0 otherwise. Given that the dependent variable is dichotomous and military assistance occurs infrequently, I estimated the model using rare events logit.

12. One might expect a robust domestic nuclear industry would provide a greater capacity to provide military assistance. However, the historical record does not support this alternative conjecture.

13. See, for example, Duane Bratt, *The Politics of CANDU Exports* (Toronto: University of Toronto Press, 2006), 119–20.

14. Some have argued that countries can learn based on the emergence of new facts, which can lead them to redefine their interests. See, for example, Robert Jervis, *Perception and Misperception in International Politics* (Princeton: Princeton University Press, 1976); Joseph Nye, "Nuclear Learning and the US Soviet Security Regimes," *International Organization* 41, no. 3 (1987): 371–402; Dan Reiter, *The Crucible of Beliefs: Learning, Alliances, and World Wars* (Ithaca: Cornell University Press, 1996); and Michael Horowitz, "The Spread of Nuclear Weapons and International Conflict: Does Experience Matter," *Journal of Conflict Resolution* 52, no. 2 (2009): 234–57.

15. The NSG founders include Canada, France, West Germany, Japan, the Soviet Union, the United Kingdom, and the United States.

16. For an overview of the NSG's history and guidelines, see Ian Anthony, Christer Ahlstrom, and Vitaly Fedchenko, *Reforming Nuclear Export Controls: The Future of the Nuclear Suppliers Group* (Oxford: Oxford University Press, 2008).

17. About half of the agreements in the dataset included a supplier that was part of the NSG; Seventy-five percent of the post-1974 treaties included an NSG supplier.

18. This is largely because the NPT did not enter in to force until 1970.

19. Note, however, that there has been a decline in the number of transfers of enrichment and reprocessing facilities to nonnuclear weapons states over time. Since 1980, no NSG member has exported enrichment or reprocessing centers to a nonnuclear weapons state that did not already possess the capability to build such facilities indigenously.

20. Limited comprehensive power NCAs were signed in 0.15 percent of the non-NSG observations in the dataset and 0.16 percent of the NSG observations.

21. This number varies slightly across model specifications due to missing data for some of the independent variables.

22. Douglas Gibler and Meredith Sarkees, "Measuring Alliances: The Correlates of War Formal Interstate Alliance Data Set," *Journal of Peace Research* 41, no. 2 (2004): 211–22.

23. Faten Ghosn, Glenn Palmer, and Stuart Bremer, "The Mid3 Data Set, 1993–2001: Procedures, Coding Rules, and Description," *Conflict Management and Peace Science* 21 (2004): 133–54. These data are supplied by *EUGene*. See D. Scott Bennett and Allan Stam, "EUGene: A Conceptual Manual," *International Interactions* 26, no. 2 (2000): 179–204.

24. Substituting a variable measuring only MIDs resulting in at least one fatality produces findings that are substantively similar.

25. James Klein, Gary Goertz, and Paul Diehl, "The New Rivalry Dataset: Procedures and Patterns," *Journal of Peace Research* 43, no. 3 (2006): 331–48. Klein, Goertz, and Diehl consider two states in a dyad to be rivals if they experience at least three militarized interstate disputes over the period 1816–2001 that are fought over related issues.

26. Monty Marshall and Keith Jaggers, "Polity IV Project: Political Regime Characteristics and Transitions, 1800–2002," (Fairfax, Va.: George Mason University, 2002).

27. Sonali Singh and Christopher Way, "The Correlates of Nuclear Proliferation: A Quantitative Test," *Journal of Conflict Resolution* 48, no. 6 (2004): 859–85.

28"NPT Membership," *Inventory of International Nonproliferation Organizations and Regimes* (Monterey, Calif.: James Martin Center for Nonproliferation Studies, 2008).

29. GDP is measured in current U.S. dollars. These data were obtained by consulting data compiled by Kristian Gleditsch. See Kristian Gleditsch, "Expanded Trade and GDP Data," *Journal of Conflict Resolution* 46, no. 5 (2002): 693–711.

30. Border countries are coded as one mile. See Gary Fitzpatrick and Marilyn Modlin, *Direct-Line Distances* (Metuchen, N.J.: Scarecrow, 1986). These data were acquired using Bennett and Stam's EUGene program.

31. Gleditsch, "Expanded Trade and GDP Data."

32. See International Atomic Energy Agency, *Nuclear Power Reactors in the World*, 2006.

33. Nathaniel Beck, Jonathan Katz, and Richard Tucker, "Taking Time Seriously in Binary Time-Series—Cross-Section Analysis," *American Journal of Political Science* 42, no. 4 (1998): 1260–88.

34. This is necessary to appropriately test my democracy promotion argument. Adding the joint democracy variable to the full estimation sample, however, produces substantively similar findings.

35. Including the NPT variable in model 3 would bias the findings against the nonproliferation argument since we would see many NCAs signed prior to 1970, a period when all states would be coded 0 with respect to NPT membership.

36. I reestimated models 1, 3, and 5 with the joint democracy variable included and it remained statistically significant and positive. The other findings were also substantively similar.

37. For a discussion of this "separation" problem, see Christopher Zorn, "A Solution to Separation in Binary Response Models," *Political Analysis* 13, no. 2 (2005): 157–70.

38. In another test not reported here, I use a different estimator to fit the models. Peaceful nuclear cooperation occurs in about 2 percent of the observations in the sample, which makes it a relatively rare event. Methodologists have developed an estimator known as rare events logit to correct for biases that could result when there are dozens or even thousands more 0s in a sample than 1s. That the results are substantively similar to the findings displayed in table 3.8 when I employ this estimator. On rare events logit, see Gary King and Langche Zeng, "Explaining Rare Events in International Relations," *International Organization* 55, no. 3 (2001): 693–715.

39. As a proxy for this, I only include suppliers that have at least one nuclear power reactor in operation.

40. Joanne Gowa, *Allies, Adversaries, and International Trade* (Princeton: Princeton University Press, 1994). On the stability of alliances in a bipolar system, see George Duncan and Randolph Siverson, "Flexibility of Alliance Partner Choice in a Multipolar System: Models and Tests," *International Studies Quarterly* 26 (1984): 511–38.

41. Lars Skalnes, *Politics, Markets, and Grand Strategy* (Ann Arbor: University of Michigan Press, 2000).

42. Note, however, that when I classify all states with nuclear weapons programs as "pursuing" the bomb, this variable is positive and statistically significant.

43. Argentina, Brazil, France, and Ukraine are among the states that signed NCAs entitling them to nuclear technology, materials, or know-how in the 1990s before they ratified the NPT.

44. A neighbor is a state within 150 miles. Distance data are obtained using EUGene.

45. J. David Singer, Stuart Bremer, and John Stuckey, "Capability Distribution, Uncertainty, and Major Power War, 1820–1965," in *Peace, War, and Numbers*, ed. Bruce Russett (Beverly Hills, Calif.: Sage, 1972) pp. 19–48.

46. Susan Carter, Scott Sigmund Gartner, Michael Haines, Alan Olmstead, Richard Sutch, and Gavin Wright, "Table Db56-59 Crude Petroleum—Average Value, Foreign Trade, and Proved Reserves: 1859–2001," *Historical Statistics of the United States* (Cambridge: Cambridge University Press, 2006).

47. "The Nuclear Suppliers Group," *Inventory of International Nonproliferation Treaties and Regimes* (Monterey, Calif.: James Martin Center for Nonproliferation Studies, 2006).

CHAPTER 4

1. I consider the statistical model to have made a correct prediction if there is a correlation between at least one of the independent variables (military alliance, conflict, shared enemy, superpower enemy, and joint democracy) and the signing of a nuclear cooperation agreement in a given year.

2. This case is not included in my dataset—and therefore it is not part of my statistical analysis—because it occurred after 2000.

3. United States Department of State, "Atoms for Peace Agreement with Iran," *Department of State Bulletin* 36, no. 929 (April 15, 1957): 629.

4. Daniel Poneman, *Nuclear Power in the Developing World* (London: George Allen and Unwin, 1982), 84.

5. *Iran Nuclear Chronology* (Washington, D.C.: Nuclear Threat Initiative, 2005).

6. Poneman, *Nuclear Power in the Developing World*.

7. Hot cells are shielded rooms with remote controlled arms that are used to chemically separate material, such as plutonium, that has been irradiated in a reactor. See David Albright, "An Iranian Nuclear Bomb?" *Bulletin of the Atomic Scientists* 51, no. 4 (1995), 21–26.

8. Henry Kissinger, "U.S. Iranian Cooperation," Department of State Telegram, March 11, 1974.

9. Joseph Sisco, *Study Report: Joint U.S.-Iranian Cooperation* (Washington, D.C.: Department of State, Interdepartmental Working Group, April 25, 1974).

10. Alfred Atherton, "Commission on Cooperation with Iran," United States Department of State Action Memorandum, September 17, 1974.

11. United States Department of State, "Iran: Joint Statement," Department of State Briefing Paper, November 3, 1974.

12. Kissinger, "U.S. Iranian Cooperation."

13. Nuclear Threat Initiative, *Iran Nuclear Chronology*.

14. Section 123 of the U.S. Atomic Energy Act of 1954 requires that a nuclear cooperation agreement exist before nuclear technology can be supplied. Poneman, *Nuclear Power in the Developing World*, 87.

15. Brent Scowcroft, "Negotiation of a Nuclear Agreement with Iran," National Security Council, April 20, 1976; Robert Ingersoll, "Department of State Response to NSSM 219," United States Department of State, April 18, 1975.

16. James Keeley, "A List of Bilateral Civilian Nuclear Cooperation Agreements," University of Calgary, 2003.

17. "US, Iran Resume Atom Power Talks," *Washington Post*, August 9, 1977.

18. Poneman, *Nuclear Power in the Developing World*, 88.

19. "U.S.-Iran Peaceful Nuclear Cooperation Agreement," Department of State Telegram, October 17, 1978.

20. For arguments on why Iran's nuclear program may not be as threatening as the conventional wisdom suggests, see John Mueller, *Atomic Obsession: Nuclear Alarmism from Hiroshima to Al-Qaeda* (Oxford: Oxford University Press, 2009).

21. Sam Roe, "U.S. Cold War Gift: Iran Nuclear Pact," *Chicago Tribune*, August 24, 2006.

22. Dafna Linzer, "Past Arguments Don't Square with Current Iran Policy," *Washington Post*, March 27, 2005.

23. United States National Security Council, "Report of the NSSM 219 Working Group Nuclear Cooperation Agreement with Iran," March 19, 1975.

24. Atherton, "Commission on Cooperation with Iran."

25. Henry Kissinger, "U.S.-Iran Cooperation," Department of State Telegram, April 11, 1974.

26. Sisco, *Study Report: Joint U.S.-Iranian Cooperation*.

27. Ingersoll, "Department of State Response to NSSM 219"; see also National Security Council, "Report of the NSSM 219 Working Group."

28. National Security Council, "Report of the NSSM 219 Working Group."

29. John Miglietta, *American Alliance Policy in the Middle East, 1945–1992: Iran, Israel, and Saudi Arabia* (Lanham, Md.: Lexington Books, 2002), 44.

30. Note, however, that America's alliance with Iran is not classified as a defense pact in the Correlates of War Dataset. Douglas Gibler and Meredith Sarkees, "Measuring Alliances: The Correlates of War Formal Interstate Alliance Data Set," *Journal of Peace Research* 41, no. 2 (2004): 211–22.

31. There is very little information in the available historical record that directly relates to this agreement. In the only congressional hearing before the Joint Committee on Atomic Energy on the subject of the 1957 agreement, U.S. officials said little to justify the arrangement other than to note that it is "similar to those negotiated with 35 other countries." It is further telling that not a single congressperson asked a question on the merits of the Iran deal. Questions instead focused on the U.S.-Norway agreement, which was controversial because it was the first power agreement signed after the declassification of power reactor data by the United States. See Joint Committee on Atomic Energy, Meeting Number 85-1-2, March 28, 1957.

32. United States National Security Council, "U.S. Policy toward Iran," 1957.

33. Ibid.

34. Miglietta, *American Alliance Policy in the Middle East.*

35. Department of State, Office of the Inspector General, Foreign Service, "The Conduct of Relations with Iran," August 1976.

36. Ibid.

37. Linzer, "Past Arguments Don't Square with Current Iran Policy."

38. Ibid.

39. This is according to rivalry data produced by Klein, Goertz, and Diehl, which I use in my quantitative analysis. James Klein, Gary Goertz, and Paul Diehl, "The New Rivalry Dataset: Procedures and Patterns," *Journal of Peace Research* 43, no. 3 (2006): 331–48.

40. Department of State, Office of the Inspector General, Foreign Service, "The Conduct of Relations with Iran," August 1976; and Miglietta, *American Alliance Policy in the Middle East.*

41. United States National Security Council, "U.S. Policy toward Iran," 1957.

42. Sisco, *Study Report.*

43. United States, Department of State, "The Evolution of the U.S.-Iranian Relationship: A Survey of U.S.-Iranian Relations, 1941–1979," State Department Report, January 29, 1980.

44. Miglietta, *American Alliance Policy in the Middle East,* 42–43. A 1957 U.S. National Security Council documented stated, for example, that "Iran is a tempting and important target of Soviet expansion because of its vulnerability to overt and covert penetration." United States National Security Council, "U.S. Policy toward Iran," 1957.

45. United States Department of State, "The Evolution of the U.S.-Iranian Relationship: A Survey of U.S.-Iranian Relations, 1941–1979," State Department Report, January 29, 1980. From 1953 through 1957, the United States provided Iran with $250 million in grants and $116 million in loans. Miglietta, *American Alliance Policy in the Middle East,* 43.

46. United States National Security Council, "U.S. Policy toward Iran," 1957.

47. Sisco, *Study Report.*

48. Kissinger, "U.S. Iranian Cooperation"; Sisco, *Study Report*; Atherton, "Commission on Cooperation with Iran."

49. Gary Sick, *All Fall Down: America's Tragic Encounter with Iran* (New York: Random House, 1985), 25.

50. Linzer, "Past Arguments Don't Square with Current Iran Policy."

51. Ibid.

52. Fred Ikle, "Comments on the Study of Conditions for U.S. Nuclear Cooperation with Iran," United States Arms Control and Disarmament Agency Memo, April 16, 1975.

53. James Schlesinger, "Nuclear Cooperation Agreement with Iran," Department of Defense Memo, April 25, 1975.

54. Sisco, *Study Report.*

55. United States National Security Council, "Report of the NSSM 219 Working Group Nuclear Cooperation Agreement with Iran."

56. Ibid.

57. Sidney Sober, "Your Meeting with the Shah at Blair House," Department of State Briefing Memorandum, May 9, 1975.

58. Ikle, "Comments on the Study of Conditions"; and United States National Security Council, "Report of the NSSM 219 Working Group Nuclear Cooperation Agreement with Iran," March 19, 1975.

59. United States Department of State, "US-Iran Commission Cements Bilateral Ties," *Current Foreign Relations,* no. 11, March 12, 1975.

60. For example, see Sisco, *Study Report.*

61. Henry Kissinger, "Strategy for Your Discussions with the Shah of Iran," Department of State Memo, May 13, 1975.

62. As noted in chapter 1, Libya received military assistance from Pakistan beginning in the 1990s.

63. Mark Martel and Warren Donnelly, "Libya's Nuclear Energy Situation," *Digital National Security Archive*, 1987.

64. Wyn Bowen, "Chapter Two: Proliferation Pathways," *Adelphi Papers* 46, no. 380 (2006): 29.

65. Ibid.

66. Thomas O'Toole, "Libya Said to Buy Soviet a-Power Plant," *Washington Post*, December 12 1977.

67. "Soviet-Libyan Nuclear Cooperation Was Included in the Various Accords," *Nucleonics Week* 22, no. 18 (1981): 5.

68. Central Intelligence Agency, "The Libyan Program: A Technical Perspective," Digital National Security Archive, 1985; and William Potter, "The Soviet Union and Nuclear Proliferation," *Slavic Review* 44, no. 3 (1985): 479.

69. "Soviet-Libyan Nuclear Cooperation Was Included in the Various Accords."

70. Martel and Donnelly, "Libya's Nuclear Energy Situation"; Bowen, "Chapter Two," 33.

71. Dissatisfied with the lack of progress, Libya turned to other nuclear suppliers. Martel and Donnelly, "Libya's Nuclear Energy Situation."

72. James Philips, *Moscow's Thriving Libyan Connection* (Washington, D.C.: Heritage Foundation, 1984).

73. Ellen Laipson, "Libya and the Soviet Union: Alliance at Arm's Length," in *The Pattern of Soviet Conduct in the Third World*, ed. Walter Laqueur (New York: Praeger, 1983), 138.

74. "Libya's Premier Jalloud Visits Moscow," *Current Digest of the Soviet Press* 36, no. 20 (1974): 17–18. The Soviet Union and Libya also cooperated in the area of arms. In 1974 the two states concluded their first major arms deal and in the 1970s Moscow exported between $10 billion and $16 billion worth of arms to Libya. See Philips, *Moscow's Thriving Libyan Connection*.

75. "Kosygin Visits Libya and Tunisia," *Current Digest of the Soviet Press* 37, no. 30 (1975): 20.

76. Potter, "Soviet Union and Nuclear Proliferation."

77. "Libya's Premier Jalloud Visits Moscow," 18.

78. Gorbachev was particularly interested in rebuilding the Soviet economy, maintaining his domestic support, and reaching arms control agreements with the United States. Robert Freedman, "US-Libyan Crisis—Moscow Keeps Its Distance," *Christian Science Monitor*, April 24, 1986; Robert Freedman, "Soviet Policy towards the Middle East," *Proceedings of the Academy of Political Science* 36, no. 4 (1987): 176–97; and Stephen Engelberg, "Soviets Sold Libya Advanced Bomber, US Officials Say," *New York Times*, April 5, 1989.

79. Freedman, "US-Libyan Crisis."

80. Before being expelled from the country in 1972, the Soviet Union had nearly twenty thousand troops and advisers in Egypt. Joseph Whelan and Michael Dixon, *The Soviet Union in the Third World: Threat to World Peace?* (Washington, D.C.: Pergamon-Brassey's, 1986), 189.

81. Ibid; and Laipson, "Libya and the Soviet Union: Alliance at Arm's Length," 138.

82. Both countries were also enemies of Israel. Moscow may have been interested in constraining the influence of Israel given that it was a key U.S. ally in the Middle East, but the United States was its principal concern. They grew to be enemies of Egypt in the early 1970s as well when Cairo adopted an independent foreign policy and cultivated close relations with Washington. Philips, *Moscow's Thriving Libyan Connection*.

83. "Libyan Delegation Visits the USSR," *Current Digest of the Soviet Press* 34, no. 10 (1972): 9–10; "Libya's Premier Jalloud Visits Moscow," 17–18.

84. John Cooley, "The Libyan Menace," *Foreign Policy* 42 (1981): 74–93.

85. Both countries also noted the importance of countering "Zionist-Israeli" aggression. "Libyan Delegation Visits the USSR," 9.

86. "Kosygin Visits Libya and Tunisia," 11.

87. "Libya's Qaddafi Visits the USSR," 10.

88. "Libya's Jalloud Visits Moscow," 18.

89. It may have also constrained the United States by forcing it to consider the possibility that one of its enemies, Libya, might acquire nuclear weapons. Moscow clearly did not want its assistance to help Libya get the bomb, but third parties might not have been aware of this, especially in light of Washington's tendency to adopt worst-case thinking when analyzing Soviet behavior. See Gloria Duffy, "Soviet Nuclear Exports," *International Security* 3, no. 1 (1978): 83–111; and Potter, "Soviet Union and Nuclear Proliferation."

90. See, for example, Deborah Welch Larson, "Shortcut to Greatness: The New Thinking and the Revolution in Soviet Foreign Policy," *International Organization* 57, no. 1 (2003): 77–109.

91. Mikhail Gorbachev, *Memoirs* (London: Bantam Books, 1996).

92. Bowen, "Chapter Two: Proliferation Pathways."

93. Joseph Micallef, "A Nuclear Bomb for Libya?" *Bulletin of the Atomic Scientists* 37, no. 7 (1981): 14–15.

94. Duffy, "Soviet Nuclear Exports;" Potter, "Soviet Union and Nuclear Proliferation."

95. Potter, "Soviet Union and Nuclear Proliferation," 470.

96. In the 1970s and 1980s, the Soviet Union engaged in nuclear commerce with other states that had weak nonproliferation records including India, Argentina, and Cuba. See Potter, "Soviet Union and Nuclear Proliferation."

97. Ibid, 477.

98. Gloria Duffy, *Soviet Nuclear Energy: Domestic and International Policies* (Santa Monica, Calif.: RAND, 1979), 84.

99. Potter, "Soviet Union and Nuclear Proliferation," 481. There was, however, a limit to Moscow willingness to cooperate with a state determined to acquire nuclear weapons. The Soviet Union built the Tajoura reactor knowing that it could produce plutonium. That reactor was capable of producing 1.2 kg of weapons-grade plutonium per year, which means that it would take ten years to accumulate enough material for a nuclear bomb. But it was reluctant to supply power reactors that were capable of producing around ten to twenty nuclear weapons per year. See also Duffy, *Soviet Nuclear Energy*, 85; and Micallef, "A Nuclear Bomb for Libya?"

100. Potter, "Soviet Union and Nuclear Proliferation," p. 478.

101. Ultimately, France and Italy supplied technology that would have enabled Iraq to produce plutonium for nuclear weapons had Israel not bombed the facility in 1981.

102. Micallef, "A Nuclear Bomb for Libya;" Bowen, "Chapter Two: Proliferation Pathways," 25–46.

103. Potter, "Soviet Union and Nuclear Proliferation."

104. Duffy, "Soviet Nuclear Exports."

105. Potter, "Soviet Union and Nuclear Proliferation."

106. Robert Bothwell, *Nucleus: The History of Atomic Energy of Canada Limited* (Toronto: University of Toronto Press, 1988).

107. Escott Reid, *Envoy to Nehru* (Oxford: Oxford University Press, 1981).

108. Robert Bothwell, "Eyes West: Canada and the Cold War in Asia," in *Canada and the Early Cold War, 1943-1957,* ed. Greg Donaghy (Ottawa: Canadian Government Publishing, 1995).

109. Canada supplied the reactor and the United States provided nuclear materials (in particular, heavy water) necessary for the plant to operate.

110. Duane Bratt, *The Politics of CANDU Exports* (Toronto: University of Toronto Press, 2006), 89.

111. Ibid, 95.

112. "India Nuclear Chronology, 1960–64" (Washington D.C.: Nuclear Threat Initiative, 2003).

113. The other half was going to be fabricated in India.

114. Bratt, *Politics of CANDU Exports*, 100.

115. The term turnkey refers to facilities that are completed by the supplier so that the only thing left for the customer to do is to "turn the key."

116. K. K. Pathak, *Nuclear Policy of India: A Third World Perspective* (New Delhi: Gitanjali Prakashan, 1980).

117. Bratt, *Politics of CANDU Exports*.

118. George Perkovich, *India's Nuclear Bomb* (Berkeley: University of California Press, 1999), 131.

119. Bratt, *Politics of CANDU Exports*.

120. "India Nuclear Chronology, 1974–75" (Washington, D.C.: Nuclear Threat Initiative, 2006).

121. For more on these negotiations, see Ashok Kapur, "The Canada-India Nuclear Negotiations: Some Hypotheses and Lessons," *World Today* 34, no. 8 (1978): 315–16.

122. Perkovich, *India's Nuclear Bomb*, 28; and Bratt, *Politics of CANDU Exports*, 118–19.

123. Bratt, *Politics of CANDU Exports*, 119.

124. Mary Halloran, "Mrs. Gandhi's Bombshell: Canadian Reactions to India's Nuclear Detonation, 1974–1976," in *Canada's Global Engagements and Relations with India*, ed. Christopher Sam Raj and Abdul Nafey (New Delhi: Manak, 2007), 279–91; and Perkovich, *India's Nuclear Bomb*.

125. M. S. Rajan, "The Indo-Canadian Entente," *International Journal* 17 (1961–62): 358–84; and Bratt, *Politics of CANDU Exports*.

126. Bratt, *The Politics of CANDU Exports*, 91–92; and Rajan, "Indo-Canadian Entente."

127. Ashok Kapur, "Canada-India Nuclear Negotiations: Context and Process," in *Canadian Policy on Nuclear Cooperation with India: Confronting New Dilemmas*, ed. Karthika Sasikumar and Wade Huntley (Vancouver, Canada: Simons Centre for Disarmament and Non-Proliferation Research, 2007), 35–46.

128. Ryan Touhey, "Troubled from the Beginning: Canada's Nuclear Relations with India during the 1960s," in *Canadian Policy on Nuclear Cooperation with India*, ed. Sasikumar and Huntley, 12.

129. Greg Donaghy, "Nehru's Reactor: The Origins of Indo-Canadian Nuclear Cooperation, 1955–1959," in *Canada's Global Engagements and Relations with India*, ed. Christopher Sam Raj and Abdul Nafey (New Delhi: Manak, 2007), 267–78; and Bothwell, *Nucleus*, 353.

130. Rajan, "Indo-Canadian Entente," 379; Bothwell, *Nucleus*.

131. Bothwell, *Nucleus*.

132. In 1950, members of the Commonwealth included Australia, Canada, India, New Zealand, Pakistan, South Africa, Sri Lanka, and the United Kingdom. Lester Pearson, *Memoirs: Volume II, 1948–1957* (New York: University of Toronto Press, 1973), 107–12.

133. See Geoffrey Pearson, *Seize the Day: Lester B. Pearson and Crisis Diplomacy* (Ottawa: Carleton University Press, 1993), 49–62.

134. Shyam Bhatia, *India's Nuclear Bomb* (Ghaziabad: Vikas, 1979).

135. Pierre-Elliott Trudeau, "Statement to the Canadian Nuclear Association, Ottawa, June 17, 1975," in *Canadian Foreign Policy, 1966–1976*, ed. Arthur Blanchette (Montreal: McGill-Queen's Press, 1980), 24.

136. Ibid.

137. See Reid, *Envoy to Nehru*, 18–19; and Pearson, *Memoirs*.

138. Bothwell, *Nucleus*, 357–58.

139. Ibid, 370–71.

140. Ibid, 368. See also John English and Ryan Touhey, "Canadian-Indian Relations: A Historical Appreciation," in *Canada's Global Engagements and Relations with India*, ed. Raj and Nafey, 231–32.

141. AECL Canada, "AECL Records, 103-a-3," Vols. 5 and 6 (July and November 1966).

142. Reid, *Envoy to Nehru*, 261.

143. Ibid, 14–98.

144. Pearson, *Seize the Day*, 49.

145. Ibid.

146. Lester Pearson, *Democracy in World Politics* (Princeton: Princeton University Press, 1955), 43.

147. Ibid.

148. Reid, *Envoy to Nehru*, 18–19.

149. Pearson, *Seize the Day*, 50.

150. Pearson, *Memoirs*, 118.

151. Pearson, *Seize the Day*, 52.

152. Ibid, 53.

153. Ibid, 51.

154. Pearson, *Memoirs*, 107.

155. Reid, *Envoy to Nehru*, 278.

156. Donaghy, "Nehru's Reactor," 269.

157. In November 1956, Prime Minister Nehru condemned the Anglo-French-Israeli attack against Egypt during the Suez crisis but refrained from criticizing the Soviet Union following its invasion of Hungary, which occurred a week earlier. English and Touhey, "Canadian-Indian Relations: A Historical Appreciation."

158. A. Appadorai and M. S. Rajan, *India's Foreign Policy and Relations* (New Delhi: South Asian Publishers, 1985); and Perkovich, *India's Nuclear Bomb*, 41.

159. English and Touhey, "Canadian-Indian Relations: A Historical Appreciation," 224

160. Bothwell, *Nucleus*.

161. Bratt, *Politics of CANDU Exports*.

162. Ivan Head and Pierre Trudeau, *The Canadian Way: Shaping Canada's Foreign Policy, 1968–1984* (Toronto: McClelland and Stewart, 1995), 122–23.

163. Donaghy, "Nehru's Reactor," 269.

164. Bratt, *Politics of CANDU Exports*, 40.

165. Louis Delvoie, "Canada and India: A Roller Coaster Ride," in *Canada's Global Engagements and Relations with India*, ed. Raj and Nafey, 241.

166. Touhey, "Troubled from the Beginning," 20.

167. Sonali Singh and Christopher Way, "The Correlates of Nuclear Proliferation: A Quantitative Test," *Journal of Conflict Resolution* 48, no. 6 (2004): 859–85; and Touhey, "Troubled from the Beginning."

168. Touhey, "Troubled from the Beginning," 15.

169. "Visit of Prime Minister Shastri of India, Briefing Book," *Library and Archives of Canada (LAC)*, RG 25 Vol. 3494, File 18-1-H-IND-1965/1 (1965). See also Touhey, "Troubled from the Beginning," 20.

170. Touhey, "Troubled from the Beginning," 19. Martin also acknowledged that this sentiment likely was not shared by Prime Ministers Nehru or Shastri.

171. See Perkovich, *India's Nuclear Bomb*.

172. "Visit of Prime Minister Shastri of India, Briefing Book." This intelligence proved to be inaccurate but Canada had no way of knowing this at the time. Touhey, "Troubled from the Beginning."

173. Bratt, *Politics of CANDU Exports*, 125–26.

174. Perkovich, *India's Nuclear Bomb*.

175. Ron Finch, *Exporting Danger: A History of the Canadian Nuclear Energy Export Programme* (Montreal: Black Rose, 1986). Some scholars have claimed that "sheer naiveté" explains Canada's nuclear cooperation with India. See, for example, Robert Morrison and Edward Wonder, *Canada's Nuclear Export Policy* (Ottawa: Carleton University Press, 1978). Finch argues that this explanation was propagated in large part to absolve Canadian decision makers for their eventual contribution to nuclear proliferation.

176. Bratt, *Politics of CANDU Exports*, 95.

177. Donaghy, "Nehru's Reactor," 271.

178. Morrison and Wonder, *Canada's Nuclear Export Policy*; Bratt, *Politics of CANDU Exports*.

179. Bothwell, *Nucleus*, 369.

180. Donaghy, "Nehru's Reactor."

181. Touhey, "Troubled from the Beginning," 12.

182. Kapur, "Canada-India Nuclear Negotiations: Context and Process," 35–46.

183. Jules Leger, "Memorandum from Under-Secretary of State for External Affairs to Secretary of State for External Affairs, March 21, 1955," *Documents on Canadian External Relations* 21, no. 254.

184. Touhey, "Troubled from the Beginning," 27

185. Touhey, "Troubled from the Beginning."

186. See, for example, George Perkovich, *Faulty Promises: The US-India Nuclear Deal* (Washington, D.C.: Carnegie Endowment for International Peace, 2005); Wade Huntley and Karthika Sasikumar, eds., *Nuclear Cooperation with India: New Challenges, New Opportunities* (Vancouver, Canada: Simons Centre for Disarmament and Nonproliferation Research, 2006); and Leonard Weiss, "U.S.-India Nuclear Cooperation," *Nonproliferation Review* 14, no. 3 (2007): 429–57.

187. Robert Norris and Hans Kristensen, "Indian Nuclear Forces, 2010," *Bulletin of the Atomic Scientists* 66, no. 5 (2010): 76-81.

188. Specifically, the NNPA required full-scope safeguards as a necessary condition for the supply of nuclear technology.

189. Sharon Squassoni, "U.S. Nuclear Cooperation with India: Issues for Congress" (Washington, D.C.: Congressional Research Service, 2006).

190. United States, "Agreement for Cooperation between the Government of the United States of America and the Government of India Concerning Peaceful Uses of Nuclear Energy," (Washington, D.C.: 2007).

191. George W. Bush and Manmohan Singh, "Joint Statement between President George W. Bush and Prime Minister Manmohan Singh" (Washington, D.C.: The White House, 2005).

192. IAEA Staff Report, "IAEA Board Approves India Safeguards Agreement, August 1, 2008." Available at http://www.iaea.org/NewsCenter/News/2008/board010808.html.

193. George W. Bush, "President's Statement on Strategic Partnership with India" (Washington, D.C.: The White House, 2004).

194. Glenn Kessler, *The Confidante: Condoleezza Rice and the Creation of the Bush Legacy* (New York: St. Martin's, 2007), 55.

195. Condoleezza Rice, "U.S.-India Civilian Nuclear Cooperation Agreement," Testimony before the House International Relations Committee, Washington, D.C., April 5, 2006.

196. Nicholas Burns, "America's Strategic Opportunity with India," *Foreign Affairs* 86, no. 6 (November/December 2007).

197. Ibid.

198. Ashley J. Tellis, "U.S.-India Atomic Cooperation: Strategic and Nonproliferation Implications," testimony before the Senate Foreign Relations Committee, April 26, 2006.

199. Nicholas Burns, "The U.S. and India: An Emerging Entente?" Testimony before the House International Relations Committee, Washington, D.C., September 8, 2005.

200. Nicholas Burns, "The U.S. and India: The New Strategic Partnership," Speech at the Asia Society, New York, October 18, 2005.

201. Rice, "U.S.-India Civilian Nuclear Cooperation Agreement."

202. See Kessler, *Confidante*, 56.

203. Quoted in ibid, 52.

204. Daniel Twining, "America's Grand Design in Asia," *Washington Quarterly* 30, no. 3 (2007): 82–83.

205. Tellis, "U.S.-India Atomic Cooperation."

206. Quoted in Michael Green and Daniel Twining, "Democracy and American Grand Strategy in Asia: The Realist Principles behind an Enduring Idealism," *Contemporary Southeast Asia* 30, no. 1 (2008): 9.

207. Joseph Cirincione, Jon Wolfsthal, and Miriam Rajkumar, *Deadly Arsenals: Nuclear, Biological, and Chemical Threats* (Washington, D.C.: Carnegie Endowment for International Peace, 2005), 222.

208. Joseph Nye, "Balancing Asia's Rivals," *Korea Times*, June 11, 2008.

209. Office of the Secretary of Defense United States, "Military Power of the People's Republic of China," (Washington, D.C., 2007).

210. Condoleezza Rice, "Promoting the National Interest," *Foreign Affairs* 79, no. 1 (January/February 2000): 45–62.

211. Rice, "U.S.-India Civilian Nuclear Cooperation Agreement."

212. Robert Blackwill, "Forging Fresh Bonds," *India Times*, February 27, 2006.

213. See, for example, Perkovich, *Faulty Promises*.

214. George W. Bush, "President Discusses Strong U.S.-India Partnership in New Delhi," in (Washington, D.C.: The White House, 2006).

215. Burns, "America's Strategic Opportunity with India."

216. Kessler, *Confidante*, 57.

217. Robert Blackwill, "The India Imperative," *National Interest*, no. 80 (2005): 9–17.

218. Ashton Carter, "America's New Strategic Partner," *Foreign Affairs* 85, no. 4 (July/August 2006).

219. Quoted in Jonathan Monten, "The Roots of the Bush Doctrine: Power, Nationalism, and Democracy Promotion in U.S. Strategy," *International Security* 29, no. 4 (2005): 112.

220. See the 2006 National Security Strategy of the United States. Available at http://www.whitehouse.gov/nsc/nss/2006/sectionII.html

221. Robert Blackwill, "A New Deal for New Delhi," *Wall Street Journal*, March 21, 2005.

222. See, for example, Ashley Tellis, "What Should We Expect from India as a Strategic Partner?" in *Gauging U.S.-Indian Strategic Cooperation* Henry Sokolski (Carlisle, Penn.: Strategic Studies Institute, 2007), 248–49.

223. Burns, "The U.S. and India: An Emerging Entente?"

224. Rice, "U.S.-India Civilian Nuclear Cooperation Agreement."

225. Quoted in Green and Twining, "Democracy and American Grand Strategy in Asia," 9.

226. Ashley Tellis, "U.S., India Advance Nuclear Trade Deal," National Public Radio interview, July 9, 2008. Available at http://www.npr.org/templates/story/story.php?storyId=92357300.

227. Green and Twining, "Democracy and American Grand Strategy in Asia," 9. The 2002 National Security Strategy of the United States indicates that Washington will "implement its strategies by organizing coalitions … of states able and willing to promote a balance of power that favors freedom." White House, "National Security Strategy of the United States of America," September 17, 2002. Available at http://www.whitehouse.gov/nsc/nss.html.

228. For an overview of Indian (and Pakistani) nuclear strategy, see Vipin Narang, "Posturing for Peace: Pakistan's Nuclear Postures and South Asian Stability," *International Security* 34, no. 3 (2009/10): 38–78.

229. Jayantha Dhanapala and Daryl Kimball, *A Nonproliferation Disaster* (Washington, D.C.: Carnegie Endowment for International Peace, 2008).

230. Kessler, *Confidante*, 51.

231. Dafna Linzer, "Bush Officials Defend India Nuclear Deal," *Washington Post,* July 20, 2005.

232. See, for example, Henry Sokolski, "Backing the U.S.-India Nuclear Deal and Nonproliferation: What's Required," testimony before a Hearing of the Senate Foreign Relations Committee, *The Nonproliferation Implications of the July 18, 2005 U.S.-India Joint Statement,* Washington, D.C., November 3, 2005.

233. On the connection between civilian uranium exports and nuclear arsenal expansion more generally see Matthew Fuhrmann, "Australia's Uranium Exports and Nuclear Arsenal Expansion: Is There a Connection?" in *Australia's Uranium Trade: The Domestic and Foreign Policy Challenges of a Contentious Export,* ed. Michael Clarke, Stephan Fruhling, and Andrew O'Neil (London: Ashgate, 2011), 39–58.

234. George Perkovich, *Faulty Promises.*

235. Kessler, *Confidante*, 58.

236. Paul Richter, "In Deal with India, Bush Has Eye on China," *Los Angeles Times*, March 4, 2006.

237. Kessler, *Confidante,* p. 58.

238. Ibid, 54.

239. Robert Joseph, "The U.S. and India: An Emerging Entente?" Testimony before the House International Relations Committee, Washington, D.C., September 8, 2005.

240. Burns, "America's Strategic Opportunity with India."

241. Joseph, "The U.S. and India: An Emerging Entente?"

242. Jackie Range, David Crawford, and Paul Glader, "India's Nuclear Deal with U.S. Faces Test," *Wall Street Journal*, August 20, 2008.

243. Rice, "U.S.-India Civilian Nuclear Cooperation Agreement."

244. Alistair Scrutton, "India Ruling Party Hails Historic Nuclear Deal," *Reuters*, October 3, 2008.

245. Rice, "U.S.-India Civilian Nuclear Cooperation Agreement."

CHAPTER 5

1. Cases not predicted by my theory are those where the four hypotheses other than hypothesis 3.2 (militarized conflict) all fail to predict the onset of cooperation. All five hypotheses collectively failed to predict nuclear assistance in only one case (India–Sri Lanka).

2. Daniel Poneman, *Nuclear Power in the Developing World* (London: George Allen and Unwin, 1982).

3. The agreement stipulated that Indonesia could possess no more than 6kg of uranium enriched to 20 percent at any one time. At least 25kg of uranium enriched to 90 percent is necessary to trigger an effective nuclear explosion. Robert Cornejo, "When Sukarno Sought the Bomb: Indonesian Nuclear Aspirations in the Mid-1960s," *Nonproliferation Review* 7, no. 2 (2000): 31–43.

4. Poneman, *Nuclear Power in the Developing World.*

5. Cornejo, "When Sukarno Sought the Bomb."

6. Ibid. Note, however, that Indonesia does not qualify as a state that pursued nuclear weapons based on the criteria I adopt in this book. See Sonali Singh and Christopher Way, "The Correlates of Nuclear Proliferation: A Quantitative Test," *Journal of Conflict Resolution* 48, no. 6 (2004): 859–85.

7. On March 11, 1966, Sukarno was overthrown by General Suharto and Indonesian nuclear weapons aspirations officially ended. Cornejo, "When Sukarno Sought the Bomb." The United States signed additional nuclear cooperation agreements with Indonesia while Suharto was in power.

8. John Finney, "U.S. Continues Atom Aid," *New York Times*, September 9, 1965.

9. McGeorge Bundy, *Danger and Survival* (New York: Random House, 1988); and Cornejo, "When Sukarno Sought the Bomb."

10. Poneman, *Nuclear Power in the Developing World*, 100.

11. Frederick Bunnell, "The Central Intelligence Agency—Deputy Directorate for Plans 1961 Secret Memorandum on Indonesia: A Study in the Politics of Policy Formulation in the Kennedy Administration," *Indonesia* no. 22 (October 1976): 131–69.

12. National Security Council, United States, "U.S. Policy on Indonesia: NSC 5901" (Washington, D.C.: Digital National Security Archive, 1959).

13. Bunnell, "A Study in the Politics of Policy Formulation," 133.

14. As I argued in chapter 2, peaceful nuclear cooperation can strengthen political ties between the supplier and recipient states by signaling favorable intentions and enhancing the likelihood of future cooperation. It can also strengthen the recipient state economically by enhancing its energy production capacity and/or its scientific knowledge base.

15. Neil Sheehan, "Indonesia Seizes Third U.S. Library," *New York Times*, February 16, 1965.

16. Department of State, United States, "Background Paper on Factors Which Could Influence National Decisions Concerning Acquisition of Nuclear Weapons" (Washington, D.C., 1964).

17. Poneman, *Nuclear Power in the Developing World*.

18. Finney, "U.S. Continues Atom Aid."

19. Ernest McCrary, "Iraq's Oil Ploy to Gain Nuclear Know How," *Business Week*, December 3, 1979.

20. Peter Eisner, "Brazil and Iraq Make Nuclear Deal," *Associated Press*, January 7, 1980; and Peter Eisner, "Brazil and Iraq Reach Oil Agreement," *Associated Press*, January 10, 1980.

21. Eisner, "Brazil and Iraq Make Nuclear Deal."

22. In particular, Brazil worried that cooperation with Iraq would jeopardize the continuation of West German nuclear aid. McCrary, "Iraq's Oil Ploy to Gain Nuclear Know How"; "Iraq Pressures Brazil for Nuclear Technology," *Chemical Week*, November 21, 1979.

23. Warren Hoge, "For Brazil, an Embarrassing Tale of Intrigue," *New York Times*, June 23, 1981; Steve Weissman and Herbert Krosney, *The Islamic Bomb: The Nuclear Threat to Israel and the Middle East* (New York: Times Books, 1981).

24. "The Iraq-Brazil Nuclear Bazaar," *Middle East Defense News*, October 29, 1990.

25. "Iraq Asserts Arabs Must Acquire Atoms Arms as a Balance to Israel," *New York Times*, June 24, 1981.

26. Some observers have suggested that Brazilian-Iraqi nuclear cooperation was part of a "mutual help scheme" designed to aid the nuclear weapons programs of both countries. Proponents of this argument maintain that Brazil initiated a "short-cut" to nuclear weapons and that it needed Iraqi help to make this plan work. See "Brazil: The Nuclear Plan Mushrooms," *Latin America Weekly Report*, July 31, 1981. The empirical evidence does not lend significant support to this argument, beyond unsubstantiated references to mutual cooperation on weapons efforts in a few media outlets.

27. "Iraq Pressures Brazil for Nuclear Technology."

28. Ibid.

29. Ibid.

30. "Iraq-Brazil Nuclear Bazaar."

31. Western officials recognized this and expressed concern about the proliferation potential of Brazilian atomic aid. See Weissman and Krosney, *Islamic Bomb*, 272.

32. Judith Miller, "3 Nations Widening Nuclear Contacts," *New York Times*, June 28, 1981.

33. Eisner, "Brazil and Iraq Make Nuclear Deal;" Eisner, "Brazil and Iraq Reach Oil Agreement."

34. Hoge, "For Brazil, an Embarrassing Tale of Intrigue."

35. Quoted in Eugene Robinson, "Sanctions Cost Brazil Key Trade Partner," *Washington Post*, August 28, 1990.

36. Mark Hibbs, "British-Korean Agreement Will Allow Reprocessing at BNFL," *Nuclear Fuel*, December 9, 1991.

37. Geoffrey Lean, "Thorp Seeks Sale to South Korea," *Independent*, November 14, 1993.

38. Ibid.

39. "North Korea Critical of British and French Plutonium Sales to South and Japan," *BBC Summary of World Broadcasts*, May 25, 1993.

40. Hibbs, "British-Korean Agreement Will Allow Reprocessing"; Mark Hibbs, "Britain Would Support BNFL Bid to Eventually Reprocess ROK Fuel," *Nuclear Fuel*, February 7, 2000.

41. Hibbs, "Britain Would Support BNFL Bid to Eventually Reprocess ROK Fuel."

42. Hibbs, "British-Korean Agreement Will Allow Reprocessing."

43. Jungmin Kang and H. A. Feiveson, "South Korea's Shifting and Controversial Interest in Spent Fuel Reprocessing," *Nonproliferation Review* 8, no. 2 (2001): 70–78.

44. Hibbs, "British-Korean Agreement Will Allow Reprocessing."

45. Ibid.

46. Pearl Marshall, "U.K. Begins Review That Will Decide if Nuclear Power Has a Place in Britain," *Nuclear Fuel*, May 23, 1994.

47. The Romanians claimed that the project initiated by the Canadians would result in the construction of *five* power reactors, but only one reactor was actually being built. Jennifer Wells, "Going Critical," *Globe and Mail Report on Business Magazine*, June 1995.

48. Duane Bratt, *The Politics of CANDU Exports* (Toronto: University of Toronto Press, 2006), 179.

49. Bratt, *Politics of CANDU Exports*.

50. Raymond Garthoff, "When and Why Romania Distanced Itself from the Warsaw Pact," *Cold War International History Project Bulletin*, no. 5 (Spring 1995): 111.

51. Duane Bratt, "Is Business Booming? Canada's Nuclear Reactor Export Policy," *International Journal* 51, no. 3 (Summer 1996): 496.

52. I thank Gary Bertsch for this insight.

53. Duane Bratt, "CANDU or CANDON'T: Competing Values behind Canada's Nuclear Sales," *Nonproliferation Review* 5, no. 3 (Spring–Summer 1998): 8.

54. Ibid.

55. For example, President Richard Nixon and Soviet premier Leonid Brezhnev participated in three major summits and engaged in cultural exchanges, technical cooperation, and promoted bilateral trade.

56. Thomas Axworthy, "To Stand Not So High Perhaps but Always Alone," in *Towards a Just Society: The Trudeau Years*, ed. Thomas Axworthy and Pierre Trudeau (Markham, Canada: Viking, 1990), 44.

57. Ibid.

58. Duane Bratt, "Is Business Booming?" 487.

59. See Robert Bothwell, *Nucleus: The History of Atomic Energy of Canada Limited* (Toronto: University of Toronto Press, 1988).

60. Bratt, "CANDU or CANDON'T," 8.

61. Bratt, *Politics of CANDU Exports*. Retransfers to Pakistan were especially worrisome because that country had not yet built nuclear weapons.

62. Quoted in Bratt, "CANDU or CANDON'T," 9.

63. James Keeley, "A List of Bilateral Civilian Nuclear Cooperation Agreements," University of Calgary, 2003.

64. Nuclear Threat Initiative, *Iraq Nuclear Chronology, 1956–1979* (Washington, D.C.: 2003).

65. The name "Osiraq" emerged because the reactor was similar to Osiris, a French reactor that operated southwest of Paris, and it was supplied to Iraq. Osiraq is also known as Tammuz I.

66. In 1975, just days after the signing of the NCA with France, Saddam Hussein declared that the agreement "is the first concrete step towards the production of an Arab atomic weapon." David Styan, *France and Iraq: Oil, Arms and French Policy Making in the Middle East* (London: I. B. Tauris, 2006).

67. Eliot Marshall, "Fallout from the Raid on Iraq," *Science* 213, no. 4503 (1981): 116–20.

68. Weissman and Krosney, *Islamic Bomb*, 92.

69. For more on this attack, see Shai Feldman, "The Bombing of Osiraq Revisited," *International Security* 7, no. 2 (1982): 114–42.

70. Styan, *France and Iraq*.

71. In July 1981, French foreign minister Claude Cheysson stated: "If Iraq wants to conclude a new accord to obtain a nuclear reactor, France will be ready to supply her under the same conditions as those applied to other customers." Despite this sentiment, President Mitterrand officially ended nuclear cooperation with Iraq in 1983. Styan, *France and Iraq*, 138–39.

72. Andre Giraud, "Energy in France," *Annual Review of Energy*, vol. 8 (1983): 165.

73. *World Development Indicators* (Washington, D.C.: World Bank, 2009).

74. Robert Lieber, "Energy, Economics and Security in Alliance Perspective," *International Security* 4, no. 4 (1980): 139–63.

75. Ibid. The SS-20 was a Soviet missile that was deployed in the western part of the Soviet Union in the late 1970s. The deployment of these missiles became a concern for France because their precision and multiple-warhead capacity meant that some French targets were vulnerable.

76. *World Development Indicators* (2009).

77. Weissman and Krosney, *Islamic Bomb*, 90.

78. Pierre Lellouche, "Breaking the Rules without Quite Stopping the Bomb: European Views," *International Organization* 35, no. 1 (1981): 50. See also Giraud, "Energy in France."

79. France recognized that Iraq could renege on any petroleum commitment it made, but it was willing to accept this chance given the salience of energy security and oil imports from Iraq in particular.

80. Weissman and Krosney, *Islamic Bomb*.

81. Styan, *France and Iraq*, 132.

82. Lieber, "Energy, Economics and Security in Alliance Perspective," 150.

83. Quoted in Avi Beker, "The Arms-Oil Connection: Fueling the Arms Race," *Armed Forces and Society* 8, no. 3 (1982): 430.

84. Tim McGirk, "How Far Will Iraq Go?" *Christian Science Monitor*, September 25, 1980.

85. Lieber, "Energy, Economics and Security in Alliance Perspective."

86. Of course, this does not undermine the conclusion that France provided Iraq with the reactor primarily to obtain assurances on oil supply. Sometimes policies simply do not have their intended effects in international politics.

87. William Lowrance, "Nuclear Futures for Sale: To Brazil from West Germany, 1975," *International Security* 1, no. 2 (1976): 147–66.

88. Norman Gall, "Atoms for Brazil, Dangers for All," *Foreign Policy*, no. 23 (Summer 1976): 155–201.

89. Ibid.

90. Lowrance, "Nuclear Futures for Sale."

91. Editorial, "Nuclear Madness," *New York Times*, June 13, 1975. The United States was correct in suspecting the presence of a weapons program. The military government of Ernesto Geisel established a secret military nuclear program, although Brazil terminated it in 1990 and joined the NPT regime several years later. Joachim Krause, "German Nuclear Export Policy and the Proliferation of Nuclear Weapons—Another Sonderweg?" Paper prepared for the conference "Germany and Nuclear Nonproliferation," Berlin, February 25, 2005.

92. Mark Hibbs and Angra dos Reis, "A Quarter Century Building, Angra-2 Vows to Match German Performance," *Nucleonics Week* 41, no. 27 (2000): 6.

93. "Siemens Permanent Withdrawal from Nuclear Business Not to Affect Brazil Angra 3 NPP – Report," *SeeNews* (Latin America), September 20, 2011.

94. Mark Hibbs, "Brazil-German Nuclear Accord Extended for Up to Five More Years," *Nucleonics Week*, November 25, 2004. See also "Brazil Agrees to Replace 30-Year-Old Nuclear Accord with Germany," Agence France Presse, November 13, 2004.

95. William Freebairn, "Brazil Plans Increased Uranium Production, Conversion, Enrichment," *Nuclear Fuel*, March 7, 2011. See also "Angra 3 Restart Gets $6.1 Billion," *Nuclear Engineering International*, February 3, 2011.

96. "Siemens Permanent Withdrawal from Nuclear Business;" and Hiroko Tabuchi, "Japan is Bullish on Nuclear Power Overseas," *International Herald Tribune*, October 12, 2011.

97. "Uranium for Know-How," *Chemical Week*, July 9, 1975; Gall, "Atoms for Brazil, Dangers for All."

98. Gall, "Atoms for Brazil, Dangers for All," 174.

99. Ibid.

100. Claus Hofhansel, *Commercial Competition and National Security: Comparing U.S. and German Export Control Policies* (Westport, Conn.: Praeger, 1996).

101. Lowrance, "Nuclear Futures for Sale," 154.

102. Gall, "Atoms for Brazil, Dangers for All," 166.

103. Lowrance, "Nuclear Futures for Sale," 154.

104. The 1999 agreement was signed under the framework of the 1986 Indo-Vietnamese agreement for nuclear cooperation.

105. P. S. Suryanarayana, "India, Vietnam Sign Pact on N-Energy Cooperation," *Hindu*, January 21, 1999.

106. Amit Baruali, "Vajpayee Offers to Develop Vietnam's Infrastructure," *Hindu*, January 10, 2001.

107. "Vietnam, India Sign Joint Statement on Comprehensive Cooperation in 21st Century," *Voice of Vietnam*, May 2, 2003.

108. G. V. C. Naidu, "Whither the Look East Policy: India and Southeast Asia," *Strategic Analysis* 28, no. 2 (2004): 331–46.

109. Faizal Yahya, "India and Southeast Asia: Revisited," *Contemporary Southeast Asia* 25, no. 1 (2003): 79–103.

110. Quoted in John Ruwitch and Nguyen Nhat Lam, "Interview—Nuclear Energy to Power 10 Percent of Vietnam by 2030," *Reuters*, August 6, 2009.

111. Ibid.

112. Sandeep Joshi, "Vietnam Favors FTA with India," *Hindu*, July 7, 2007.

113. John Cherian, "Friend in Need," *Frontline*, July 14–27, 2007.

114. C. Raja Mohan, "India's Geopolitics and Southeast Asian Security," *Southeast Asian Affairs* Vol. 2008 (April 2008): 52.

115. Yogendra Singh, *India-Vietnam Relations: The Road Ahead* (New Delhi: Institute of Peace and Conflict Studies, 2007).

116. Cherian, "Friend in Need."

117. Amit Kumar, "Fresh Impetus for India-Vietnam Strategic Ties: China as the Driving Force," *South Asia Monitor* (December 2007).

118. Interestingly, both India and Vietnam publicly claim to be developing more friendly relations with China but the actions of both countries are clearly motivated to balance against its growing influence in the region. Cherian, "Friend in Need." See also Anindya Batabyal, "Balancing China in Asia: A Realist Assessment of India's Look East Strategy," *China Report* 42, no. 2 (2006): 179–97.

119. Micool Brooke, "India Courts Vietnam with Arms and Nuclear Technology," *Asia-Pacific Defence Reporter* 25, no. 5 (2000): 20.

120. Note, however, that some Indian strategic thinkers have argued that India should provide additional nuclear technology to Vietnam to impose constraints on Beijing. Bharat Karnad, *Nuclear Weapons and Indian Security: The Realist Foundations of Strategy* (New Delhi: Macmillan, 2002).

121. On India's cultivation of ties with Southeast Asian states as a hedging strategy, see Walter C. Ladwig, "Delhi's Pacific Ambition: Naval Power, 'Look East,' and India's Emerging Influence in the Asia-Pacific," *Asian Security* 5, no. 2 (2009): 90.

122. Weissman and Krosney, *Islamic Bomb*.

123. Ibid, 98.

124. Marshall, "Fallout from the Raid on Iraq."

125. Weissman and Krosney, *Islamic Bomb*, 265.

126. Joseph Nye, "Energy Nightmares," *Foreign Policy* no. 40 (1980): 132–54; Lellouche, "Breaking the Rules without Quite Stopping the Bomb"; and Leonard Spector, "Silent Spread," *Foreign Policy* no. 58 (1985): 53–78;

127. Weissman and Krosney, *Islamic Bomb*, 97.

128. Ibid, 265.

129. Richard Gardner, *Mission Italy: On the Front Lines of the Cold War* (Lanham, Md.: Rowman and Littlefield, 2005), 255; and Weissman and Krosney, *Islamic Bomb*.

130. Richard Burt, "Italy Selling Nuclear Know-How to Iraq," *New York Times*, March 18, 1980.

131. Weissman and Krosney, *Islamic Bomb*, 97–98.

132. Ibid.

133. Carlo Mancini and Giuseppe Maria Borga, "Italy's Policies and Practices," in *The Nuclear Suppliers and Nonproliferation*, ed. Rodney Jones, Cesare Merlini, Joseph Pilat, and William Potter (Lexington, Mass.: Lexington Books, 1985), 58.

134. Weissman and Krosney, *Islamic Bomb*, 97.

135. Not surprisingly, the French were equally quick to blame the Italians. Ibid, 101–2.

136. William Potter, "The Soviet Union and Nuclear Proliferation," *Slavic Review* 44, no. 3 (1985), 468–88; William Potter, Djuro Miljanic, and Ivo Slaus, "Tito's Nuclear Legacy," *Bulletin of the Atomic Scientists* 56, no. 2 (2000): 63–70; and Department of Foreign Affairs and International Trade Canada, "Relations between the Soviet Union and Yugoslavia" (Ottawa, 1956).

137. Andrew Koch, "Yugoslavia's Nuclear Legacy: Should We Worry?" *Nonproliferation Review* 4, no. 2 (1997): 123–28.

138. Potter, Miljanic, and Slaus, "Tito's Nuclear Legacy."

139. Yugoslavia was not pursuing nuclear weapons based on the criteria employed in this book but it was at least exploring the development of nuclear weapons. See Singh and Way, "Correlates of Nuclear Proliferation."

140. Potter, Miljanic, and Slaus, "Tito's Nuclear Legacy."

141. Koch, "Yugoslavia's Nuclear Legacy."

142. Yugoslavia resumed the nuclear weapons program in 1974, following the Indian nuclear test, only to suspend it again in 1987. Potter, Miljanic, and Slaus, "Tito's Nuclear Legacy."

143. This policy was altered in the late 1950s when China used Soviet nuclear aid to acquire nuclear weapons. Gloria Duffy, *Soviet Nuclear Energy: Domestic and International Policies* (Santa Monica, Calif.: RAND, 1979).

144. George Ginsburgs, "Soviet Atomic Energy Agreements," *International Organization* 15, no. 1 (1961): 48–65.

145. Paul Zinner, "Soviet Policies in Eastern Europe," *Annals of the American Academy of Political and Social Science* 303 (1956): 152–65.

146. United States Central Intelligence Agency, "Yugoslavia and Its Future Orientation" (Washington, D.C.: National Intelligence Council, 1955).

147. Zinner, "Soviet Policies in Eastern Europe."

148. Ibid.

149. Department of Foreign Affairs and International Trade Canada, "Relations between the Soviet Union and Yugoslavia" (Ottawa, 1956).

150. Ginsburgs, "Soviet Atomic Energy Agreements"; and Zinner, "Soviet Policies in Eastern Europe."

151. Central Intelligence Agency, "Yugoslavia and Its Future Orientation."

152. Branko Peselj, "Communist Economic Offensive Soviet Foreign Aid: Means and Effects," *Law and Contemporary Problems* 29, no. 4 (1964): 983–99.

153. Central Intelligence Agency, "Yugoslavia and Its Future Orientation."

154. Jacques Hymans, "Proliferation Implications of Civilian Nuclear Cooperation," *Security Studies,* 20, no. 1 (2011): 89.

CHAPTER 6

1. The United States does not produce a substantial amount of electricity at oil-fired power plants, but many countries rely on oil for a significant percentage of their electricity production.

2. For an overview of the relevant arguments, see Jeff Colgan, "Oil and Revolutionary Governments: Fuel for International Conflict," *International Organization* 64, no. 4 (2010): 661–94.

3. For example, Steve Chan, "The Consequences of Expensive Oil on Arms Transfers," *Journal of Peace Research* 17, no. 3 (1980): 235–46; and Lewis Snider, "Arms Exports for Oil Imports? The Test of a Nonlinear Model," *Journal of Conflict Resolution* 28, no. 4 (1984): 665–700.

4. Some scholars have suggested that Soviet trade with Libya, which I discussed in chapter 4, was also part of a strategy to obtain oil and meet growing domestic needs. See Robert Freedman, *Soviet Policy toward the Middle East since 1970* (New York: Praeger, 1982), 165.

5. See, for example, Joseph Sisco, *Study Report: Joint U.S.-Iranian Cooperation* (Washington, D.C.: Department of State, Interdepartmental Working Group, April 25, 1974).

6. Kissinger recognized that the Europeans and Japan were considering similar policies vis-à-vis Iran and he wanted to offer Tehran an "alternative." Henry Kissinger, "U.S. Iranian Cooperation," Department of State Telegram, March 11, 1974.

7. Ibid.

8. Sisco, *Study Report.*

9. Given that the evidence cited above comes from declassified documents, it is unlikely that it merely reflects an effort to sell the policy to the public. If this were the case, we would

expect policymakers to connect atomic assistance and oil in public speeches but not necessarily in private correspondence with other officials. U.S. elites recognized that this issue was complicated but they genuinely believed that providing the shah with nuclear power *could* lead to concessions on energy prices and this motivated their pursuit of peaceful nuclear cooperation.

10. Suzanne Goldenberg, "Obama Breaks with Bush Oil Bosses and Puts Environment at Top of Agenda," *Guardian*, December 16, 2008.

11. "India-Kazakhstan Agreement Likely on Uranium Supply, Nuclear Cooperation," *BBC Worldwide Monitoring*, January 12, 2009; "Kazakh, Japanese Leaders Sign Up to Nuclear Cooperation," *RIA Novosti*, August 28, 2006; and "Kazakhstan Plans Civil Nuclear Cooperation with UK," *BBC Worldwide Monitoring*, July 16, 2006.

12. Borzou Daragahi, "U.S. and UAE Forge Nuclear Cooperation Deal," *Los Angeles Times*, December 16, 2008; "U.S. Unveils Deals with Saudi on Nuclear Power, Oil Protection," Agence France Presse, May 16, 2008; and Lisa Bryant, "French President Sarkozy Signs Nuclear, Military Accords in Gulf," *Voice of America*, January 15, 2008.

13. Ann MacLachlan, "France, Libya Initial Nuclear Pact, Emphasize Infrastructure-Building," *Nucleonics Week*, December 13, 2007; and Bryant, "French President Sarkozy Signs Nuclear, Military Accords in Gulf."

14. "Russia-Venezuela Nuclear Accord," *BBC News*, November 27, 2008; and "Bahrain, Russia Sign Peaceful Nuclear Energy Memorandum of Understanding," *BBC Worldwide Monitoring*, December 3, 2008.

15. "S. Korea Promotes Sale of Indigenous Nuclear Reactor to India," Yonhap (South Korea), July 25, 2007; "Iran, Nigeria to Share Peaceful Nuclear Technology," Associated Press, August 28, 2008; and "Argentina, Algeria Sign Nuclear Cooperation Agreements," Xinhua, November 11, 2008.

16. Rory Carroll and Luke Harding, "Russia to Build Nuclear Reactor for Chavez," *Guardian*, November 19, 2008.

17. See Daragahi, "U.S. and UAE Forge Nuclear Cooperation Deal."

18. "Kazakh, Japanese Leaders Sign Up to Nuclear Cooperation."

19. "Bahrain, Russia Sign Peaceful Nuclear Energy Memorandum of Understanding."

20. "US Unveils Deals with Saudi on Nuclear Power, Oil Protection."

21. Ibid.

22. Allegra Straton, "France Signs 10bn Trade Deal with Libya," *Guardian*, December 10, 2007.

23. Jela De Franceschi, "Debating France's Nuclear Diplomacy," *Voice of America*, April 21, 2008; and Matthew Fuhrmann, "Oil for Nukes—Mostly a Bad Idea," *Christian Science Monitor*, February 29, 2008.

24. Mycle Schneider, "The Reality of France's Aggressive Nuclear Power Push," *Bulletin of the Atomic Scientists* (online), June 3, 2008. Available at http://www.thebulletin.org/web-edition/op-eds/the-reality-of-frances-aggressive-nuclear-power-push.

25. Laurent Pirot, "French Offer Saudi Nuclear Energy Help," Associated Press, January 13, 2008; and "Nuclear Power: France Offers to Help Saudi Arabia with Development," *Greenwire*, January 14, 2008.

26. Elaine Ganley, "France, Libya Sign Deals on Armaments, Nuclear Reactor," *USA Today*, December 10, 2007.

27. Ibid.

28. Hugh Schofield, "Sarkozys Lead French Libya Push," *BBC News*, July 24, 2007.

29. Katrin Bennhold, "France Looks for Deals with Libya's Military," *International Herald Tribune*, February 7, 2005.

30. Dong-Joon Jo and Erik Gartzke, "Determinants of Nuclear Weapons Proliferation," *Journal of Conflict Resolution* 51, no. 1 (2007): 1–28; and Sonali Singh and Christopher Way, "The Correlates of Nuclear Proliferation: A Quantitative Test," *Journal of Conflict Resolution* 48, no 6 (2004): 859–85.

31. See, for example, Gordon Corera, *Shopping for Bombs: Global Insecurity, Nuclear Proliferation, and the Rise and Fall of the A.Q. Khan Network* (Oxford: Oxford University Press, 2006).

32. In addition to the changes mentioned here I include all of the explanatory variables in the statistical model and I do not distinguish between democratic and nondemocratic suppliers.

33. Fearon and Laitin define a country as an oil exporter if its oil exports exceed one-third of its total export revenues in a particular year. They rely on data from the World Bank to construct this measure. See James Fearon and David Laitin, "Insurgency, Ethnicity, Identity, and Civil War," *American Political Science Review* 97, no. 1 (2003): 75–90.

34. Susan Carter, Scott Sigmund Gartner, Michael Haines, Alan Olmstead, Richard Sutch, and Gavin Wright, "Table Db56-59 Crude Petroleum—Average Value, Foreign Trade, and Proved Reserves: 1859–2001," *Historical Statistics of the United States* (Cambridge: Cambridge University Press, 2006).

35. The coefficient on the variable measuring the supplier state's nuclear power reactors per capita is negative and statistically significant, consistent with the export pressure argument; without the oil variable in the model, this variable was positively correlated with nuclear assistance.

CHAPTER 7

1. On the political effects of nuclear weapons see, for example, Thomas Schelling, *Arms and Influence* (New Haven: Yale University Press, 1966); Kenneth Waltz, "Nuclear Myths and Political Realities," *American Political Science Review* 84, no. 3 (1990): 731–45; Robert Powell, *Nuclear Deterrence Theory: The Search for Credibility* (Cambridge: Cambridge University Press, 1990); Erik Gartzke and Dong-Joon Jo, "Bargaining, Nuclear Proliferation, and International Disputes," *Journal of Conflict Resolution* 52, no. 2 (2009): 209–33; Michael Horowitz, "The Spread of Nuclear Weapons and International Conflict: Does Experience Matter," *Journal of Conflict Resolution* 52, no. 2 (2009): 234–57; and Kyle Beardsley and Victor Asal, "Winning with the Bomb," *Journal of Conflict Resolution* 52, no. 2 (2009): 278–301.

2. There is, however, a limit to the coercive effects of nuclear weapons. Although nuclear weapons may be useful for deterrence, they do not necessarily allow states to blackmail their enemies into making changes to the existing status quo (i.e., compellence). On nuclear blackmail, see Richard Betts, *Nuclear Blackmail and Nuclear Balance* (Washington, D.C.: Brookings, 1987); and Todd Sechser and Matthew Fuhrmann, "Crisis Bargaining and Nuclear Blackmail," unpublished manuscript, University of Virginia, Charlottesville.

3. George Quester, *The Politics of Nuclear Proliferation* (Baltimore: Johns Hopkins University Press, 1973); William Potter, *Nuclear Power and Nonproliferation: An Interdisciplinary Perspective* (Cambridge, Mass.: Oelgeschlager, Gunn and Hain, 1982); and Sonali Singh and Christopher Way, "The Correlates of Nuclear Proliferation: A Quantitative Test," *Journal of Conflict Resolution* 48, no. 6 (2004): 859–85.

4. See Stephen Schwartz, *Atomic Audit: The Costs and Consequences of U.S. Nuclear Weapons since 1940* (Washington, D.C.: Brookings, 1998).

5. See, for example, Etel Solingen, *Nuclear Logics: Contrasting Paths in East Asia and the Middle East* (Princeton: Princeton University Press, 2007). Iran has suffered economically for its alleged pursuit of nuclear weapons; the United Nations Security Council (UNSC) has passed several resolutions sanctioning the country.

6. Michael C. Horowitz, *The Diffusion of Military Power: Causes and Consequences for International Politics* (Princeton: Princeton University Press, 2010), 99.

7. Matthew Fuhrmann and Sarah Kreps, "Targeting Nuclear Programs in War and Peace: A Quantitative Empirical Analysis, 1942–2000," *Journal of Conflict Resolution* 54, no. 6 (2010): 831–59; Sarah Kreps and Matthew Fuhrmann, "Attacking the Atom: Does Bombing

Nuclear Facilities Affect Proliferation?" *Journal of Strategic Studies* 34, no. 2 (2011): 161–87; and Dan Reiter, "Preventive Attacks against Nuclear, Biological, and Chemical Weapons Programs: The Track Record," in *Hitting First, Preventive Force in US Security Strategy*, by William Walton Keller and Gordon R. Mitchell (Pittsburgh, Penn.: Pittsburgh University Press, 2006).

8. United States Office of Technology Assessment (OTA), *Technologies Underlying Weapons of Mass Destruction* (Washington, D.C.: U.S. Government Printing Office, 1993).

9. Joseph Cirincione, Jon Wolfsthal, and Miriam Rajkumar, *Deadly Arsenals: Nuclear, Biological, and Chemical Threats* (Washington: Carnegie Endowment for International Peace, 2005).

10. Roberta Wohlstetter, "The Buddha Smiles: U.S. Peaceful Aid and the Indian Bomb," in *Nuclear Policies: Fuel without the Bomb*, ed. Albert Wohlstetter, Victor Gilinsky, Robert Gillette and Roberta Wohlstetter (Cambridge, Mass.: Ballinger, 1978), 57–72.

11. Recognizing the proliferation consequences of using plutonium in civilian nuclear reactors, many countries ceased doing so in the 1970s and 1980s. Cirincione, Wolfsthal, and Rajkumar, *Deadly Arsenals*; and M. D. Zentner, G. L. Coles, and R. J. Talbert, *Nuclear Proliferation Technology Trends Analysis* (Richland, Wash.: United States Department of Energy, Pacific Northwest National Laboratory, 2005).

12. Donald MacKenzie and Graham Spinardi, "Tacit Knowledge, Weapons Design, and the Uninvention of Nuclear Weapons," *American Journal of Sociology* 101, no. 1 (1995): 44–99. Neutronics is the study of the behavior of neutrons in fissile materials. It is important from a weapons design standpoint because expertise in this field helps ensure that bombs explode rather than fissile and that a critical mass is not prematurely formed during the assembly of fissile material.

13. OTA, *Technologies Underlying Weapons of Mass Destruction*, 153.

14. Stephen Meyer, *The Dynamics of Nuclear Proliferation* (Chicago: University of Chicago Press, 1984), 143; and Ted Greenwood, Harold Feiveson, and Theodore Taylor, *Nuclear Proliferation: Motivations, Capabilities, and Strategies for Control* (New York: McGraw Hill, 1977), 150.

15. Meyer, *Dynamics of Nuclear Proliferation*, 143.

16. See, for example, Morton Halperin, *Bureaucratic Politics and Foreign Policy* (Washington, D.C.: Brookings, 1974). For an excellent overview of how bureaucratic politics can influence proliferation, see Scott Sagan, "Why Do States Build Nuclear Weapons? Three Models in Search of a Bomb," *International Security* 21, no. 3 (1996–97): 63–65. Scientists do not always push leaders down the nuclear path, but in many cases they do. In Germany, for example, scientists lobbied leaders not to develop the bomb. I thank Etel Solingen for this insight.

17. Itty Abraham, "Contra-Proliferation: Interpreting the Meanings of India's Nuclear Tests in 1974 and 1988," in *Inside Nuclear South Asia*, ed. Scott Sagan (Stanford: Stanford University Press, 2009), 109.

18. Scientists typically overplay the indigenous contribution to a technological breakthrough, even when foreign assistance plays a critical role. See ibid, 109–10.

19. See, for example, Sagan, "Why Do States Build Nuclear Weapons?" 63–64; Peter Liberman, "The Rise and Fall of the South African Bomb," *International Security* 26, no. 2 (2001): 45–86; and Peter Lavoy, "Nuclear Myths and the Causes of Nuclear Proliferation," *Security Studies* 2, nos. 3–4 (1993): 192–212.

20. Quoted in Matthew Bunn, *Civilian Nuclear Energy and Nuclear Weapons Programs: The Record* (Cambridge, Mass.: Belfer Center for Science and International Affairs, 2001).

21. Sagan, "Why Do States Build Nuclear Weapons?" 64.

22. Dong-Joon Jo and Erik Gartzke, "Determinants of Nuclear Weapons Proliferation," *Journal of Conflict Resolution* 51, no. 1 (2007): 1–28. See also Singh and Way, "Correlates of Nuclear Proliferation."

23. Gary King and Langche Zeng, "Logistic Regression in Rare Events Data," *Political Analysis* 9, no. 2 (1998): 137–63.

24. Readers may notice that the substantive effects reported in the book—which are based on an updated dataset—differ from the effects reported in some of my earlier work even though there is little difference in terms of statistical significance. See Matthew Fuhrmann, "Spreading Temptation: Proliferation and Peaceful Nuclear Cooperation Agreements," *International Security* 34, no. 1 (2009): 28.

25. For a discussion on the utility of statistics in this regard see Singh and Way, "Correlates of Nuclear Proliferation," 860–61.

26. Singh and Way, "Correlates of Nuclear Proliferation."

27. It would be misleading, for instance, to code Russia as having a nuclear weapons program for every year since 1945. Note, however, that countries can reenter the dataset if they end their weapons program (e.g., Brazil post-1990). See Alexander Montgomery and Scott Sagan, "The Perils of Predicting Proliferation," *Journal of Conflict Resolution* 52, no. 2 (2009): 302–28.

28. Richard Betts, "Paranoids, Pygmies, Pariahs, and Nonproliferation Revisited," in *The Proliferation Puzzle: Why Nuclear Weapons Spread (and What Results)*, ed. Zachary Davis and Benjamin Frankel (Portland, Ore.: Frank Cass, 1993); and Bradley Thayer, "The Causes of Nuclear Proliferation and the Nonproliferation Regime," *Security Studies* 4, no. 3 (1993): 463–519.

29. Singh and Way, "Correlates of Nuclear Proliferation," 863.

30. For research that emphasizes the role of norms in international security see, for example, Nina Tannenwald, *The Nuclear Taboo: The United States and the Non-Use of Nuclear Weapons since 1945* (Cambridge: Cambridge University Press, 2007); and Richard Price, *The Chemical Weapons Taboo* (Ithaca: Cornell University Press, 1997).

31. Sagan, "Why Do States Build Nuclear Weapons?" 73–82; and Jo and Gartzke, "Determinants of Nuclear Weapons Proliferation," 171.

32. See William Long and Suzette Grillot, "Ideas, Beliefs, and Nuclear Policies: The Cases of South Africa and Ukraine," *Nonproliferation Review* 7, no. 1 (2000): 24–40; and Maria Rost Rublee, *Nonproliferation Norms: Why States Choose Nuclear Restraint* (Athens; University of Georgia Press, 2009).

33. See Shahram Chubin. "The Middle East," in *Nuclear Proliferation after the Cold War*, in Mitchell Reiss and Robert Litwak (Cambridge, Mass.: Ballinger, 1994), 33–66; and Ali Sheikh, "Pakistan," in *Nuclear Proliferation after the Cold War*, ed. Reiss and Litwak, 191–206.

34. See Jack Snyder, *From Voting to Violence: Democratization and Nationalist Conflict* (New York: W. W. Norton, 2000).

35. Solingen, *Nuclear Logics*.

36. Singh and Way, "Correlates of Nuclear Proliferation," 864.

37. Ibid.

38. To make this calculation I set the NCA variable to 0 and all variables to their sample means. This calculation is based on model 2 (see appendix 7.1).

39. The squared term for GDP per capita is statistically significant but GDP per capita is insignificant.

40. See, for example, Singh and Way, "Correlates of Nuclear Proliferation"; Jo and Gartzke, "Determinants of Nuclear Weapons Proliferation"; and Fuhrmann, "Spreading Temptation."

41. I thank an anonymous reviewer for this insight.

42. Ariel Levite, "Never Say Never Again: Nuclear Reversal Revisited," *International Security* 27, no. 3 (Winter 2002–03): 59–88.

43. Matthew Kroenig, *Exporting the Bomb: Technology Transfer and the Spread of Nuclear Weapons* (Ithaca: Cornell University Press, 2010), 172.

44. The marginal effect is the probability of weapons program onset in the "treatment" group (i.e., high levels of assistance) minus the probability of program initiation in the "control" group (i.e., low levels of assistance).

45. Specifically, the probability rises from 0.000012 to 0.000049.

46. At high levels of conflict, the probability of weapons program onset approaches 1 with increases in peaceful aid, but countries that face so many security threats are also likely to proliferate in the absence of assistance. This is why the marginal effect displayed in figure 7.3 declines slightly after fourteen disputes.

47. J. W. de Villiers, Roger Jardine, and Mitchell Reiss, "Why South Africa Gave Up the Bomb," *Foreign Affairs* 72, no. 5 (November–December 1993), 99.

48. Helen Purkitt and Stephen Burgess, *South Africa's Weapons of Mass Destruction* (Bloomington: Indiana University Press, 2005), 27. See also Mitchell Reiss, *Without the Bomb: The Politics of Nuclear Nonproliferation* (New York: Columbia University Press, 1988), 179.

49. Recall from chapter 3 that profit motives shed light on why suppliers provide nuclear assistance but the principal goal is to gain political leverage and influence.

50. Zdnek Cervenka and Barbara Rogers, *The Nuclear Axis: Secret Collaboration between West Germany and South Africa* (New York: Times Books, 1978), 164.

51. Purkitt and Burgess, *South Africa's Weapons of Mass Destruction*, 32.

52. Ronald Walters, *South Africa and the Bomb: Responsibility and Deterrence* (Lexington, Mass.: Lexington Books, 1987), 23.

53. Purkitt and Burgess, *South Africa's Weapons of Mass Destruction*, 35.

54. Reiss, *Without the Bomb*, 182–83.

55. Ibid, 182.

56. David Albright, "South Africa and the Affordable Bomb," *Bulletin of the Atomic Scientists* 50, no. 4 (July–August 1994), 40.

57. Quoted in Frank V. Pabian, "South Africa's Nuclear Weapons Program: Lessons for U.S. Nonproliferation Policy," *Nonproliferation Review* 3, no. 1 (1995): 2

58. Reiss, *Without the Bomb*, 182.

59. Purkitt and Burgess, *South Africa's Weapons of Mass Destruction*, 35.

60. See Cervenka and Rogers, *Nuclear Axis*.

61. Purkitt and Burgess, *South Africa's Weapons of Mass Destruction*, 35.

62. Albright, "South Africa and the Affordable Bomb," 39.

63. Walters, *South Africa and the Bomb*, 26.

64. Liberman, "The Rise and Fall of the South African Bomb," 65.

65. Ibid, 64.

66. Purkitt and Burgess, *South Africa's Weapons of Mass Destruction*, 38.

67. Mitchell Reiss, *Bridled Ambition: Why Countries Constrain Their Nuclear Capabilities* (Baltimore: Johns Hopkins University Press, 1995), 29. The military opposed programs with applications for nuclear weapons as late as 1977 due to cost concerns. See Verne Harris, Sello Hatang, and Peter Liberman, "Unveiling South Africa's Nuclear Past," *Journal of Southern African Studies* 30, no. 3 (2004): 462.

68. Reiss, *Bridled Ambition*, 29.

69. Walters, *South Africa and the Bomb*, 26.

70. Quoted in Reiss, *Without the* Bomb, 187.

71. Liberman, "The Rise and Fall of the South African Bomb," 64.

72. Quoted in ibid.

73. Purkitt and Burgess, *South Africa's Weapons of Mass Destruction*, 41.

74. Albright, "South Africa and the Affordable Bomb," 37.

75. Waldo Stumpf, "South Africa's Nuclear Weapons Program: From Deterrence to Dismantlement," *Arms Control Today* 25, no. 10 (December–January, 1995–96), 4; and Liberman, "Rise and Fall of the South African Bomb." Donald Sole, the former South African ambassador

to the United States, also highlights the import role of scientists in convincing Vorster to authorize the weapons program. See Donald Sole, "The South African Nuclear Case in Light of Recent Revelations," in *New Horizons and Challenges in Arms Control and Verification*, ed. James Brown (Amsterdam: V. U. University Press, 1994), 71–80.

76. There is some disagreement about precisely when South Africa initiated the nuclear weapons program. See Purkitt and Burgess, *South Africa's Weapons of Mass Destruction*, 41; and Albright, "South Africa and the Affordable Bomb," 43.

77. Purkitt and Burgess, *South Africa's Weapons of Mass Destruction*, 41–42.

78. Quoted in Liberman, "Rise and Fall of the South African Bomb," 64.

79. Note that the military leadership was not involved in this decision. In fact, had the military been involved, they would have advised Vorster not to build nuclear weapons. See Liberman, "Rise and Fall of the South African Bomb," 66–67; Sagan, "Why Do States Build Nuclear Weapons?" 70.

80. Reiss, *Bridled Ambition*, 29.

81. For a discussion of security threats contributing to the weapons program, see Purkitt and Burgess, *South Africa's Weapons of Mass Destruction*, 52–55.

82. Quoted in Sagan, "Why Do States Build Nuclear Weapons?" 69.

83. Ibid.

84. Harris, Hatang, and Liberman, "Unveiling South Africa's Nuclear Past."

85. Nehru went on to acknowledge that nuclear technology could be used for military purposes but emphasized that this was not the purpose of commencing a nuclear program. Quoted in George Perkovich, *India's Nuclear Bomb: The Impact on Global Proliferation* (Berkeley: University of California Press, 1999), 20.

86. Ibid.

87. Roberta Wohlstetter, "The Buddha Smiles: U.S. Peaceful Aid and the Indian Bomb," in *Nuclear Policies: Fuel without the Bomb*, ed. Albert Wohlstetter, Victor Gilinsky, Robert Gillette and Roberta Wohlstetter (Cambridge, Mass.: Ballinger, 1978), 57–72.

88. Shyam Bhatia, *India's Nuclear Bomb* (Ghaziabad, India: Vikas, 1979).

89. Duane Bratt, *The Politics of CANDU Exports* (Toronto: University of Toronto Press, 2006), 89.

90. Perkovich, *India's Nuclear Bomb*, 30.

91. Ibid, 64.

92. Nuclear Threat Initiative, "India Nuclear Chronology, 1960–1964," August 2003. Available at http://www.nti.org/e_research/profiles/India/Nuclear/2296_2346.html.

93. Bratt, *Politics of CANDU Exports.*

94. Abraham, "Contra-Proliferation," 109.

95. Ashok Kapur, *Pokhran and Beyond: India's Nuclear Behaviour* (Oxford: Oxford University Press, 2001), 160.

96. Ibid.

97. Ibid, 159. Bhabha occasionally argued that mastering nuclear technology would greatly enhance economic development and it was cost effective compared to the alternatives, but these statements were not viewed as sincerely as remarks made by Nehru. See Perkovich, *India's Nuclear Bomb*, 20.

98. Perkovich, *India's Nuclear Bomb*, 35.

99. Ibid.

100. Reiss, *Without the Bomb*, 215.

101. Singh and Way, "Correlates of Nuclear Proliferation." Additional datasets code 1964 as the first year of Indian nuclear weapons pursuit. See, especially, Philipp Bleek, "Assessing Reactive Proliferation: Why Nuclear Dominoes Rarely Fall," PhD diss., Georgetown University, 2010.

102. Reiss, *Without the Bomb*, 216.

103. Raja Ramanna, *Years of Pilgrimage: An Autobiography* (New Delhi: Viking, 1991), 74. See also Sagan, "Why Do States Build Nuclear Weapons?" 66.

104. Sagan, "Why Do States Build Nuclear Weapons?" 66.

105. K. Rangaswami, "Leaders Reject Demand for Atom Bomb," *Hindu*, November 9, 1964.

106. Quoted in Perkovich, *India's Nuclear Bomb*, 65.

107. Quoted in Reiss, *Without the Bomb*, 213.

108. Perkovich, *India's Nuclear Bomb*, 84.

109. Ibid, 65.

110. Ibid, 71.

111. Ibid, 83.

112. Brahma Chellaney, "India," in *Nuclear Proliferation after the Cold War*, ed. Reiss and Litwak, 165–90; and Kapur, *India's Nuclear Option*; and Kapur, *Pokhran and Beyond*. It is important not to overstate the impact of China on Indian nuclear decision making. See George Perkovich, "The Nuclear and Security Balance," in *The India-China Relationship: What The United States Needs to Know*, ed. Francine Frankel and Harry Harding (New York: Columbia University Press, 2004), 183.

113. "India Urged to Produce Atom Bomb," *Times of India*, October 26, 1964.

114. Bhumitra Chakma, "Toward Pokhran II: Explaining India's Nuclearisation Process," *Modern Asia Studies* 39, no. 1 (2005): 201.

115. For arguments on the inadequacies of realism in explaining the Indian case, see, for example, Sagan, "Why Do States Build Nuclear Weapons?" 65–66; and Jacques Hymans, *The Psychology of Nuclear Proliferation: Identity, Emotions, and Foreign Policy* (Cambridge: Cambridge University Press, 2006), 171–203.

116. Quoted in Chris Smith, *India's Ad Hoc Arsenal: Direction or Drift in Defense Policy?* (Oxford: Oxford University Press, 1994), 181.

117. Quoted in Perkovich, *India's Nuclear Bomb*, 67.

118. See ibid, 70.

119. It is difficult to reach a stronger conclusion given that the Chinese nuclear test coincided with critical technological advances, such as the separation of plutonium from the Trombay reprocessing plant.

120. J. P. Jain, *Nuclear India*, vol. 2 (New Delhi: Radiant, 1974), 179.

121. It is unclear based on the available evidence whether Gandhi authorized Sarabhai's decision to end the explosives program. But she did not explicitly endorse the program in her initial years in office, as Prime Minister Shastri did in 1964. See Perkovich, *India's Nuclear Bomb*, 121.

122. Ibid, 123.

123. P. K. Iyengar, "Twenty Years after Pokhran," *India Express*, May 18, 1994, 11.

124. It could have also come from the British-supplied Apsara research reactor or the indigenously built Zerlina reactor, which went critical in 1961. See Arjun Makhijani, Howard Hu, and Katherine Yih, *Nuclear Wastelands: A Global Guide to Nuclear Weapons Production and Its Health and Environmental Effects* (Cambridge: MIT Press, 1995).

125. Perkovich, *India's Nuclear Bomb*, 150.

126. Francine Frankel, *India's Political Economy: The Gradual Revolution* (Princeton: Princeton University Press, 1978), 493.

127. Indian Institute of Public Opinion, *Monthly Public Opinion Surveys* 16, nos. 11–12 (August-September): 53.

128. K. Subrahmanyam, "Options for India," *Institute for Defense Studies and Analysis Journal* 3, no. 1 (1970):102–18.

129. Perkovich, *India's Nuclear Bomb*, 157, 174.

130. Quoted in ibid, 169.

131. Ibid, 176.

132. Ibid, 175.

133. Singh and Way also code the initiation of a weapons program based on the development of dedicated military facilities. Singh and Way, "Correlates of Nuclear Proliferation," 866.

134. Singh and Way code India as launching a weapons program on two separate occasions (1964 and 1980).

135. Technology, materials, and knowledge generally cannot be taken away once they are provided. For example, if a country received such assistance in 1970, this aid would still be around in 1980 (and in subsequent years).

136. I am interested in whether countries *receiving* peaceful nuclear assistance are more likely to proliferate. Some agreements have one supplier and one recipient state. Others authorize both countries to receive assistance. In the latter case, both parties to the agreement are coded as recipients.

137. Faten Ghosn, Glenn Palmer, and Stuart Bremer, "The MID3 Data Set, 1993–2001: Procedures, Coding Rules, and Description," *Conflict Management and Peace Science* 21, no. 2 (2004): 133–54.

138. When testing conditional hypothesis (e.g., hypothesis 7.2), it is necessary to include an interaction term along with the two constituent variables. Thomas Brambor, William Clark, and Matt Golder, "Understanding Interaction Models: Improving Empirical Analysis," *Political Analysis* 14, no. 1 (2006): 63–82.

139. Singh and Way, "Correlates of Nuclear Proliferation."

140. Higher values indicate greater levels of democracy. See Monty Marshall and Keith Jaggers, "Polity IV Project: Political Regime Characteristics and Transitions, 1800–2002" (Fairfax, Va.: George Mason University, 2002).

141. Singh and Way take their GDP data from version 6.1 of the Penn World Tables. A. Heston, R. Summers, and B. Aten, *Penn World Table Version 6.1* (Philadelphia: University of Pennsylvania, Center for International Comparisons, 2002).

142. I code this variable based on the James Martin Center for Nonproliferation Studies' *Inventory of International Nonproliferation Organizations and Regimes* (Monterey, Calif.: Monterey Institute for International Studies, 2009).

143. Nathaniel Beck, Jonathan Katz, and Richard Tucker, "Taking Time Seriously in Binary Time-Series-Cross-Section Analysis," *American Journal of Political Science* 42, no. 4 (1998): 1260–88.

144. King and Zeng, "Logistic Regression in Rare Events Data."

145. Rare events logit is appropriate when the dependent variable has thousands of times fewer ones than zeros. The dependent variable used here is well within the ratio of ones to zeros that is suitable for rare events logit; weapons program onset occurs in 0.24 observations in the dataset.

146. One could argue that there is a threshold effect involving NCAs whereby making a few agreements increases the risk of proliferation but many agreements makes states more likely to foreswear the bomb. In other words, there is a "tipping point" with respect to peaceful nuclear cooperation. To test for this, I add a squared term of the NCA variable to the models displayed in table 7.3. The results did not support this conjecture.

147. GDP per capita squared is statistically significant in both models but I excluded it from figure 7.2 because GDP per capita was not significant. The "inverted-U" relationship only works if both of these variables achieve statistical significance.

148. In another test not reported here, I removed variables that were statistically insignificant in model 2. Nuclear cooperation agreements, militarized disputes, and rivalry all remained positive and statistically significant. The statistical significance of GDP per capita squared washed away, however.

149. This is a standard practice for dealing with endogeneity that is widely implemented in political science and economics. For other examples, see Paul Collier, Anke Hoeffler, and Dominic Rohner, "Beyond Greed and Grievance: Feasibility and Civil War," *Oxford Economic Papers* 61, no. 1 (2009): 1–27; and Gartzke and Jo, "Bargaining, Nuclear Proliferation, and Interstate Disputes."

150. See G. S. Maddala, *Introduction to Econometrics, Third Edition* (New York: John Wiley and Sons, 2005), 354.

151. G. Maddala, *Limited Dependent and Qualitative Variables in Econometrics* (Cambridge: Cambridge University Press, 1983); and Omar Keshk, "CDSIMEQ: A Program to Implement Two-Stage Probit Least Squares," *Stata Journal* 3, no. 2 (2003): 157–67.

152. It is important to include states with weapons programs in the estimation sample—as long as they have not successfully produced the bomb—because nonnuclear weapons states might seek atomic assistance.

153. The dependent variable mirrors the measurement of the nuclear weapons pursuit variable in chapter 3.

154. In other words, this relationship is not driven by the factors that motivate states to seek assistance; nuclear hedging does not take place except in a handful of isolated cases.

155. Brambor, Clark, and Golder, "Understanding Interaction Models."

156. If zero is included in the 95 percent confidence interval, then the marginal effect of atomic assistance is statistically insignificant at that particular level of conflict. Brambor, Clark, and Golder, "Understanding Interaction Models."

157. Jo and Gartzke, "Determinants of Nuclear Weapons Proliferation."

158. I update their data to extend through 2000.

159. Dong-Joon Jo and Erik Gartzke, "Codebook and Data Notes for 'Determinants of Nuclear Weapons Proliferation: A Quantitative Model," September 2006. Available at http://dss.ucsd.edu/~egartzke/data/jo_gartzke_0207_codebk_0906.pdf. This is a reasonable coding decision for their research, but it potentially creates problems for my purposes because they define "nuclear activities" as "nuclear reactor construction or purchase" and other procurement activities that may be conducted exclusively for civilian purposes

160. These countries include Australia, Egypt, Syria, Yugoslavia, Taiwan, Romania, and Sweden.

161. The United States and Russia were included in the tests presented in tables 7.1, 7.3, and 7.5 but excluded from table 7.4, 7.6, and 7.7. I used lagged variables in the latter three tables. Since these states were coded as pursuing nuclear weapons in 1945—the same year that my dataset begins—they were dropped from the estimation sample when I used a dependent variable in year *t+1*.

162. I also replicate models 6–17 using these alternate dependent variables and the results are similar.

CHAPTER 8

1. See, for example, Matthew Bunn, *Civilian Nuclear Energy and Nuclear Weapons Programs: The Record* (Cambridge, Mass.: Belfer Center for Science and International Affairs, 2001).

2. Donald MacKenzie and Graham Spinardi, "Tacit Knowledge, Weapons Design, and the Uninvention of Nuclear Weapons," *American Journal of Sociology* 101, no. 1 (1995): 44–99.

3. M. D. Zentner, G. L Coles, and R. J. Talbert, *Nuclear Proliferation Technology Trends Analysis*(Richland, Wash.: United States Department of Energy, Pacific Northwest National Laboratory, 2005).

4. The findings from the previous chapter likewise give us no reason to believe that such a relationship exists.

5. Many of these findings are consistent with previous research. See, for example, Sonali Singh and Christopher Way, "The Correlates of Nuclear Proliferation: A Quantitative Test," *Journal of Conflict Resolution*48, no. 6 (2004): 859–85; Matthew Fuhrmann, "Spreading Temptation: Proliferation and Peaceful Nuclear Cooperation Agreements," *International Security* 34, no. 1 (2009): 78–41; and Dong-Joon Jo, and Erik Gartzke, "Determinants of Nuclear Weapons Proliferation," *Journal of Conflict Resolution* 51, no. 1 (2007): 167-194.

6. Recall that this is consistent with what I found in chapter 7 regarding the relationship between peaceful aid and nuclear weapons pursuit.

7. I focus on this NCA type because it was closely associated with weapons acquisition in the preceding analysis, whereas other treaty types were statistically insignificant. The other agreements—intangible NCAs, material NCAs, and comprehensive research NCAs—do not raise the likelihood of bomb acquisition even at high levels of conflict (see appendix 8.1).

8. John Redick, "Nuclear Illusions: Argentina and Brazil," Henry L. Stimson Center, Occasional Paper no. 25 (December 1995), 5.

9. Leonard Spector, *Nuclear Exports: The Challenge of Control* (Washington, D.C.: Carnegie Endowment for International Peace, 1990), 5.

10. Mitchell Reiss, *Bridled Ambition: Why Countries Constrain Their Nuclear Capabilities* (Washington, D.C.: Woodrow Wilson Center Press, 1995), 47.

11. Leonard Spector, *Nuclear Ambitions: The Spread of Nuclear Weapons, 1989–1990* (Boulder, Co.: Westview Press, 1990), 253–59; and Redick, "Nuclear Illusions," 12.

12. Spector, *Nuclear Ambitions*, 244.

13. Stephen Meyer, *The Dynamics of Nuclear Proliferation* (Chicago: University of Chicago Press, 1984), 129.

14. Julio Carasales, "The So-Called Proliferator That Wasn't: The Story of Argentina's Nuclear Policy," *Nonproliferation Review* 6, no. 4 (1999): 56.

15. David Myers, "Brazil: Reluctant Pursuit of the Nuclear Option," *Orbis* 27, no. 4 (1984): 881–911.

16. T. V. Paul, *Power versus Prudence: Why States Forgo Nuclear Weapons* (Montreal: McGill-Queens University Press, 2000), 105.

17. Faten Ghosn, Glenn Palmer, and Stuart Bremer, "The MID3 Data Set, 1993–2001: Procedures, Coding Rules, and Description," *Conflict Management and Peace Science* 21, no. 2 (2004): 133–54.

18. Redick, "Nuclear Illusions," 14.

19. Paul, *Power versus Prudence*, 101.

20. Reiss, *Bridled Ambition*, 52.

21. Jacques Hymans, *The Psychology of Nuclear Proliferation: Identity, Emotions, and Foreign Policy* (Cambridge: Cambridge University Press, 2006), 141–70.

22. The Falklands War was the most significant episode of conflict during this period. Yet Argentina experienced minimal conflict during the 1980s compared to proliferators such as Iran, Iraq, and Pakistan. On Argentina's pursuit of nuclear weapons, see Singh and Way, "The Correlates of Nuclear Proliferation."

23. Meyer, *Dynamics of Nuclear Proliferation*, 130.

24. Michael Barletta, "The Military Nuclear Program in Brazil," Stanford University, Center for International Security and Cooperation, August 1997.

25. Quoted in ibid, 16. Emphasis added.

26. Jean Krasno, "Brazil's Secret Nuclear Program," *Orbis* 38, no. 3 (1994): 425–38.

27. The transition to democracy in both countries also influenced the move towards nuclear restraint. See, for example, Paul, *Power versus Prudence*, 99–112.

28. See Avner Cohen, *Israel and the Bomb* (New York: Columbia University Press, 1998).

29. Ibid.

30. Gary Milhollin, "Heavy Water Cheaters," *Foreign Policy* 69 (Winter 1987–88): 100–119; Astrid Forlan, "Norway's Nuclear Odyssey: From Optimistic Proponent to Nonproliferator," *Nonproliferation Review* 4, no. 2 (Winter 1997): 1–16; and "UK Helped Israel Get Nuclear Bomb," *BBC News*, August 4, 2005.

31. Milhollin, "Heavy Water Cheaters."

32. United States National Security Council, "Peaceful Uses of Atomic Energy," NSC 5725, November 22, 1957. Accessed via the Digital National Security Archive.

33. James Keeley, "A List of Bilateral Civilian Nuclear Cooperation Agreements," University of Calgary, 2003.

34. Ibid.

35. Post-1960, France used civil power reactors to produce plutonium for anywhere between sixty-three and 250 nuclear weapons. See "French Nuclear Facilities," *Nuclear Weapon Archive*, May 2001, Available at: http://nuclearweaponarchive.org/France/FranceFacility. html; Albert Donnay and Martin Kuster, "France," in *Nuclear Wastelands: A Global Guide to Nuclear Weapons Production and Its Health and Environmental Effects*, ed. Arjun Makhijani, Howard Hu, and Katherine Hiy (Cambridge: MIT Press, 1995), 462; and Mary Davis, *Nuclear France: Materials and Sites* (Paris: WISE-Paris, 2002). Available at http://www.francenuc.org/en_sources/sources_plut_e.htm.

36. Quoted in David Albright, "French Military Plans for Superphenix?" *Bulletin of the Atomic Scientists* 40, no. 9 (1984): 30.

37. Alexander Zhebin, "A Political History of Soviet–North Korean Nuclear Cooperation," in *The North Korean Nuclear Program: Security, Strategy, and New Perspectives from Russia*, ed. James Clay Moltz (New York: Routledge, 2000). The Soviet Union is believed to have trained more than three hundred North Korean nuclear scientists during the period of cooperation.

38. Faye Flam, "American Scientists Explain North Korean Nuclear Test," *Philadelphia Inquirer*, October 10, 2006.

39. Singh and Way, "Correlates of Nuclear Proliferation"; Mitchell Reiss, *Bridled Ambition: Why Countries Constrain Their Nuclear Capabilities* (Washington, D.C.: Woodrow Wilson Center Press, 1995); Peter Liberman, "The Rise and Fall of the South African Bomb," *International Security* 26, no. 2 (2001): 45–86; Verne Harris, Sello Hatang, and Peter Liberman, "Unveiling South Africa's Nuclear Past," *Journal of Southern African Studies* 30, no. 3 (2004): 457–76; and Waldo Stumpf, "South Africa's Nuclear Weapons Program: From Deterrence to Dismantlement," *Arms Control Today* 25, no. 10 (December–January, 1995–96): 3–8.

40. *International Institute of Nuclear Science and Engineering Classbook* (Argonne, Ill.: Argonne National Laboratory, 1961); and author's correspondence with Walter Kato, Cambridge, Mass., November 20, 2008.

41. United States Central Intelligence Agency, "The Chinese Communist Atomic Energy Program," NIE Number 13-2-1960, December 13, 1960, accessed via the Digital National Security Archive. See also John Wilson Lewis and Xue Litai, *China Builds the Bomb* (Stanford: Stanford University Press, 1988).

42. MacKenzie and Spinardi, "Tacit Knowledge, Weapons Design, and the Uninvention of Nuclear Weapons," 74; and Keeley, "Bilateral Civilian Nuclear Cooperation Agreements."

43. Lewis and Litai, *China Builds the Bomb*, 104–36.

44. Iraq, Libya, and South Korea all signed fewer than the average number of NCAs among states with weapons programs. Iran signed 30 percent more NCAs than the average state pursuing the bomb but it concluded far fewer agreements than proliferators such as India.

45. Sam Roe, "U.S. Cold War Gift: Iran Nuclear Pact," *Chicago Tribune*, August 24, 2006.

46. See, for example, Gordon Corera, *Shopping for Bombs: Nuclear Proliferation, Global Insecurity, and the Rise and Fall of the A. Q. Khan Network* (Oxford: Oxford University Press, 2006).

47. Nazila Fathi, David Sanger, and William Broad, "Iran Says It Is Making Nuclear Fuel, Defying UN," *New York Times*, April 12, 2006.

48. Zentner, Coles, and Talbert, *Nuclear Proliferation Technology Trends Analysis*, 20.

49. See Etel Solingen, *Nuclear Logics: Contrasting Paths in East Asia and the Middle East* (Princeton, NJ: Princeton University Press, 2007), 149.

50. See, for example, Matthew Fuhrmann and Sarah Kreps, "Targeting Nuclear Programs in War and Peace: A Quantitative Empirical Analysis, 1942–2000," *Journal of Conflict Resolution* 54, no. 6 (2010): 831–59.

51. See, for example, Dan Reiter, "Preventive Attacks against Nuclear Programs and the 'Success' at Osiraq," *Nonproliferation Review* 12, no.2 (July 2005): 355–71; Dan Reiter, *Preventive War and Its Alternatives: The Lessons of History* (Carlisle, Penn.: Strategic Studies Institute, 2006); Jeremy Tamsett, "The Israeli Bombing of Osiraq Reconsidered: Successful Counterproliferation?" *Nonproliferation Review* 11, no. 3 (September 2004): 70–85; and Sarah Kreps and Matthew Fuhrmann, "Attacking the Atom: Does Bombing Nuclear Facilities Affect Proliferation?" *Journal of Strategic Studies* 34, no. 2 (2011): 161–87.

52. Mahdi Obeidi and Kurt Pitzer, *The Bomb in My Garden: The Secrets of Saddam's Nuclear Mastermind* (New York: Wiley and Sons, 2004), 103.

53. Quoted in Richard Kokoski, *Technology and the Proliferation of Nuclear Weapons* (Oxford: Oxford University Press, 1995), 143. See also David Albright and Mark Hibbs, "Iraq's Bomb: Blueprints and Artifacts," *Bulletin of the Atomic Scientists* 48, no. 1 (January–February 1992); and Kreps and Fuhrmann, "Attacking the Atom."

54. Reiss, *Without the Bomb*, 89.

55. Keeley, "Bilateral Civilian Nuclear Cooperation Agreements."

56. U.S. Central Intelligence Agency, "Prospects for Further Proliferation of Nuclear Weapons," DCI NIO 1945/74, September 4, 1974. Accessed via the Digital National Security Archive.

57. Reiss, *Without the Bomb*, 106–7.

58. Ibid, 91.

59. Ibid; Solingen, *Nuclear Logics;* Rebecca Hersman and Robert Peters, "Nuclear U-Turns: Learning from South Korean and Taiwanese Rollback," *Nonproliferation Review* 13, no. 3 (2006): 539–53.

60. Hersman and Peters, "Nuclear U-Turns," 542.

61. United States, Central Intelligence Agency, "Pakistan's Nuclear Program," National Intelligence Estimate, April 26, 1978. Accessed via the Digital National Security Archive.

62. Ashok Kapur, *Pakistan's Nuclear Development* (London: Croom Helm, 1987), 75.

63. Ibid, 156.

64. United States, Department of State, "The Pakistani Nuclear Program," June 23, 1983. Accessed via the Digital National Security Archive.

65. Andrew Koch and Jennifer Topping, "Pakistan's Nuclear-Related Facilities," Center for Nonproliferation Studies Fact Sheet, 1997.

66. United States, Department of State, "Apprehensions Regarding Pakistan's Nuclear Intentions," September 3, 1975. Accessed via the Digital National Security Archive.

67. Central Intelligence Agency, "Pakistan's Nuclear Program."

68. *International Institute of Nuclear Science and Engineering Classbook*.

69. United States, Department of Energy, "International School Focused on Peaceful Uses of Nuclear Energy," October 12, 1996. Available at http://www.anl.gov/Media_Center/News/History/news961012.html.

70. Shahid Rehman, *Long Road to Chagai* (Islamabad: Print Wise Publication, 1999), 36–37.

71. See, for example, Ashok Kapur, *Pakistan's Nuclear Development* (London: Croom Helm, 1987); Corera, *Shopping for Bombs*; and Hassan Abbas, *Causes That Led to Nuclear Proliferation from Pakistan to Iran, Libya, and North Korea*, PhD diss., Fletcher School of Law and Diplomacy, Tufts University, 2008.

72. Author's correspondence with Walter Kato, Cambridge, Mass., November 20, 2008.

73. Kapur, *Pakistan's Nuclear Development*, 169.

74. Ibid, 136.

75. Ibid, 169; and Central Intelligence Agency, "Pakistan's Nuclear Program."

76. See especially Corera, *Shopping for Bombs*.

77. Carnegie Endowment for International Peace, "A. Q. Khan Nuclear Chronology, *Nonproliferation Issue Brief* 8, no. 8 (September 7, 2005).

78. Ibid.

79. James Markham, "Bonn Checks Report of Smuggling of Atomic Technology to Pakistan," *New York Times*, May 5, 1987; Corera, *Shopping for Bombs*, 23; and Shelby McNichols, "Chronology of Pakistani Nuclear Development," Center for Nonproliferation Studies, July 2000.

80. Corera, *Shopping for Bombs*, 27.

81. Ibid, 22.

82. David Albright and Kevin O'Neill, "ISIS Technical Assessment: Pakistan's Stock of Weapon Grade Uranium," Institute for Science and International Security, June 1998; Christopher Clary, "Dr. Khan's Nuclear WalMart," *Disarmament Diplomacy*, March–April 2004; David Albright and Mark Hibbs, "Pakistan's Bomb: Out of the Closet," *Bulletin of the Atomic Scientists* 48 (July–August 1992): 39; Corera, *Shopping for Bombs*, 49; and Kapur, *Pakistan's Nuclear Development*, 208.

83. Quoted in George Perkovich, "Nuclear Power and Nuclear Weapons in India, Pakistan, and Iran," in *Nuclear Power and the Spread of Nuclear Weapons: Can We Have One without the Other?* ed. Paul Leventhal, Sharon Tanzer, and Steven Dolley (Washington, D.C.: Brassey's, 2002), 194.

84. Samar Mubarakmand, Capital Talk Special, Geo-TV, May 3, 2004. Available at http://www.pakdef.info/forum/showthread.php?t=9214.

85. Joseph Cirincione, with Jon B. Wolfsthal and Miriam Rajkumar, *Deadly Arsenals, Tracking Weapons of Mass Destruction* (Washington, DC: Carnegie Endowment for International Peace, 2002), 307; and Kenneth Timmerman, *Weapons of Mass Destruction: The Cases of Iran, Syria, and Libya* (Los Angeles: Simon Wiesenthal Center, August 1992), 89.

86. Leonard Spector, *Nuclear Proliferation Today* (Cambridge, Mass.: Ballinger, 1984), 152.

87. Gloria Duffy, "Soviet Nuclear Exports," *International Security* 3, no. 1 (1978): 83–111.

88. See Ian Traynor, "Nuclear Chief tells of Black Market in Bomb Equipment," *Guardian*, January 26, 2004.

89. Iran and North Korea were also beneficiaries of Khan's network, as I noted in chapter 1. See, for example, Corera, *Shopping for Bombs*.

90. In 2002 Libya moved this facility to al Fallah due to security concerns. See Wyn Bowen, "Libya and Nuclear Proliferation: Stepping Back from the Brink," *Adelphi Papers*, No. 380 (2006), 40.

91. *Technology Transfer to the Middle East* (Washington, D.C.: U.S. Congress, Office of Technology Assessment, 1984), 386.

92. Corera, *Shopping for Bombs*, 109.

93. Bowen, "Libya and Nuclear Proliferation," 44.

94. International Atomic Energy Agency, "Implementation of the NPT Safeguards Agreement in the Libyan Arab Jamahiriya," GOV/2008/39, September 12, 2008.

95. Bowen, "Libya and Nuclear Proliferation."

96. Spector, *Nuclear Ambitions*, 182.

97. Quoted in Solingen, *Nuclear Logics*, 224.

98. Ibid, 223–24.

99. Meghan O'Sullivan, *Shrewd Sanctions: Statecraft and State Sponsors of Terrorism* (Washington, D.C.: Brookings Institution Press, 2003), 204.

100. The perceived absence of strategic incentives to possess nuclear weapons clearly had an effect on Qaddafi's decision to terminate the bomb program in 2003. See Scott Macleod and Amany Radwan, "10 Questions for Muammar Gaddafi," *Time*, January 30, 2005; Solingen, *Nuclear Logics*, 216; and Malfrid Braut-Hegghammer, "Libya's Nuclear Turnaround: Perspectives from Tripoli," *Middle East Journal* 62, no. 1 (Winter 2008): 71.

101. I code this variable by consulting data compiled by Singh and Way, "Correlates of Nuclear Proliferation." I make two modifications to the Singh and Way data. First, I code Israel as acquiring nuclear weapons in 1967, four years earlier than Singh and Way. I made this alternation on the basis of Avner Cohen's definitive account of Israeli nuclear history, which states: "In 1966–67 Israel completed the development stage of its first nuclear weapon, and on the eve of the Six-Day War it already had a rudimentary, but operational, nuclear weapons capability." Avner Cohen, *Israel and the Bomb* (New York: Columbia University Press, 1998), 1. Second, I code India as acquiring nuclear weapons in 1988; Singh and Way code New Delhi as acquiring nuclear weapons on separate occasions in 1974 and 1988. Updated data provided by Christopher Way indicate that it is more appropriate to use 1988 as the single year of acquisition for India. Personal communication with Christopher Way, October 31, 2006.

102. The lack of variation between the NPT variable and bomb production creates problems for statistical modeling, which is why other published quantitative studies of bomb acquisition often exclude it.

103. I take this variable from Jo and Gartzke's study of nuclear proliferation, "Determinants of Nuclear Weapons Proliferation."

104. Nathaniel Beck, Jonathan Katz, and Richard Tucker, "Taking Time Seriously in Binary Time-Series-Cross Section Analysis," *American Journal of Political Science* 42, no. 4 (1998): 1260–88.

105. Theoretically, we would expect the explanatory variables to precede the dependent variable.

106. Gary King and Langche Zeng, "Logistic Regression in Rare Events Data," *Political Analysis* 9, no 2 (1998): 137–63.

107. The calculations used to produce figure 8.1 were based on this model 2.

108. The countries included here are defined as exploring nuclear weapons by Singh and Way. See Singh and Way, "Correlates of Nuclear Proliferation."

109. When I remove the two nuclear assistance variables from model 2, industrial capacity becomes statistically significant.

110. I replicated model 2 with these variables removed and the results were substantively similar. Most important, nonsafety NCAs remained positive and statistically significant.

111. Thomas Brambor, William Clark, and Matt Golder, "Understanding Interaction Models: Improving Empirical Analysis," *Political Analysis* 14, no. 1 (2006): 63–82. See also Bear Braumoeller, "Hypothesis Testing and Multiplicative Interaction Terms," *International Organization* 58, no. 4 (2004): 807–20.

112. In other tests, comprehensive research NCAs were negative and significant, indicating that these agreements lower the likelihood of nuclear weapons acquisition in the absence of militarized conflict. This is also not a robust finding, however.

113. I conduct additional robustness tests not discussed here. The United States was dropped from the initial analysis due to the use of a leading dependent variable (the dataset be-

gins in 1945, the same year that the United States acquired the bomb). I replicated the models with the United States included and NCAs remained positive and statistically significant.

114. See National Intelligence Council, *Global Trends 2015: A Dialogue about the Future with Nongovernment Experts.* NIC 2000–02, December 2000, 36; Robert Shuey, "Nuclear, Biological, and Chemical Weapons and Missiles: The Current Situation and Trends," Congressional Research Service, RL30699, August 10, 2001; and Mary Beth Nikitin, "North Korea's Nuclear Weapons," Congressional Research Service, RL34256, February 12, 2009.

115. Jo and Gartzke, "Determinants of Nuclear Weapons Proliferation."

CHAPTER 9

1. See, for example, Robert Keohane, *After Hegemony: Cooperation and Discord in the World Political Economy* (Princeton: Princeton University Press, 1984).

2. Xinyuan Dai, *International Institutions and National Policies* (Cambridge: Cambridge University Press, 2007), 33.

3. Unlike the institution envisioned by Dwight D. Eisenhower, the IAEA does not assert strict control over nuclear activities or "own" materials. See U.S. Congress, Office of Technology Assessment, *Nuclear Safeguards and the International Atomic Energy Agency*, OTA-ISS-615 (Washington, D.C.: U.S. Government Printing Office, June 1995), 27.

4. Ibid.

5. See Roger K. Smith, "Explaining the Non-Proliferation Regime: Anomalies for Contemporary International Relations Theory," *International Organization* 41, no. 2 (Spring 1987): 253–81; and Joseph Nye, "Maintaining the Non-Proliferation Regime," *International Organization* 35, no. 1 (Winter 1981): 15–38.

6. See, for example, Stephen Krasner, ed., *International Regimes* (Ithaca: Cornell University Press, 1983).

7. The NSG was never explicitly tied to the NPT. For further details, see Ian Anthony, Christer Ahlstrom, and Vitaly Fedchenko, *Reforming Nuclear Export Controls: The Future of the Nuclear Suppliers Group* (Oxford: Oxford University Press, 2008).

8. Jack Boureston and Charles Ferguson, "Strengthening Nuclear Safeguards: Special Committee to the Rescue?" *Arms Control Today* 35, no. 10 (December 2005): 17–22; and Pierre Goldschmidt, "The IAEA Safeguards System Moves Into the 21st Century," *IAEA Bulletin* 41, no. 4 (December 1999): 1–20.

9. Safeguards did not apply to undeclared facilities, meaning that it remained theoretically possible for a state to operate covert reactors that could not be inspected by the IAEA.

10. Theodore Hirsch, "The IAEA Additional Protocol: What It Is and Why It Matters," *Nonproliferation Review* 11, no. 3 (2004): 143

11. Ibid.

12. As part of the AP, states must report on manufacturing activities that are not directly linked to the production or use of nuclear materials but are relevant to building nuclear weapons.

13. The number of AP members has increased steadily since Australia became the first country to bolster its safeguards commitment in December 1997. As of September 2009, a total of ninety-two countries had ratified the AP. Iran, Saudi Arabia, Syria, and others with possible nuclear weapons ambitions are among the states yet to ratify AP agreements.

14. See Jacques Hymans, *The Psychology of Nuclear Proliferation* (Cambridge: Cambridge University Press, 2006), 6.

15. On the importance of ratification in international law, see, for example, Oona Hathaway, "Why Do Countries Commit to Human Rights Treaties?" *Journal of Conflict Resolution* 51, no. 4 (2007): 588–621.

16. It is theoretically possible for a state make another international commitment foreswearing nuclear weapons even if they do not ratify the NPT. Regional Nuclear Weapon Free Zone (NWFZ) treaties, for instance, require states to make several nonproliferation commitments. Such zones currently exist in Africa, Central Asia, Latin America, Southeast Asia, and the South Pacific.

17. Karthika Sasikumar and Christopher Way, "Paper Tiger or Barrier to Proliferation? What Accessions Reveal about NPT Effectiveness," unpublished manuscript, San Jose State University.

18. See, for example, Maria Rost Rublee, *Nonproliferation Norms: Why States Choose Nuclear Restraint* (Athens: University of Georgia Press, 2009).

19. Nina Tannenwald, *The Nuclear Taboo: The United States and the Non-Use of Nuclear Weapons since 1945* (Cambridge: Cambridge University Press, 2007). For another argument on nuclear nonuse, see T. V. Paul, *The Tradition of Non-Use of Nuclear Weapons* (Stanford: Stanford University Press, 2009).

20. The marginal effect is the change in probability of nuclear weapons program initiation that results when the comprehensive NCA variable increases from its mean to one standard deviation above the mean, holding all other factors constant. The number of militarized interstate disputes is set to ten. All other continuous variables are set at the mean and dummy variables are set at the mode.

21. I define "modest" assistance as signing at least three NCAs entitling a state to nuclear technology, materials, or know-how.

22. I define a "high conflict environment" as experiencing an average of at least two militarized interstate disputes over a five-year period. This value translates to roughly one standard deviation above the mean.

23. For an analysis of why some proliferators enter the NPT and cheat while others remain outside the treaty, see Matthew Fuhrmann and Jeffrey D. Berejikian, "Disaggregating Noncompliance: Abstention versus Predation in the Nuclear Nonproliferation Treaty," *Journal of Conflict Resolution* (forthcoming).

24. There is a general debate in political science on whether treaties "screen" or "constrain." See, for example, Beth Simmons, "International Law and State Behavior: Commitment and Compliance in International Monetary Affairs," *American Political Science Review* 94, no. 4 (2000): 819–35; and Jana von Stein, "Do Treaties Constrain or Screen? Selection Bias in Treaty Compliance," *American Political Science Review* 99, no. 4 (2005): 611–22.

25. See Hymans, *Psychology of Nuclear Proliferation*, 6.

26. I calculate the marginal effect by generating the predicted probability of weapons program onset when the NCA variable is set to its mean and all other factors are set at their mean (for continuous variables) or mode (for dummy variables). Then I calculate the same probability when the NCA variable is set to one standard deviation above the mean and take the difference between the two probabilities.

27. Roughly 2 percent of the observations in the dataset experience this high level of nuclear assistance.

28. Richard K. Betts, "Universal Deterrence or Conceptual Collapse? Liberal Pessimism and Utopian Realism," in *The Coming Crisis: Nuclear Proliferation, U.S. Interests, and World Order*, ed. Victor Utgoff (Cambridge: MIT Press, 2000), 69.

29. Betts suggested that there may not be a country that would have pursued nuclear weapons if the NPT was never created. Betts, "Universal Deterrence or Conceptual Collapse?," 69.

30. See, for example, Rublee, *Nonproliferation Norms*; Mitchell Reiss, *Without the Bomb: The Politics of Nuclear Nonproliferation* (New York: Columbia University Press, 1988), 109–37; and John Endicott, *Japan's Nuclear Option: Political, Technical, and Strategic Factors* (New York: Praeger, 1975).

31. See, for example, Christopher Walker, "Israel Tries to Halt Syrian Reactor Deal," *The Times* (London), July 15, 1995; and Magnus Normark, Anders Lindblad, Anders Norqvist, Bjorn Sandstrom, and Louise Waldenstrom, *Syria and WMD: Incentives and Capabilities* (Stockholm: Swedish Defense Research Agency, June 2004).

32. Recall that the substantive interpretation of the findings presented in chapter 7 is similar if I code Syria as a proliferator (see appendix 7.1).

33. Recall from chapter 7 that this is the criterion I use to identify the onset of nuclear weapons programs. This is consistent with Sonali Singh and Christopher Way, "The Correlates of Nuclear Proliferation: A Quantitative Test," *Journal of Conflict Resolution* 48, no. 6 (2004): 859–85.

34. "Background Briefing with Senior U.S. Officials on Syria's Covert Nuclear Reactor and North Korea's Involvement," Washington, D.C., April 24, 2008. Available at http://www.dni.gov/interviews/20080424_interview.pdf.

35. See Chris Schneidmiller, "Syria Unlikely to Resume Nuclear-Weapon Program, Expert Says," *Global Security Newswire*, April 7, 2009.

36. See, for example, David Sanger and Mark Mazzetti, "Israel Struck Syrian Nuclear Project, Analysts Say," *New York Times*, October 14, 2007.

37. Quoted in Leonard Spector and Avner Cohen, "Israel's Airstrike on Syria's Reactor: Implications for the Nonproliferation Regime," *Arms Control Today* 38 (July–August 2008). Available at http://www.armscontrol.org/act/2008_07-08/SpectorCohen.

38. Julien Barnes-Dacey, "Can Syria Avoid Sanctions with a UN Nuclear Inspection?" *Christian Science Monitor*, June 25, 2008.

39. See International Atomic Energy Agency, "Implementation of the NPT Safeguards Agreement in the Syrian Arab Republic," GOV/2008/60, November 19, 2008. This is the first of four reports (as of September 2009) issued by the Board of Governors on Syria's compliance with its safeguards agreement.

40. Such materials are often referred to as "anthropogenic."

41. Davis Albright and Paul Brannan, "IAEA Report on Syria," Washington, D.C.: Institute for Science and International Security, June 5, 2009.

42. International Atomic Energy Agency, "Implementation of the NPT Safeguards Agreement in the Syrian Arab Republic," GOV/2008/60, November 19, 2008.

43. International Atomic Energy Agency, "Implementation of the NPT Safeguards Agreement in the Syrian Arab Republic," GOV/2009/56, August 28, 2009.

44. Barnes-Dacey, "Can Syria Avoid Sanctions with a UN Nuclear Inspection?"

45. Hirsch, "IAEA Additional Protocol," 143.

46. The agency could call for "special inspections" of undeclared facilities in Syria but there are ambiguities in the statutory language that provides this authority. See James Acton, Mark Fitzpatrick, Pierre Goldschmidt, "The IAEA Should Call for a Special Inspection in Syria," Carnegie Endowment for International Peace, February 26, 2009: http://www.carnegieendowment.org/publications/index.cfm?fa=view&id=22791; and Fiona Simpson, "IAEA Special Inspections after Israel's Raid on Syria," *Bulletin of the Atomic Scientists*, February 10, 2008.

47. George Jahn, "IAEA Chief Baffled over Lack of Syria Nuclear Info," Associated Press, November 29, 2008.

48. "U.N. Nuclear Agency to Study Claims of Secret Syrian Reactor," Reuters, April 26, 2008.

49. Gregory Schulte, "Damascus Deception," *Foreign Policy*, September 2, 2009. Available at http://www.foreignpolicy.com/articles/2009/09/02/stuck_on_damascus.

50. Murhaf Jouejati, "Syrian Motives for Its WMD Programs and What to Do about Them," *Middle East Journal* 59, no. 1 (Winter 2005): 57.

51. Ibid.

52. M. Zuhair Diab, "Syria's Chemical and Biological Weapons: Assessing Capabilities and Motivations," *Nonproliferation Review 5*, no. 1 (1997): 107.

53. Benedict Brogan, "We Won't Scrap WMD Stockpile Unless Israel Does, Says Asad," *Daily Telegraph*, January 6, 2003.

54. Kurt Campbell and Tsuyoshi Sunohara, "Japan: Thinking the Unthinkable," in *The Nuclear Tipping Point: Why States Reconsider Their Nuclear Choices*, by Kurt Campbell, Robert Einhorn and Mitchell Reiss (Washington, D.C.: Brookings Institution Press, 2004), 243.

55. As I noted in chapter 7, the presence of outliers does not undermine my theory because the argument is probabilistic, not deterministic.

56. See, for example, George Quester, *The Politics of Nuclear Proliferation* (Baltimore: Johns Hopkins University Press, 1973); John Endicott, "The 1975–76 Debate over Ratification of the NPT in Japan," *Asian Survey* 17, no. 3 (1977): 275–92; Roger Gale, "Nuclear Power and Japan's Proliferation Option," *Asian Survey* 18, no. 11 (1978): 1117–33; T. V. Paul, *Power versus Prudence: Why States Forgo Nuclear Weapons* (Montreal: McGill-Queens University Press, 2000); Campbell and Sunohara, "Japan: Thinking the Unthinkable"; Mike Mochizuki, "Japan Tests the Nuclear Taboo," *Nonproliferation Review* 14, no. 2 (2007): 303–28; Llewelyn Hughes, "Why Japan Will Not Go Nuclear (Yet): International and Domestic Constraints on the Nuclearization of Japan," *International Security* 31, no. 4 (2007): 67–96; Etel Solingen, *Nuclear Logics: Contrasting Paths in East Asia and the Middle East* (Princeton: Princeton University Press, 2007); and Rublee, *Nonproliferation Norms*.

57. Yuri Kase, "Japan's Nonnuclear Weapons Policy in the Changing Security Environment: Issues, Challenges, and Strategies," *World Affairs* 165, no. 3 (2003): 126; and Hughes, "Why Japan Will Not Go Nuclear."

58. Rublee, *Nonproliferation Norms*; and John Endicott, "Commentary," in *The Nuclear Dimension of the U.S.-Japan Alliance*, by Morton Halperin (San Francisco: Nautilus Institute, 1999).

59. Quoted in Marie Rost Rublee, "Taking Stock of the Nuclear Nonproliferation Regime: Using Social Psychology to Understand Regime Effectiveness," *International Studies Review* 10, no. 3 (2008): 437.

60. Endicott, "Commentary," in *Nuclear Dimension of the U.S.-Japan Alliance.*

61. Ali Reza Moaiyeri, "Statement at the Second Session of the Preparatory Committee for the 2010 Review Conference of the Parties to the Treaty on the Nonproliferation of Nuclear Weapons," Geneva, Switzerland, April 29, 2008.

62. Article X of the treaty permits states to withdraw in the event that they provide ninety days notice. See Solingen, *Nuclear Logics*, 63–65.

63. Howard French, "Nuclear Arms Taboo Is Challenged in Japan," *New York Times*, June 9, 2002.

64. Lawrence Scheinman, "Tokai Reprocessing Plant Options Paper," U.S. Department of State, Confidential Memorandum, August 1, 1977.

65. Campbell and Sunohara, "Japan: Thinking the Unthinkable," 223.

66. Solingen, *Nuclear Logics*, 65.

67. Paul, *Power versus Prudence*, 48–49.

68. Campbell and Sunohara, "Japan: Thinking the Unthinkable," 224.

69. Matake Kamiya, "Nuclear Japan: Oxymoron or Coming Soon?" *Washington Quarterly* 26, no. 1 (2002–03): 68; and Hughes, "Why Japan Will Not Go Nuclear," 78.

70. Campbell and Sunohara, "Japan," 227–28.

71. Ibid.

72. Jacques Hymans, "The Effects of Institutional Change on Japanese Nuclear Policies," paper presented at the Annual Meeting of the American Political Science Association, Toronto, September 3–6, 2009.

73. Hughes, "Why Japan Will Not Go Nuclear," 84.

74. See, for example, Peter Katzenstein and Nobuo Okawara, *Japan's National Security: Structures, Norms, and Policy Responses in a Changing World* (Ithaca: Cornell University Press, 1993), 128–29.

75. Hughes, "Why Japan Will Not Go Nuclear," 89–90.

76. Other scholars reach a similar conclusion. See, for example, Scott Sagan, "Realist Perspectives on Ethical Norms and Weapons of Mass Destruction," in *Ethics and Weapons of Mass Destruction: Religious and Secular Perspectives*, ed. Sohail Hashmi and Steven Lee (Cambridge: Cambridge University Press, 2004), 88.

77. Solingen, *Nuclear Logics*, 77.

78. Ibid, 78.

79. See, for example, Mochizuki, "Japan Tests the Nuclear Taboo."

80. Campbell and Sunohara, "Japan," 228.

81. Moreover, China did not ratify the NPT until the early 1990s.

82. "Further Measures to be Taken for Strengthening the Treaty on the Non-Proliferation of Nuclear Weapons," working paper submitted by Japan, 2005 Review Conference of the Parties to the Treaty on the Non-Proliferation of Nuclear Weapons, May 17, 2005.

83. Rublee, "Taking Stock of the Nuclear Nonproliferation Regime," 444.

84. Campbell and Sunohara, "Japan," 240.

85. Tetsuro Fukuyama, "Statement at the 2010 Review Conference of the Parties to the Treaty on the Non-Proliferation of Nuclear Weapons," New York, May 4, 2010. Available at http://www.mofa.go.jp/announce/svm/state100504.html.

86. Joseph Cirincione, "The Asian Nuclear Reaction Chain," *Foreign Policy*, no. 118 (2000): 126.

87. For a discussion on nuclear weapons and prestige, see, for example, Scott Sagan, "Why Do States Build Nuclear Weapons? Three Models in Search of a Bomb," *International Security* 21, no. 3 (Winter 1996–97): 73–82.

88. Rublee, "Taking Stock of the Nuclear Nonproliferation Regime," 441.

89. Hisane Maski, "Japanese Nukes: Voicing the Unthinkable," *Asia Times*, November 16, 2006.

90. The NPT prohibits all NNWS from building nuclear weapons, so in one sense post-1970 acquisitions of the bomb are an outlier. But here I am primarily interested in the NPT's attempts to prevent states from using nuclear assistance to build nuclear weapons.

91. Israel technically assembled its first nuclear weapon prior to 1970 but after the IAEA instituted a safeguards regime that covered power reactors in 1965.

92. See, for example, David Leigh, "How the UK Gave Israel the Bomb," *Guardian*, August 4, 2005.

93. As I discussed in chapter 8, Pakistan ultimately did not use plutonium from this reactor to build its first nuclear weapon but it is clear that safeguards played little role in this outcome.

94. According to Singh and Way, North Korea initiated a weapons program in 1980. See Singh and Way, "Correlates of Nuclear Proliferation."

95. Jean du Preez and William Potter, "North Korea's Withdrawal from the NPT: A Reality Check," James Martin Center for Nonproliferation Studies, April 9, 2003.

96. Daryl Kimball and Peter Crail, "Chronology of U.S.–North Korean Nuclear and Missile Diplomacy," Washington, D.C.: Arms Control Association, 2009.

97. Jeffrey Richelson, *Spying on the Bomb: American Nuclear Intelligence from Nazi Germany to Iran and North Korea* (New York: W. W. Norton, 2007), 519.

98. For more on the Agreed Framework see, for example, Curtis Martin, "Rewarding North Korea: Theoretical Perspectives on the 1994 Agreed Framework," *Journal of Peace Research* 39, no. 1 (2002): 51–68; Jonathan Pollack, "The United States, North Korea, and the End of the Agreed Framework," *Naval War College Review* 56, no. 3 (2003): 11–49; and Scott Sagan, "How to Keep the Bomb from Iran," *Foreign Affairs* 85, no. 5 (2006): 45–59.

99. Richelson, *Spying on the Bomb*, 530.

100. International Atomic Energy Agency, "Implementation of the Safeguards Agreement between the Agency and the Democratic People's Republic of Korea Pursuant to the Treaty on the Non-Proliferation of Nuclear Weapons," GC(47)/19, August 13, 1.

101. Andrew Semmel, "Statement to the Second Session of the Preparatory Committee for the 2005 NPT Review Conference," Geneva, Switzerland: http://usmission.ch/press2003/0507semmel.htm

102. John Redick, "Nuclear Illusions: Argentina and Brazil," Henry L. Stimson Center, Occasional Paper No. 25, December 1995, 2–3.

103. Leonard Spector, *Nuclear Ambitions: The Spread of Nuclear Weapons 1989–90* (Boulder, Colo.: Westview Press, 1990), 244.

104. Paula DeSutter, *Denial and Jeopardy: Deterring Iranian Use of NBC Weapons* (Washington, D.C.: National Defense University Press, 1997), 44.

105. International Atomic Energy Agency, "Implementation of the NPT Safeguards Agreement in the Islamic Republic of Iran," GOV/2003/75, November 10, 2003.

106. Solingen, *Nuclear Logics*, 171. See also, David Albright, "An Iranian Bomb?" *Bulletin of the Atomic Scientists* 51, no. 4 (1995): 20–26.

107. International Atomic Energy Agency, "Implementation of the NPT Safeguards Agreement in the Islamic Republic of Iran," GOV/2003/40, June 6, 2003.

108. International Atomic Energy Agency, "Implementation of the NPT Safeguards Agreement in the Islamic Republic of Iran," GOV/2006/15, February 27, 2006.

109. Ibid.

110. Catherine Philip, Francis Elliot, and Giles Whittell, "How Secrecy over Iran's Qom Nuclear Facility Was Finally Blown Away," *The Times* (London), September 26, 2009.

111. See, for example, Richard Koloski, *Technology and the Proliferation of Nuclear Weapons* (New York: Oxford University Press, 1995), 97–146; Solingen, *Nuclear Logics*; and Anthony, Ahlstrom, and Fedchenko, *Reforming Nuclear Export Controls*.

112. For an alternate perspective, see Dan Reiter, "Preventive Attacks against Nuclear Programs and the 'Success' at Osiraq," *Nonproliferation Review* 12, no. 2 (2005): 355–71.

113. Quoted in Solingen, *Nuclear Logics*, 149.

114. Shai Feldman, "The Bombing of Osiraq—Revisited," *International Security* 7, no. 2 (1982): 114.

115. Quoted in Solingen, *Nuclear Logics*, 153.

116. Hirsch, "IAEA Additional Protocol," 142.

117. Solingen, *Nuclear Logics*, 151.

118. *Comprehensive Report of the Special Advisor to the DCI on Iraq's WMD* (Washington, D.C.: Government Printing Office, 2004). This report is commonly referred to as the "Duelfer Report."

119. Ibid.

120. *Gulf War Air Power Survey*, vol. 2, "Planning and Command and Control" (Washington, D.C.: Government Printing Office, 1993), 485.

121. International Atomic Energy Agency, "Implementation of the NPT Safeguards Agreement of the Socialist People's Libyan Arab Jamahiriya," GOV/2004/12, February 20, 2004, 6.

122. Ibid, 3.

123. Ibid, 8.

124. As I discussed above, safeguards are not intended to be foolproof. The logic is that by raising the probability of being caught, safeguards will deter violations in the first place.

125. Wyn Bowen, "Libya and Nuclear Proliferation: Stepping Back from the Brink," *Adelphi* Papers, no. 380 (2006): 33. See also Solingen, *Nuclear Logics*, 219.

126. International treaties do not bind countries until they are ratified. Thus, South Korea officially cheated on its NPT commitments between 1975 and 1978.

127. Solingen, *Nuclear Logics*, 86.

128. Jungmin Kang, Peter Hayes, Li Bin, Tatsujiro Suzuki, and Richard Tanter, "South Korea's Nuclear Surprise," *Bulletin of the Atomic Scientists* 61, no. 1 (2005): 41.

129. Reiss, *Without the Bomb*, 85–86; and Solingen, *Nuclear Logics*, 91.

130. Reiss, *Without the Bomb*; Solingen, *Nuclear Logics*; and Rebecca Hersman and Robert Peters, "Nuclear U-Turns: Learning from South Korean and Taiwanese Rollback," *Nonproliferation Review* 13, no. 3 (2006): 539–53.

131. Singh and Way, "Correlates of Nuclear Proliferation."

132. Thomas Brambor, William Clark, and Matt Golder, "Understanding Interaction Models: Improving Empirical Analysis," *Political Analysis* 14, no. 1 (2006): 63–82.

133. Figure 9.2 was generated based on the estimates from model 2.

134. I exclude safety NCAs because there is not a theoretical reason to expect that they would be linked with nuclear weapons program onset.

135. Note that this finding is sensitive to model specification.

136. Brambor, Clark, and Golder, "Understanding Interaction Models."

137. Note that the results are substantively similar when I recalculate figure 9.2 based on the estimates from models 1 and 3.

138. Figure 9.5 was based on the estimates from model 4, which employed the nuclear cooperation variable that only measured comprehensive power NCAs.

139. I replicated models 4 and 5 using the alternate proliferation data compiled by Jo and Gartzke to construct the dependent variable. Dong-Joon Jo, and Erik Gartzke, "Determinants of Nuclear Weapons Proliferation," *Journal of Conflict Resolution* 51, no. 1 (2007): 167–94. The findings were less favorable to the NPT argument. The variable measuring whether a state had ratified the NPT was statistically insignificant across the whole range of the comprehensive power NCA variable. And NCAs remained statistically related to nuclear weapons program initiation for members of the NPT.

CONCLUSION

1. Quoted in Stephen Brooks, *Producing Security: Multinational Corporations, Globalization, and the Changing Calculus of Conflict* (Princeton: Princeton University Press, 2005).

2. Interestingly, those NCAs that are best explained by my theory of nuclear cooperation are also the most proliferation-prone treaties. Recall that comprehensive power agreements raise the risk of both nuclear weapons pursuit and bomb acquisition. Intangible NCAs contribute to weapons pursuit but not acquisition. Comprehensive research NCAs contribute to weapons programs but only in the presence of militarized conflict.

3. This conclusion is admittedly preliminary. Future research should devote greater attention to the role of time horizons in international politics, particularly in the context of peaceful nuclear assistance. Political scientists are beginning to devote more attention to the role of time horizons. See, for example, Philip Streich and Jack Levy, "Time Horizons, Discounting, and Inter-Temporal Choice," *Journal of Conflict Resolution* 51, no. 2 (2007): 199–226.

4. There is a tendency to focus on either the supply side or the demand side. Scholars who focus on both tend to treat them independently without examining how they could interact.

5. See, for example, Chaim Braun and Christopher Chyba, "Proliferation Rings: New Challenges to the Nuclear Nonproliferation Regime," *International Security* 29, no. 2 (Fall 2004): 5–49; Alexander Montgomery, "Ringing in Proliferation: How to Dismantle an Atomic Bomb Network," *International Security* 30, no. 2 (Fall 2005): 153–87; Gordon Corera, *Shopping for Bombs: Nuclear Proliferation, Global Insecurity, and the Rise and Fall of the A. Q. Khan Network* (Oxford: Oxford University Press, 2006); Matthew Kroenig, *Exporting the Bomb: Technology Transfer and the Spread of Nuclear Weapons* (Ithaca: Cornell University

Press, 2010); and David Albright, *Peddling Peril: How the Secret Nuclear Trade Arms America's Enemies* (New York: Free Press, 2010).

6. For a recent analysis on this topic see Michael Horowitz, *The Diffusion of Military Power: Causes and Consequences for International Politics* (Princeton: Princeton University Press, 2010).

7. See, for example, John Alic, Lewis Branscomb, Harvey Brooks, Ashton Carter, and Gerald Epstein, *Beyond Spinoff: Military and Commercial Technologies in a Changing World* (Boston: Harvard Business School Press, 1992).

8. See, for example, Mark Lorell, Julia Lowell, Michael Kennedy, and Hugh Levaux, *Cheaper, Faster, Better? Commercial Approaches to Weapons Acquisition* (Santa Monica: RAND, 2000).

9. See, for example, John Mearsheimer, "The False Promise of International Institutions," *International Security* 19, no. 1 (1994–95): 5–49.

10. See, for example, Robert Keohane, *After Hegemony: Cooperation and Discord in the World Political Economy* (Princeton: Princeton University Press, 1984); and Robert Keohane and Lisa Martin, "The Promise of Institutionalist Theory," *International Security* 20, no. 1 (1995): 39–51.

11. See, for example, Abram Chayes and Antonia Handler Chayes, *The New Sovereignty: Compliance with International Regulatory Agreements* (Cambridge: Harvard University Press, 1995).

12. Oran Young, "The Effectiveness of International Institutions: Hard Cases and Critical Variables," in *Governance without Government: Order and Change in World Politics*, ed. James Rosenau and Ernst-Otto Czempiel (Cambridge: Cambridge University Press, 1992), 183; see also Jonas Tallberg, "Paths to Compliance: Enforcement, Management, and the European Union," *International Organization* 56, no. 3 (Summer 2002): 613–14.

13. See, Barbara Koremenos, Charles Lipson, and Duncan Snidal, "The Rational Design of International Institutions," *International Organization* 55, no. 4 (Autumn 2001): 761–99.

14. See, for example, Koremenos, Lipson, and Snidal, "Rational Design of International Institutions"; and Ronald Mitchell, "Regime Design Matters: International Oil Pollution and Treaty Compliance," *International Organization* 48, no. 3 (Summer 1994): 425–58.

15. See, for example, Virginia Page Fortna, *Peace Time: Cease-Fire Agreements and the Durability of Peace* (Princeton: Princeton University Press, 2004).

16. See, for example, Xinyuan Dai, *International Institutions and National Policies* (Cambridge: Cambridge University Press, 2007); and Beth Simmons, *Mobilizing for Human Rights: International Law in Domestic Politics* (Cambridge: Cambridge University Press, 2009).

17. Kenneth Abbott and Duncan Snidal, "Hard and Soft Law in International Governance," *International Organization* 54, no. 3 (2000): 421–56.

18. See, for example, Jana von Stein, "Do Treaties Constrain or Screen? Selection Bias in Treaty Compliance," *American Political Science Review* 99, no. 4 (2005): 611–22.

19. Joshua Williams and Jon Wolfsthal, "The NPT at 35: A Crisis of Compliance or a Crisis of Confidence" (New York: United Nations Association of the United States of America, 2005).

20. Promising research examining the determinants of NPT ratification is already under way. See Karthika Sasikumar and Christopher Way, "Paper Tiger or Barrier to Proliferation? What Accessions Reveal about NPT Effectiveness," unpublished manuscript, San Jose State University.

21. The Statute of the IAEA is available at http://www.iaea.org/About/statute_text.html#A1.12.

22. This is routinely cited as a weakness of the safeguards regime. See, for example, U.S. Congress, Office of Technology Assessment, *Nuclear Safeguards and the International*

Atomic Energy Agency, OTA-ISS-615 (Washington, D.C.: U.S. Government Printing Office, June 1995), 33.

23. See James Acton, "The Problem with Nuclear Mind Reading," *Survival* 51, no. 1 (2009): 119–42.

24. It is plausible that the NPT contributes to state-driven nonproliferation policies, even though the institution itself lacks robust enforcement capacity. This is a fruitful area for future research.

25. Office of Technology Assessment, *Nuclear Safeguards and the International Atomic Energy Agency,* 31.

26. Note, however, that these states can reach limited (i.e., facility-specific) safeguards agreements with the IAEA.

27. See Matthew Fuhrmann and Jeffrey D. Berejikian, "Disaggregating Noncompliance: Abstention versus Predation in the Nuclear Nonproliferation Treaty," *Journal of Conflict Resolution* (forthcoming).

28. For more details, see Ian Anthony, Christer Ahlstrom, and Vitaly Fedchenko, *Reforming Nuclear Export Controls: The Future of the Nuclear Suppliers Group* (Oxford: Oxford University Press, 2007), 24.

29. Ibid, 20.

30. Ibid.

31. Quoted in Gert Brieger, "Do the Sick No Harm," *New York Times,* November 22, 1987. See Albert Wohlstetter, Thomas Brown, Gregory Jones, David McGarvey, Henry Rowen, Vince Taylor, and Roberta Wohlstetter, *Swords from Plowshares: The Military Potential of Civilian Nuclear Energy* (Chicago: University of Chicago Press, 1979).

32. Countries that would have struggled to build nuclear weapons at the time they did in the absence of civilian nuclear assistance include India, Israel, North Korea, Pakistan, and South Africa.

33. See, for example, Steven Miller and Scott Sagan, "Nuclear Power without Nuclear Proliferation?" *Daedalus* 138, no. 4 (2009): 10; and Matthew Fuhrmann, "Splitting Atoms: Why Do Countries Build Nuclear Power Plants?" *International Interactions* 38, no. 1 (2012).

34. Quoted in "Jordan's King Abdullah II Wants His Own Nuclear Program," Associated Press, November 19, 2007.

35. John Mueller, *Atomic Obsession: Nuclear Alarmism from Hiroshima to Al-Qaeda* (Oxford: Oxford University Press, 2009).

36. Marty M. Natalegawa, "Statement on behalf of the NAM States Party to the Non-Proliferation of Nuclear Weapons Treaty," 2010 NPT Review Conference, New York, May 3, 2010. Available at http://www.un.org/en/conf/npt/2010/statements/pdf/nam_en.pdf.

37. Quoted in Larry Rohter and Juan Forero, "Venezuela's Leader Covets a Nuclear Energy Program," *New York Times,* November 27, 2005.

38. Mustafa Kibaroglu and Baris Caglar, "Implications of a Nuclear Iran for Turkey," *Middle East Policy* 15, no. 4 (2008): 68.

39. One policy pursued by the United States and other countries to limit the proliferation potential of peaceful nuclear cooperation is to reduce the use of highly enriched uranium (HEU) in civilian reactors. This measure certainly could not hurt, but it is likely to be more effective in addressing concerns about nuclear terrorism than concerns about nuclear proliferation.

40. Countries might withdraw from the NPT, for instance, if the international community were to provide the IAEA with extreme monitoring and enforcement powers.

41. The flip side of this recommendation is that countries should encourage defections among key scientists in proliferating countries. The United States has implemented a "Brain-drain Program" to lure scientists out of Iran. This policy could prove to be fruitful given the

historical importance of knowledge in building nuclear weapons. See David Sanger, "Beyond Iran Sanctions, Plans B, C, D, and ..." *New York Times*, June 10, 2010.

42. Hillary Clinton, "Remarks at the Review Conference of the Nuclear Nonproliferation Treaty," New York, May 3, 2010.

43. Others have also advocated this policy. See, for example, Henry Sokolski and Victor Gilinski, "Locking Down the NPT," *Bulletin of the Atomic Scientists*, Web edition, June 17, 2009. Available at http://www.thebulletin.org/web-edition/op-eds/locking-down-the-npt.

44. Miles Pomper, "US International Nuclear Energy Policy: Change and Continuity," Nuclear Energy Futures Paper No. 10, January 2010, 10.

45. Quoted in Pierre Goldschmidt, "Nuclear Renaissance and Non-Proliferation," lecture at the 24th Conference of the Nuclear Societies, Israel, February 19–21, 2008, 7.

46. Clinton, "Remarks at the Review Conference of the Nuclear Nonproliferation Treaty."

47. Ibid.

48. Paul Kerr, "ElBaradei: IAEA Budget Problems Dangerous, *"Arms Control Today* 37, no. 6 (2007). Available at http://www.armscontrol.org/act/2007_07-08/IAEABudget.

49. The IAEA can receive voluntary contributions from member states.

50. This is consistent with a recommendation issued by an independent commission on the future of safeguards. *Report of the Commission of Eminent Persons on the Future of the Agency*, International Atomic Energy Agency, Board of Governors, GOV/2008/22-GC(52)/ INF/4, May 23, 2008.

Index

Note: Page numbers followed by n indicate notes.